150575921 3

AF743042

University of Bristol Library
Tyndall Avenue, Bristol BS8 1TJ

Date due for return
UNLESS RECALLED EARLIER

NOT FOR LOAN
- 1 APR 1998

Used 2/12

- 6 MAR 2012

UNIVERSITY OF BRISTOL
WITHDRAWN

KNOWLEDGE AND INQUIRY

POZNAŃ STUDIES IN THE PHILOSOPHY OF THE SCIENCES AND THE HUMANITIES

VOLUME 51

EDITORS

Tomasz Banaszak (assistant-editor)
Jerzy Brzeziński
Robert Egiert (assistant-editor)
Andrzej Klawiter
Krzysztof Łastowski
Leszek Nowak (editor-in-chief)
Katarzyna Paprzycka (Pittsburgh)
Marcin Paprzycki (Odessa, TX)
Piotr Przybysz (assistant-editor)

ADVISORY COMMITTEE

Joseph Agassi (Tel-Aviv)
Étienne Balibar (Paris)
Wolfgang Balzer (München)
Mario Bunge (Montreal)
Nancy Cartwright (London)
Robert S. Cohen (Boston)
Francesco Coniglione (Catania)
Andrzej Falkiewicz (Wrocław)
Dagfinn Føllesdal (Oslo)
Bert Hamminga (Tilburg)
Jaakko Hintikka (Boston)
Jerzy Kmita (Poznań)
Władysław Krajewski (Warszawa)
Theo A.F. Kuipers (Groningen)
Witold Marciszewski (Warszawa)
Ilkka Niiniluoto (Helsinki)
Günter Patzig (Göttingen)
Marian Przełęcki (Warszawa)
Jan Such (Poznań)
Jerzy Topolski (Poznań)
Ryszard Wójcicki (Warszawa)
Georg H. von Wright (Helsinki)

The address: **prof. L. Nowak, Cybulskiego 13, 60-247 Poznań, Poland.**
Fax: (061) 477-079 or (061) 471-555
E-mail: epistemo@vm.amu.edu.pl

KNOWLEDGE AND INQUIRY:

Essays on Jaakko Hintikka's Epistemology and Philosophy of Science

Edited by

Matti Sintonen

Rodopi

Amsterdam - Atlanta, GA 1997

B 4715.H54 KNO

∞ The paper on which this book is printed meets the requirements of "ISO 9706:1994, Information and documentation - Paper for documents - Requirements for permanence".

ISSN 0303-8157
ISBN: 90-420-0080-5 (bound)
ISBN: 90-420-0069-4 (paper)
©Editions Rodopi B.V., Amsterdam - Atlanta, GA 1997
Printed in The Netherlands

UNIVERSITY OF BRISTOL
UNIVERSITY OF BRISTOL LIBRARY
WITHDRAWN

CONTENTS

IV: GROWTH OF KNOWLEDGE: EXPLANATION AND DISCOVERY

Poznań Studies in the Philosophy of the Sciences and the Humanities
1997, Vol. 51, pp. 1–2

Matti Sintonen

FROM THE SCIENCE OF LOGIC TO THE LOGIC OF SCIENCE

Twentieth century analytic philosophy of science has both benefited and suffered from its close alliance and cooperation with logic. This cooperation reached its highest peak during the heyday of logical empirism, but has been in the decline ever since. There has been a growing feeling that the science of logic has been and is unable to reach the truly interesting problems in the logic of science. Along with this skeptical current there is also another important development, viz., the growing dissatisfaction with general philosophy of science, or logic of science at the most general level of concept and theory development. Together these developments have led to the rise of historical, and also is part, sociological studies of science on one hand, and to the rise of "special" philosophies of science. As a result of this, the science of logic and the logic of science were sundered a part.

One prominent dissident on this philosophical scene has been Jaakko Hintikka. Hintikka has written on almost all areas of philosophy, from logic to philosophy of science, and from Wittgenstein to the history of philosophy. But despite this variety of topics there is in his philosophy the underlying conviction that the resources of modern logic have not been exhausted, quite the contrary. Where skeptics have yielded to the temptations of purely historical case studies, or turned into epistemology at large, Hintikka has consistently favoured a different strategy: although the emergence of historical and related studies is a welcome development, it does not make the logic of science obsolete. And although the rise of special philosophies of science, from philosophy of quantum physics to philosophy of economics is a welcome turn, it does not make general philosophy of science outdated. Rather, what we must do is to develop new logical tools which meet these demands.

If there is a single theme that characterizes Hintikka's approach to, and impact on, epistemology and philosophy of science it surely is the deep commitment to developing new logical tools for epistemologists and philosophers of science. This started with Hintikka's work on distributive normal form which provided important insights for inductive logic. One of the most recent breakthroughs on this front is the coming of age of the interrogative model of inquiry. Although its roots are in the early logic of questions and epistemic logic it is now rapidly growing into a fullblown philosophy of science and knowledge acquisition.

This collection of essays dedicated to the various historical and systematic aspects of Hintikka's epistemology and philosophy of science by his friends and collegues all around the world demostrates the scope and depth of his contributions. But they also vindicate his conviction that the rendering apart of the science of logic and the logic of science is not final and has been detrimental to the development of the latter. In fact, as can be seen from this collection of essays, Hintikka's new logical tools throw light on important topics which were thought to be beyond "logical reconstruction", such as explanation and discovery.

I

HISTORICAL PERSPECTIVES

Poznań Studies in the Philosophy of the Sciences and the Humanities
1997, Vol. 51, pp. 5–18

Zev Bechler

HINTIKKA ON PLENITUDE IN ARISTOTLE

I have known for a long time now that Jaakko Hintikka's work on Aristotle was the most decisive of the influences on my whole thinking, and in my recent book on Aristotle (*Aristotle's Theory of Actuality*, SUNY Press, 1995) I pointed this out. But I never knew to what extreme degree I followed him until I sat down to write this paper and reread his works as a preparation. It now became obvious that what I did was hardly more than dotting and crossing some of Hintikka's i's and t's. And it gave me a deep sense of peace – I was fortunate enough to be the follower of such great intellect.

1. Plenitude and the Necessity of Eternal Motion

In what follows I use the term "consistency-potentiality" for what Hintikka called "partial-potentiality". A potentiality is partial only when it comprises only part of all the conditions necessary for the actualization of the end. But it is a potentiality nevertheless because it is consistent with all we know to assume that it will actualize. Some writers call this brand "logical possibility", and Aristotle called it possibility *haplōs*, i.e., without taking into account the exact conditions that must be supplied for it to actualize. By contrast, potentiality is "genuine" only when all the necessary conditions for its actualization are at hand. It is agreed all around that Aristotle held that what is actual is prior in every sense (ontological, epistemic, and temporal) to what is only potential. It can be easily shown that such priority of the actual entails, together with the previous definitions of the two potentialities, that only actualities are real and that potentialities possess no reality whatsoever. The thesis that only actualities are real I call "actualism" and I hold that Aristotle's actualism fully determined the details of his philosophy of nature. One major consequence of Aristotle's actualism was his peculiar brand of determinism. For reasons which will become clear in a while, this should be called logical determinism for it holds that all that happens, happens by logical necessity, i.e., it would involve a contradiction if what actually happened were assumed *ex hypothesi* not to have happened. A closely related thesis, similarly determined by Aristotle's actualism, was his doctrine of plenitude.

Hintikka was the first to argue in detail that, contrary to many appearances, Aristotle in fact subscribed to the thesis of plenitude, according to which all possibilities are actualized eventually.[1] In fact, it may be easily seen that Aristotle's actualism entails the plenitude thesis in its strongest version and it entails it trivially, in two stages. No potentiality can be said to exist and yet not actualize. This is trivially true for consistency-potentials since Aristotle regarded them as non-real things. It is also true for genuine potentials, since these actualize necessarily and instantaneously: when all the necessary conditions are present, it is logically necessary for the result to actualize. In sum, potentialities actualize, and all states are necessary while they exist because all genuine potentialities necessarily actualize. This is the plenitude thesis with a deterministic vengeance. Though at this stage not all possibilities are actualized eventually, nevertheless all the genuine ones are actualized instantaneously and all – and not only eternal – states are necessary, as we shall see now. Though at this first stage consistency-potentials are taken as nonreal entities, at the second stage we shall see that the same is true even when they are assumed to possess some reality.[2]

A major difficulty about this version of plenitude is that it conflates three categories, the "always or usual", the "possible", and the "necessary". Aristotle, however, seems to hold the commonsense usage that what is "always or usually" (*Phys* 198b6) is what has or may have exceptions and hence is different from the "necessary". Also, the "possible" is what may but also may not happen, and so it too is obviously different from "necessary". The principle of plenitude, however, entails that all three are strictly coextensive and as a result also cointensive: If whatever is possible eventually actualizes, the possible is coextensive with what happens necessarily (since it cannot *not* happen) and so with what always happens. I'll employ only Hintikka's major argument, which shows that a weak version of the plenitude principle follows from Aristotle's argument that eternity entails necessity. Aristotle declared that

> What is 'of necessity' coincides with what is 'always', since that which 'must be' cannot possibly 'not be'. Hence a thing is eternal if its 'being' is necessary, and if it is eternal, its 'being' is necessary. (*DGC* 338a1–3, Joachim)

And yet he explicitly declared also that things could exist forever and yet not be necessary (*Met* 1088b18–25; and *AP* 19a25–7). What, then, is the element that distinguishes between such two kinds of enduring things? Most of Aristotle's argument that if something exists forever it must also be assumed to exist necessarily occurs in DC 1,12, and it says that a state *x* is eternal because there is no consistency-potentiality for either non-*x* or for *x*. From the necessity of eternal existence he then inferred that an eternal existent is ungenerated, that is, that a generated existent cannot be an eternal existent, just as an ungenerated existent cannot stop existing and so is undestructible. As proof he argued that if something exists forever but is also assumed to have the potentiality for stopping to exist, this entails a contradiction:

Assumption 1: State *x* exists forever.
Assumption 2: It has consistency-potentiality to become non-*x*.
Consistency-Test: It is permissible, or consistent, to infer that whatever has potentiality for something, actualizes it.
Conclusion 1: From Assumption 2 and the Consistency-Test we may infer that *x* actually became non-*x*, but this contradicts Assumption 1.

Hence Conclusion 1 shows that the conjunction of Assumption 2 and the Consistency-Test contradicts Assumption 1. But the Consistency-Test is presumably analytically true for consistency-potentials, for it merely says what it means for them to be consistent. Hence it is Assumptions 1 and 2 that are to blame, and since 1 states a fact, 2 must be false. This is what Aristotle says here:

> Thus if anything which exists for an infinite time is destructible, it must have the potentiality of not being [Assumption 2]. It exists, then, for an infinite time, but we may suppose this potentiality of not being to be realized [Consistency-Test]. Then it will both be and not be, in actuality, at the same time [Conclusion 1 cum Assumption 1]. That is, a false conclusion will result because the premise [Assumption 2] laid down was false. . .. Hence everything which exists forever is absolutely (*haplōs*) indestructible [Assumption 1]. (*DC* 281b20–25, Guthrie, with "potentiality" for Guthrie's "power", as also in all following translations from Guthrie)

Aristotle says that if something is forever in some state, it follows that it is necessarily so, that is, there can be no contingently eternal things. But to say that the eternity of *x* entails lack of potentiality for non-*x* is to imply that if there is potentiality for non-*x* then *x* is temporary, and so *x* will cease eventually and non-*x* will actualize eventually, and this is true for every *x*. This is the weak version of the plenitude principle and Hintikka was the first to point out this implication. Hintikka's argument may be put thus:

A: The eternity of x entails the necessity of x.
B: Hence, nonnecessity of x entails the noneternity of x.
C: That is, the possibility of non-x entails its eventual actualization and so it is necessary for actualization to occur at some time.

We see that the reason why Hintikka's argument is true, and why the plenitude principle binds Aristotle, is that it is the consequence of his actualism; potentiality necessarily depends on its actualization. Where no actualization is possible, as in eternal states, no potentiality exists. And consequently, where potentiality does exist, there its actuality must exists. Potentiality necessarily actualizes simply because genuine potentiality can be attributed only after it became an actual thing. It is therefore an elementary truth of actualism that potentiality pertains only to temporary states and to all such states, but then it is as trivially true that all potentials actualize eventually, for this is what to be temporary means. Hence, by necessity, all potentials actualize eventually. But notice that this cannot be an informative truth, for it also means that any imaginable possibility that never did and never will actualize is an inconsistent possibility and so is no possibility at all.

2. *The Necessity of All States*

If Aristotle's commitment to the plenitude principle is in fact nothing more than a consequence of actualism, it would follow that consistency-potentials are exactly (all and only) those which actualize, and then they will be identical with genuine potentials. At this second stage it can be seen that Aristotle's argument for the necessity of eternal states applies with equal force to temporary, apparently contingent states. Consequently, it will follow that not only is it impossible for a potential to remain unactualized forever, but rather, not surprisingly for us now, it is impossible for a potential to remain unactualized for any limited time. This will mean that all states, and not just the eternal ones, are necessary while they last. Allowing for some awkwardness, temporary states are contingently necessary states, whereas eternal ones are necessarily necessary. For it follows that there is no difference between Aristotle's pair of Assumptions 1+2 and the following Assumptions $a+b$:

Assumption a: State x exists during the whole finite time-interval t.
Assumption b: x can change during the same time-interval t.
Conclusion A: From Assumption b and the Consistency-Test we may infer that x actually changed during the time-interval t.

This conclusion, however, contradicts Assumption a. Hence the infinite duration of the state in Assumption 1 is irrelevant: for whatever time a state exists, while it exists, there is no potentiality of its not existing. It then follows that the only potentiality it has concerning its present state is to endure in it, and since obviously this is not a mere consistency-potentiality, it is a genuine potentiality. But this means that every state exists necessarily. If Aristotle is committed to the necessity of eternal states, he must also accept the Megarian thesis that things are possible only when they are actual, and this in spite of his protests. This consequence follows from the Consistency-Test exactly as for the case of infinite duration: To be a consistency-potential for non-x is to entail no contradiction if the potential for non-x actualizes, but, given that during this time-interval x holds, to assume that non-x also holds then, is self-contradictory. Hence, during the time x exists, no consistency-potentiality for non-x exists, and therefore while x exists it exists by logical necessity. Given that the man will stand during the next hour, it is inconsistent to say that he nevertheless can sit during the next hour, and so he does not have the consistency-potentiality for sitting. This means that the consistency-potential for non-x is time-indexed to apply after x stops.

We have thus learned a lesson in the delicate art of evaluating the alleged consistency of proposed potentials; that is, these must be qualified so as not to clash with any actual state, and this means that they must be identical with some actual state. The sitting man has consistency-potentiality to stand only after sitting, but not before:

> A man has at the same time the potentiality of sitting and that of standing, in the sense that when he has the one he also has the other; but this does not mean that he is able to sit and stand simultaneously, but only successively. (*DC* 281b16–18, Guthrie)

Assume the man is now standing. Then assuming what the first clause says, leads to the contradiction described in Conclusion *A*. But this he denies in the second clause because, as he had already stated,

> If certain things have the potentiality both of being and of not being, an outside limit must be set to the time of their being and their not being, the time, I mean, for which the thing can (*dunaton*) be or not be. (*DC* 281a28–30, Guthrie).

This is an explicit limitation on the temporal aspect of possibilities. So even though it may be said that while he stands the man has the potentiality to sit, this is a time-indexed potentiality, since it is the potentiality to sit-*after*-standing. Thus, even if this potentiality can be said to overlap the time of standing, it is irrelevant for the state of standing for it cannot be the potentiality to sit-while-standing. This is the meaning of the clause "the time for which the thing *can* be or not be," (rather than "is and is not"). This interpretation gains plausibility in the next sentence. Aristotle goes to show that in case we do not set such a time limit to a given state but assume it to be infinite, the potentiality for another state must be for a state after an infinite time had passed, which is a contradictory concept:

> For if the time is not of a certain definite length, but is always more than any given time, and there is none laid down which must exceed it, then the same thing will have the potentiality of being for an infinite time and not being for another infinite time; which is impossible. (*DC* 281a35–281b2, Guthrie)

That is, if while in *x* the object has potentiality for some non-*x*, then it is for non-*x*-after-a-certain-definite-time. Since for an eternal *x* this would mean potentiality for non-*x*-after-eternity, this non-*x* will have to actualize exactly after eternity. But since this is impossible, then if all potentiality must actualize, both *x* and non-*x* will exist simultaneously:

> if a thing has more than one potentiality for an infinite time, there is no "successively"; it must actualize the other potentiality simultaneously. (*DC* 281b18, Guthrie)

It is easy to see that such time-indexing of consistency-potentiality enormously strengthens the plenitude ontology for finite and determined states. For this same argument must hold for finite times, and consequently the potentiality for non-*x* while in *x* is in fact for non-*x*-after-*x*-stops. Either this or both states would be actual simultaneously. Therefore, the potentiality for non-*x* during the time for which it is given that the state is *x*, is self-contradictory, not a consistency-potentiality. Thus Hintikka's argument for Aristotle's plenitude accords with these consequences, derived from Aristotle's own argument; temporary states, exactly as eternal states, do not have any consistency-potentiality for their alternatives. Both kinds, and so all states, exist necessarily while they exist; that is, no unactualized potentials are possible. Consequently, lack of potentiality cannot be taken as what distinguishes eternal from temporary states. Though Aristotle says that things are eternal because they exist "in actuality", meaning in pure actuality with no admixture of potentiality, this applies equally to temporary, contingent, states within his ontology of potentiality.

3. Kinēsis Is No Escape

Now, the whole train of reasoning up to this point can be seen to be independent of Aristotle's physics. It is a purely logical argument, depending merely on the thesis that genuine potentiality is a state that includes all the necessary conditions for the actualization of some effect (the "end"), and that whatever is not a genuine potentiality, is no potentiality at all. Given this framework, I don't see how complete logical determinism and plenitude can be avoided in any way. That is, I don't see how it is possible to avoid the consequence that given all the necessary conditions, the outcome follows at once and necessarily. And if plenitude is the thesis that all genuine potentiality actualize, then I don't see how plenitude is avoidable as well. Consequently, there must be some error in arguing that Aristotle did find a way out of this logical determinism.

Hintikka argued that Aristotle's way out, and a valid way out at that, was in his definition of *kinēsis*. Hintikka wrote that *kinēsis* is the only case "in which a genuine potentiality can exist unfulfilled according to Aristotle" (1977, p. 59). Hintikka's reasoning is that since *kinēsis* is by Aristotle's definition the actuality of some potentiality as such, this potentiality keeps being a potentiality during the whole course of the *kinēsis* and so is real or actual as potentiality without actualizing into its end during that *kinēsis*. But I can't see that this potentiality for the end is a genuine one at all during the *kinēsis*: Clearly, not all the necessary conditions exist during the *kinēsis* for the actualization of the end. This is, indeed, why there is *kinēsis* – the end is not actual yet, and this entails that not all the necessary conditions for it are at hand yet. The potentiality is not a genuine potentiality for that end, but non-genuine potentialities are not the issue of determinism or plenitude at all. Indeed, Aristotle says just this in clear words – what the potentiality (whose actuality this building *kinēsis* is) is genuine for is the buildable, not the house (201b10ff.). And being a genuine potentiality for the buildable entails that it is the buildable that exists in actuality right from the moment building has begun. The building *kinēsis* is the actuality of the potentiality as such (for the house) and such actuality is the buildable (and not the house). It is just this that Aristotle says in *Phys* 201a28–9, where he identifies motion with the *entelecheia* of what exists potentially "when it is already fully real and operative, not as itself but as movable". The last words refer to the buildable, which is what is fully real and operative in the building *kinēsis* and is the actuality of the potentiality (for the house) as such. The "thing" whose "attribute is *kinēsis* just when it is fully real in this way" in 201b6–7 is the same potentiality (for the house) but "as such", i.e., the buildable. In other words, the buildable, and in general what Aristotle calls "the movable", denotes an actuality, not a potentiality of any kind. During a *kinēsis*, what is fully and completely real is that special entity, the movable.

Moreover, since qua movable it does not change during the whole *kinēsis*, it follows that *kinēsis* is a steady state, not a change at all. And since it seems to be the case (by 225b15ff) that *kinēsis* does not come-into-being nor passes-away gradually, neither does the movable. Thus, the buildable springs into full blooming being the moment the building *kinēsis* begins, and disappears as suddenly the moment it ends (and a house similarly springs then into full being). The difference between the two states, building and house, must therefore be only in the different natures of its respective actualities – the buildable and the house – but in what this difference consists cannot be said, since change, which is the apparently obvious characteristic of *kinēsis*, has now been eliminated once the entity "movable" is taken as a fully blown actuality. The potentiality as such, i.e., the movable, does not change during its entire actuality.

What all this means is that *kinēsis* is definitely *not* a feasible locus of escape from logical determinism. Rather, it is a regular and standard state in which potentiality, in so far as it is genuine is for the movable and so actualizes instantaneously and necessarily, but in so far as it is potentiality for the end product (house) is not genuine and so cannot possibly actualize. So again, the only way out of logical determinism remained for Aristotle via the notion of non-genuine potentiality, and I think that this is a main reason why he kept such a notion under the concept of absolute (*haplōs*) potentiality even though it had for him no physical significance at all.

Since *kinēsis* cannot serve as a locus of actual potentiality but only of fully actualized potentiality, the notion of actual or real potentiality can find no place within Aristotle's conceptual scheme. And if strict actualism is followed, this means that no potentiality is a real thing or state within it.

4. Actualism and Anti-Informationism

It follows from this that Aristotle's potentialities cannot act, i.e., they cannot be "powers" or "forces" or "tendencies" or whatever one may name such pushing causes. The reason is clear: non-genuine potentialities do not act by definition (and this contributes to the claim that they are mere conceptual entities and are not real entities), whereas genuine potentialities never have time in which to act for they diappear at the same instant they become genuine. Hence when Aristotle says that potentiality is *archē kinēseōs* he cannot mean that it is an initiating or a pushing force or cause.

And since the *phusis* or nature of any given substance is just as inefficient an *archē kinēseōs* as non-genuine potentiality (as Aristotle points out in *Phys VIII*, 1 and *II*, 1, see Hintikka 1977, p. 64), neither are natures such initiating or pushing forces and causes.

But since natures and potentialities are the only *archai* of natural motions in Aristotle's system of nature, it follows that Aristotle's system of nature contains no powers or forces by which natural motions can be explained.

In other words, Aristotle's system of nature cannot consistently contain physically necessitating factors which determine the outcomes of any given situations. Consequently, the logical determinism that is entailed by the non-reality of all potentiality, is accompanied by a correlative physical indeterminism. Consequently, all scientific explanation, i.e., the explanation which exhibits the necessity of the explanandum, must be a strictly logical explanation, that is, it cannot contain elements which are merely contingently true. In other words, its premises must be logically true propositions, and Aristotle took care of this by postulating that the premises of the paradigm of scientific syllogism, the Barbara, must be definitions. But since the essence which is stated in a proper definition is nothing separate from the subject, a definition is not a predicative statement: "In a definition one thing is not predicated of another" (*An Po* 90b35). That is, a definition is an identity and it is because of this, and only as such, that it can be necessarily true. But as such it also contains no objective information about the world. Consequently, Aristotle's actualism entails that scientific explanation must be objectively non-informative, and this exactly because his actualism entailed that whereas all potentiality is necessarily non-real, all scientific explanation of natural motion must be formulated in terms of just such potentiality, e.g., essences and natures.

Anti-actualism may now be described as the thesis that some potentialities are fully real. That means that some potentialities may be actual and yet not actualize into their ends. I argued in my *Newton's Physics and the Conceptual Structure of the Scientific Revolution* (Kluwer, 1991) that two paradigmatic examples of such anti-actualism are Plato's ontology of separate Ideas and geometric atoms in space, and Newton's ontology of separate laws and forces and separate infinitely many actual geometric forms in space. Because they are real and yet not necessarily manifested as (or transformed into) their actualizations, their necessity is physical and not logical, the premises of scientific explanations by their means (e.g., Newton's laws of motion) are predicative propositions "predicating one thing of another" and so are objectively informative. Consequently the possibility of error is a merit only of anti-actualism. Moreover, I suspect that all objectively informative explanation must be based on anti-actualistic ontology.

In fact objective informativity and contingency are one and the same: The laws of nature are contingent in the sense that they are not logically necessary truths and so could have been other than they are had God chosen so. Contingency next pertains to every event in the derived sense that it is necessitated by contingent laws and in the direct sense that given these laws the initial conditions of the world at the moment of its creation could be differently chosen by God. So Newton's world, as a paradigm of an anti-actualistic world, is completely deterministic in being fully determined by the initial conditions of the world and its laws, but is also fully contingent in the above senses.

Within such a world, genuine potentialities are those of events or world-states which are determined by the world's initial conditions and laws. These potentialities can be fully calculated by a Laplace demon at the initial instant of the world and so they are all actual potentialities, i.e., they are real and yet do not actualize at once. For some of them, e.g., the big crunch, it will be billions of years before they actualize. All other potentialities can be classified according to the two species of contingencies – such as pertain to a world with identical laws but different initial conditions and such as pertain to a world with different laws. As a rule, such potentialities are never to be actualized in this world and may be called non-genuine or other-world potentialities.

It thus turns out that weak plenitude holds equally within Aristotle's actualistic world and Newton's anti-actualistic world: In both worlds, all genuine potentialities actualize. But the crucial difference between them is that only in Aristotle's world a super-plenitude holds as well – all potentialities, and not just genuine ones, actualize and, more important still, they actualize instantaneously, i.e., at the instant they become potentialities. In effect, Aristotle's actualistic world does not contain potentialities at all, whereas Newton's potentialistic world contains an actual infinity of them and they are all genuine yet unactualized.

Consequently, plenitude as such is a condition that is independent of the ontology. Genuine potentiality necessarily actualizes equally in a world such as Newton's, where causality and determinism are fully physical, as well as in a world in which there is no physical causality. In such an indeterministic world no set of initial conditions constitutes a complete set of sufficient conditions for any given effect and the only way to make it complete is by including in it the effect itself as one of the conditions. Aristotle referred to something close when he posited that the absence of interfering factors must be included in the very concept of genuine potentiality (*Met* 1048a1, 17–21). Causality and determinism thus become strictly logical force: Though nothing necessitates any definite effect before it happens (hence physical indeterminism), the effect becomes logically necessary once it happened since now it can be included in the set of prior conditions which now becomes a complete set. This is the stuff anti-informationism is made of: scientific explanations cannot and must not be objectively informative since complete explanation can be only given by means of logically enforcing causes.

5. Einstein, Bohr and Plenitude

Most gratifying, I found Hintikka's thesis surprisingly fertile even when applied to the development of scientific thought up to our own time. Let me round up these notes with a sketchy example of the role of plenitude in the history of quantum theory.

When after 1927 quantum theory needed a justification and defence as a scientific explanation, Bohr developed his actualistic philosophy of nature. Its central aim was to argue that a physically indeterministic world can be explained completely by forgoing physical causality and adopting logical causality instead. And when Einstein attacked the theory because of its incompleteness (Einstein 1935) what he argued was that certain genuinely predictable (i.e., potentially predicted) magnitudes are denied actuality (or reality) in the theory. These are pairs of conjugate values, each member of which is predictable. From the genuinity of this separate potentiality of prediction, i.e., the predictability of the magnitude, Einstein inferred the reality of the magnitude itself, even if the potentiality is not actualized, i.e., even if the value is not actually predicted.

It was thus a distinctly potentialistic ontology that grounded Einstein's argument, and Bohr was quick to point out this "misunderstanding": The value of a magnitude becomes genuinely predictable, Bohr argued in effect, if and only if it is actually predicted and not an instant before. Consequently there are no realities corresponding to what is not predicted in actuality, and consequently it is only the actual prediction that instantaneouly creates the reality of what is predicted. Thus quantum theory became a complete theory by a kind of an actualistic plenitude principle: All genuine predictability is actualized by an actual prediction, and since actual prediction creates its predicted magnitude, all genuine predictability denotes an actual magnitude and value.

Moreover, Bohr made it clear that the force by which such a potentiality is actualized is strictly a logical and not a physical force ("mechanical disturbance"), i.e., "of course,... there is... no question of mechanical disturbance of the system during the last critical stage of the measuring procedure. But even at this stage there is essentially the question of *an influence on the very conditions which define the possible types of predictions (sic)*" (Bohr 1935, p. 138).

And just as Aristotle included the absence of interfering factors in the definition of the genuine potentiality, so Bohr added that the "very conditions which define the possible types of predictions", i.e., which define the genuinity of a given predictability, "constitute an inherent element of the description of any phenomenon to which the term 'physical reality' can be properly attached" (*ibid.*).

Bohr called this reasoning "complementarity" (*ibid.*, p. 139) but in fact it denotes an actualistic principle of plenitude: "Physical reality" is just whatever is actually measured and so whatever is genuinely measurable, thus all genuine reality is genuinely predictable and actually predicted. Since the measurable, the measured and physical reality thus denote an identical extension, this plenitude meant that quantum theory is necessarily complete: There is nothing in actuality that is not predictable.

Aristotle's program of anti-informationism, which lay as mere consistency-potentiality since Newton's time, thus gradually actualized in the Copenhagen interpretation and at present there is little doubt that it actually carried the day. (See Bechler 1995a.)

The Cohn Institute for the History & Philosophy of Science & Ideas
Tel-Aviv University
Tel-Aviv 69 978, Israel

NOTES

1 Hintikka's persistent and convincing argument that Aristotle held the principle of plenitude can be found in papers collected in his (1973). Hintikka's thesis was attacked in Mulhern (1969) and M. Kneale (1974), and these were rebutted by Hintikka (1977). Though Hintikka's thesis and arguments determined much of my views, I cannot see how to avoid the centrality of consistency-potentials in Aristotle, but Hintikka sees this as negligible (1977, p. 26). More important yet, I conclude that the notion of potentiality as striving and force has no place in Aristotle but Hintikka states the contrary (*ibid.*, pp. 25–6) and views the "dynamic" interpretation as essential to him, though he does not spell out the details. Without consistency-potentials I see no explanation of such difficult passages as *DGC* 316a23, *Met* 1003a2, 1071b13, *DI* 19a12, and the most problematic *An Po* 74b37–75a35.

2 Hintikka (1973, pp. 97–102) has defended his theory convincingly and in detail with relation to some texts which appear to refute it, such as *Met* 1003a2 and 1071b13. The "coat passage" in *DI* 19a12 was explained by a somewhat fatal weakening of the

thesis, urging that it applies in Aristotle only to kinds and not to individuals, since only the former are eternal (*ibid.*, pp. 94, 100–1). This is unsatisfactory since the notion that the kind coat includes some individual coats that will be torn up before wearing out is ludicrous: The kind does not say anything about the accidental career of any of its individuals. However, Hintikka was unhappy about this solution and expressed his doubt explicitly (1973, p. 174). His analysis of *An Po* 75a31 (*ibid.*, p. 101) is also unsatisfactory, since it ignores the text's implication that even though we may not know that a property is accidental, it may nevertheless be such even though it is eternal. However, none of these defences will hold if Aristotle states that something (a kind, possibly) has the potentiality for *x*-ing, even though *x*-ing is a super-task. The "kind defence" would not apply since no individual of the kind did or will ever *x*. But this is what Aristotle says in the following text (which is not mentioned by Hintikka):

> Therefore, supposing it is of a nature to be divisible throughout, by a series of similar bisections or on any other principle, nothing impossible will have been achieved if it has actually been divided, since, even if it has been divided into innumerable parts innumerable times, there is no impossibility, *though perhaps no one would carry out this division.* (*DGC* 316a23, Forster)

REFERENCES

Aristotle

De Generatione et Corruptione

Categories & De Interpretatione

On the Heavens

Metaphysics

Physics

Posterior Analytics

Prior Analytics

Bechler, Z. (1991). *Newton's Physics and the Conceptual Structure of the Scientific Revolution*. Dordrecht: Kluwer.

Bechler, Z. (1995). *Aristotle's Theory of Actuality*. Albany, New York: State University of New York.

Bechler, Z. (1995a). *Our Justification of Emptiness* (forthcoming).

Bohr, N. (1935). Can quantum mechanical description of physical reality be considered complete? In: S. Toulmin (Ed.), *Physical Reality*. NY: Harper & Row 1970.

Einstein, A., B. Podolsky, N. Rosen (1935). Can quantum mechanical description of physical reality be considered complete? In: S. Toulmin (Ed.), *Physical Reality*, 1970.

Hintikka, J. (1957). Necessity, universality and time in Aristotle, *Ajatus* **20**.

Hintikka, J. (1966). Aristotelian Infinity, *Phil. Rev.* **75**.

Hintikka, J. (1967). Time, truth and knowledge in Aristotle and other Greek philosophers, *Amer. Phil. Quart.* **4**.

Hintikka, J. (1967a). A.O. Lovejoy on plenitude in Aristotle, *Ajatus* **29**.

Hintikka, J. (1972). On the different ingredients of an Aristotelian science, *Nous* **6**.

Hintikka, J. (1973). *Time and Necessity*. Oxford.
Hintikka, J. (1974). *Knowledge and the Known*. Reidel.
Hintikka, J. (1974a). Reply to Dorothea Frede, *Synthese* **28**.
Hintikka, J. (ed.) (1977). *Acta Philosophica Fennica* **29**.
Hintikka, J. (1980). Aristotelian induction, *Rev. Inter. de Phil.* **34**.
Hintikka, J. (1980a). Parmenides' Cogito argument, *Ancient Phil.* **1**.
Kneale, M. (1974). Review of Hintikka 1973, *Phil. Quart.* **24**.
Mulhern, K. (1969). Aristotle on Universality & Necessity, *Logique et Analyse* **12**.

Poznań Studies in the Philosophy of the Sciences and the Humanities
1997, Vol. 51, pp. 19–39

Marja-Liisa Kakkuri-Knuuttila

WHAT CAN THE SCIENCES OF MAN LEARN FROM ARISTOTLE?

This article considers Aristotle's significance for *research* in human science today. The emphasis on the first part of the phrase means that, in spite of the considerable relevance of Aristotle's social philosophy to current research in human science, I am not going to look at his socio-theoretical views, but rather his thoughts on conducting research.[1] I would like to introduce some aspects that have been less apparent outside actual research on Aristotle himself. As recent philosophers and human scientists have concentrated on the substance of Aristotle's ethical and socio-theoretical views, his ideas on the formal aspects of thought have largely been ignored. I am not only referring here to his theory of science, but also to some of his fundamental insights into the theory of dialectic and rhetoric as well. In the following, I will first look at some features of Aristotle's views on dialectic, rhetoric and philosophy relevant to current human and social studies. I will then show with the help of a paradigmatic case that the Aristotelian model of rhetorical speech does better justice to the traditional narrative historical research than the empiricist nomothetic ideal of science. I wish also to argue that Kuhn's conceptions of interparadigmatic choice correspond to the Aristotelian view of decision-making in rhetorical situations. In the last section, I intend to point out a gap in post-positivist philosophy of science of particular relevance to a common type of social research. There is so far no epistemological view that would sufficiently take the nature of research material into account in theory formation, especially the significance of the common sense knowledge of both the investigator and his or her objects of study. I will claim that the Aristotelian concept of *endoxa* and the method of *saving the phenomena* in ethics provide fundamental epistemological solutions, the details of which are worth closer study. I am thus going to prove that Aristotle provides two rather different types of research model, both of which opposed to the nomothetic view of science; one deriving from his rhetorical theory, and the other from his way of reasoning in ethics.

1. The Main Concepts of Rhetoric

Aristotle's *Rhetoric* has been a rich source of inspiration, and it has certainly not lost any of its vitality. The treatise includes two interesting trichotomies fundamental to its systematic presentation, the most familiar of which is the division of rhetorical means of influence into *logos, ethos*, and *pathos*. As is well-known *logos* refers to the substance of the argument, *ethos* to the speaker's means of showing to be himself trustworthy, and *pathos* to the means of appealing to the emotions of the audience.[2] These components are all relevant to studying scientific discourse as well, and it seems to me that rhetorical research into scientific discourse would greatly benefit from a more conscious use of this conventional trichotomy of means of influence.[3]

Another structural trichotomy in Aristotle's *Rhetoric* is his division of speeches into political, forensic and eulogistic, which complies with the conventions of public communication of his time. Perhaps not so surprisingly, this division is still applicable in spite of the considerable external changes that have taken place in our communication culture particularly through the rise of mass media. The most significant difference between the various speeches consists in the speech's hearer, or rather in the context of oration, which is reflected in the structure of the main argument. When a political speaker attempts to persuade his audience to adopt or reject some action, he reasons that it will either promote or hinder their happiness. The main conclusion of a forensic speech is that the accused is either guilty or not guilty of a criminal offence. The main argument of an eulogy, on the other hand, is either to praise or blame the object of celebration for virtue or vice.[4]

The three types of communication, political, forensic and eulogies, still exist, though often in combination, and the awareness of their basic structure facilitates the rhetorical analysis of various kinds of texts. To take an example, the majority of the leading articles in newspapers appear to represent political speech.[5] They present a burning issue with a suggested solution that seems suitable. Besides, there are also leading articles that follow the pattern of eulogies, giving reasons for either blame or praise, the eulogies of politicians being an excellent case in point. In my rhetoric courses, I have analyzed ministerial speeches at corporate functions, speeches which combine two structural principles. They are festal speeches, designed to pour praise on the object of celebration on one hand, but with the inclusion of an imperative element on the other, in which the minister, being a political actor, presents a platform for more extensive political influence.

2. *Topoi and non-formal logic*

Rhetoric and dialectic have one concept in common, the *topos*, which, for Aristotle, is the vehicle for discovering arguments. To be more precise, the *topos* is a reasoned instruction for formulating an argument for a given conclusion, the inferential rule yielding the reason for the instruction.[6] The *topoi* in dialectic are *general* in character, which means that their rules of inference apply to several kinds of objects.[7] Many of them deal with the relationship between species and genus. The following provides a good instance of a different sort of rule of inference for a dialectical *topos*:

> If what is more probable does not happen, then that what is less probable won't happen either.
> If what is less probable happens, then what is more probable will happen as well.[8]

The inferential principles for rhetorical *topoi* are usually *particular*, and only apply to one genus of speech. The following mainly applies to forensic speeches:

> If one has the capacity and motivation to do something, then one will do it,[9]

and the following statements mainly apply to political speeches:

> Whatever promotes happiness, should be done.
> Whatever creates harm, should be refrained from.
> One's countrymen should not offend a state stronger than their own, or a state with which it is advantageous to trade.
> What is rare is a greater good than what is plentiful.[10]

These instances illustrate three things. Firstly, the *topoi* deal with common sense knowledge, since rhetorical and dialectical discussions are not scholarly but part of our everyday life. Secondly, most of our everyday reasoning is not *deductive*, i.e. logically valid, where the truth of our premises would warrant the truth of our conclusions, but could rather be called *inductive,* because our conclusions are supported by our premises to some extent. Thirdly, the ideal of formalisation which has inspired modern research in logic ever since Leibniz, does not seem applicable to the examples above. It is a simple fact that the statements above hardly form any inspiring axiomatic systems. When Perelmān and Toulmin invented their informal logic in the late 1950s, they were hardly aware of the fact that Aristotle had come up with the same as much as 2300 years before them.[11]

How useful are the *topoi* from Aristotle's *Topics* and *Rhetoric* for modern human science? Since the species genus structure no longer forms the ideal for presentation, the Aristotelian classification system and the *topoi* relevant to it that will evidently be inadequate in modern human science. Each school or orientation contains its own categories of explanation, so we may consider Aristotle's textbooks as yielding a model of disclosing the *topoi* characteristic of each. What, then, are the *topoi* of for instance survey research, interactionism, discourse analysis, or ethnomethodology like? Are there *topoi* to be discerned at different levels of generality, analogous to the *topoi* in dialectic and rhetoric? Besides the *topoi* used in substantial explanation, we could look for *topoi* common in scientific writing, used to validate the choice of subject or methods, or any other aspect the author has to justify in addition to the results themselves.[12] The next subsection presents a historical study, where the main explanation is based on one of the *topoi* in the *Rhetoric*.

3. Historical narrative, structured like a speech

Aristotle's handbook on the art of rhetoric includes a great number of inference rules for deliberation in various decision-making situations, such as whether someone has done a certain deed or whether a deed ought to be done and in what conditions, be they either factual or concern the character of the agent. The rhetorical *topoi* thus appear to apply especially well to conventional research in history. This is further confirmed by my analysis of a historical study, titled *Tuomari Olaus Petrin vaa'assa (The Judge assessed by Olaus Petri)* by one of our highly productive Finnish historians, Heikki Ylikangas.[13]

In my analysis, I explicated the structure of the Ylikangas article, looking for patterns of explanation and argument. The main problem that Ylikangas had set out to solve is the identity of the author of the jurisdictional instructions appended to the Swedish and Finnish body of laws and statutes since the late 16th century. Ylikangas is particularly concerned with the question of what these instructions were originally intended for.

Ylikangas accepts the traditional view that they were written by a well-known theologian and religious reformer, Olaus Petri, born the son of a blacksmith in the 1490s and died 1552. Olaus Petri studied in Wittenberg, at the university of the reformers Martin Luther and Philip Melanchthon and was thus bound to have been influenced by their circles. After his return, Olaus Petri became an adviser to the Swedish King Gustav Wasa, and one of his most intimate and trusted friends until their disgreement about regal power turned them into mortal enemies. Olaus Petri went on criticizing the King partly openly, and eventually so vehemently that he was sentenced to death by the King in 1540; the City of Stockholm however bought his release from the sentence.[14]

Ylikangas summarizes with one of the inference models from the *Rhetoric* mentioned above: if one has the capacity and motivation to do something, one will do so. This principle is expressly stated in Ylikangas' article,[15] which is first of all an attempt to point out that Olaus Petri is a person with sufficient knowledge and intellectual resources to write the jurisdictional instructions, having moreover sufficient motivation to do so. The *curriculum vitae* of Olaus Petri given above is vital for the premises of this main argument. Petri was familiar with the abuses in jurisdictional practice manifested in these instructions, from his time as secretary of the Stockholm Council. The motive, on the other hand, comes from the pressures on him as a representative of the Estate of the Clergy having supported the King in his attempts to reduce the authority of the Church. Thus Ylikangas interprets the instructions essentially as an attack on the King, as a part of Olaus' attempt to wash some blood off his hands.[16]

The preceding analysis reveals some interesting aspects of the applicability of the so-called *covering law* model as the ideal for narrative historical writing. Superficially, the Ylikangas study is a narrative of Olaus Petri's life and political activities. However, it is an extremely selective story, the plot of which could be given as a chain of arguments. Only aspects relevant to the main argument are presented, rendering support to the two premises:

Olaus Petri was able to write the instructions.
Olaus Petri was motivated to write the instructions.

These imply on the basis of the aforementioned *topos* that

Olaus Petri wrote the instructions.

Thus the article has a strictly argumentative structure rather than being a mere chronological presentation of events, or an arbitrary literary story.[17] Perceiving this naturally involves a reader who can place the facts in their *context*,[18] and accept the *enthymematic* structure of the presentation.[19] The plot need not necessarily proceed directly, and seeking support for the main premises need not mean abandoning a critical approach. Any scholarly study naturally ramifies in weighing alternative interpretations and commenting on earlier research.

The short presentation above is intended as a prelude to my assessment of the value of the well-known Hempel–Oppenheim theory of explanation for this type of historical studies.[20] According to Hempel, the covering law model also applies to explanations produced by historical research. A phenomenon is explained by suggesting it as an individual instance of a general law; why its description, logically or plausibly, is derived from given laws, along with the description of its initial conditions. My study certainly employs general rules as explanatory principles, but they are intrinsically different from the regularities presupposed in the covering law model; they are not laws or hypotheses generalized by a historian with plenty of historical evidence, as required by Hempel.[21]

Instead of general laws, the explanatory principles employed by Ylikangas are firstly probabilistic principles at the level of *common sense knowledge*, thus corresponding to the inference rules of Aristotle's *rhetorical topoi*.[22] Secondly, most of the inferential rules themselves are not expressly stated, in agreement with rhetorical practice.[23] Furthermore, since the explanation does not consist of one inferential step only, but of a number of associations between minor explanations and arguments, it is pertinent to call this approach *the stepwise model of explanation*.[24]

On the whole, it can be stated that the study in question shows a surprising agreement with Aristotle's notion of rhetorical structure. Firstly, its arguments are basically structured like a speech, consisting as it does of a main argument or explanation, and justification of its premises; the inferences are presented in an elliptic form, and the underlying principle of the *topos* is usually not expressly stated. Besides arguing for one's own thesis, the rhetor also has to challenge the thesis of his/her opponent, and its defence, or possible defence of it, which is exactly what a researcher is supposed to do.[25] Furthermore, the familiarizing introduction to a speech is analogous to the introduction of a study with its presentation of the problem and literature survey.

Hempel certainly recognizes this type of historical explanation. He points out that the commonly used hypotheses are often associated with individual or socio-psychological factors which are commonly known from everyday experience. Admittedly these factors are usually not universal, but statistical principles. However, it is difficult in many cases to verbalize them as hypotheses to be accurately tested, so that they are no more than mere explanatory schemes. Hempel also suggests the necessity of several small explanations in stating that a researcher has to resort to generalisations when drawing conclusions concerning past events on the basis of the current material.[26]

Hempel's ideal of historical research relies on positivism, formulating general laws and testing them. The conventional research described here fails these requirements to a great extent, and is therefore inferior. Personally, I am opposed to this sort of normative approach. It seems philosophically more interesting to follow Aristotle's approach to various forms of communication. Applied to this context, it means exploring the underlying forms of rationality for different types of research.[27] Consequently, conventional historical research corresponds primarily to the type of rationality which Aristotle associates with rhetoric, as opposed to dialectic and philosophy. To be more precise, historical research represents forensic oratory rather than political or festal. Historical research solves problems about past events, such as whether a person did a certain thing and why he did it.

The conventional expositions in historical research are designed for a certain *audience*, not only for fellow scholars, but for a wider forum as well, let us say, for anyone with the basic education.[28] Therefore the chief explanations do not plead theories only familiar to a restricted number of experts. The underlying *topoi* are part of common knowledge. The historian explicates the deliberation and motivation of his human subjects in basically the same frame of reference as his audience or readers evaluate their own behaviour. From this point of view, it is not practical to criticize conventional research in history for failing to challenge common sense reasoning strongly, but rather to take it as given, reflecting it in one's line of argument and explanation. This makes perfectly practicable history, though it is not the only form of historiography.[29]

This in no way implies that conventional historical research might not supplement common sense thinking by working out new *topos* rules, or by using methods from other fields.[30] To give an example, the study discussed above uses lines of argument which are essentially based on systematic scholarly interpretation of the contents of the jurisdictional instructions. As far as a historian uses methods of source criticism or methods from other fields of research, these modes of argument remain subordinate to the common sense reasoning concerning the goals, motives and character of the historical actor. The point of view that I am advocating does not imply either that analysis should give up a critical approach, since there is always a possibility of criticism in disclosing inferential structures. Examining the adequacy of his/her premises is an essential part of a historian's everyday life, but judging the validity of individual *topos* principles is a gateway to philosophical activity.

4. The rhetorical and the scientific solution

Relativism is one of the burning issues in current discussions of science, strongly influenced by *The Structure of Scientific Revolutions* by Thomas Kuhn. Kuhn establishes that the choice of research paradigm is not purely based on rational argument, but is influenced by socio-psychological and aesthetic aspects as well. This results primarily from the fact that the criteria used in argument involve value judgements, which makes it difficult to come to an agreement on the application of these criteria. Secondly, it results from the incommensurability of the paradigm theories, which results from the fact that observations are paradigm dependent.[31]

The contrast between the conventional, rational view of science, and the wider notion that observes so-called non-rational aspects, is in some respects similar to Aristotle's distinction between philosophy and rhetoric. Thus it might be illuminating for our current discussion to take a look into the latter. One of the fundamental distinctions that Aristotle makes is that it is possible in philosophy to produce agreement based on argument, on certain conditions in any case, while in rhetoric, the final decision is made by voting. In the following, I shall concentrate on some basic features relevant to my argument, since I intend to go into the details of Aristotle's view of philosophical research in section 3.

According to Aristotle, philosophical research presupposes a host of extensive discussions with plenty of diversions on major issues, the divergent views conflicting with each other. These contributions to the discussion are called *endoxa*, translated in English as *the opinions of reliable people*. As Aristotle states at the beginning of his *Topics*, *endoxa* consist of views agreeable to all or the great majority of reliable people; or agreeable to all the wise, philosophers in other words, or the majority of them; or views presented by the most reliable of them.[32] Somewhat later he writes that *endoxa* comprehend expert views as well.[33] Thus, *endoxa* cover the researcher's opinions and common sense thinking, including views developed in various kinds of professional activities. These make up the actual basis of dialectical discussion as well as of philosophical research.

Accordingly, there is no essential difference between what we call philosophy and the more specialized fields of science. They comprise one and the same project inasmuch as they are based on the available views of their subject matter with all their gaps and conflicts. Research aims at working these problems out, and solving them in a way that maintains the truth of the *endoxa*. This research program is called *saving the phenomena* and, in propitious cases, it results in an overall, systematic theory of its object.[34] By the end of the study, it is in principle possible to reach an agreement on the benefits of the new theory compared with the old ones, presupposing that their supporters are willing to follow the new line of argument. This is what Aristotle is aiming at. A minimal condition for its realisation is that the supporters of *endoxa* be ready to engage in a debate, and prepared to accept that the other schools may be partly right as well.

A philosophical study is a project completed in finite time, because it deals with general features of reality. New arguments in favour of a given thesis or against it cannot be expressed *ad infinitum*. Having examined Aristotle's criteria for adequately dealing with a problem, I have come to the conclusion that he adopts the same criterion for philosophy as can be identified in his *Topics* for dialectic. This principle could be called *the last argument wins*, which implies that a belief system is acceptable if no irrefutable counterargument can be offered against any of its beliefs.[35]

Rhetorical decision-making is essentially different from that described above, even if it involves debate in the same sense as in philosophy. One speaker is defending a thesis, and the other argues against it. However, rhetoric does not deal with general aspects of some field of reality, but with an individual incident to be decided upon,[36] whereas in philosophy, the choice is to be made between different theories. Individual incidents are so infinitely complex that not all their aspects can be fully taken into account, and it is always possible to present new, relevant ones. The decision is to be made by voting, since there is no hope that either of the parties to the debate could refute all the arguments of the other. Thus, in a rhetorical debate, both sides are left with some irrefutable arguments, and it is unlikely that the matter could be solved on the grounds of some purely rational principles, such as *the last argument wins* described above.

Here we may see a connection with Kuhn's relativism. Kuhn seems to consider science as Aristotle does his rhetoric. Both present rational arguments which, however, are insufficient to solve complex situations of choice. Discussions based on rational reasoning become so complicated that some of the questions are bound to be excluded. As for Kuhn, the choice is about questions such as whether to maintain Newton's paradigm, or whether to adopt Einstein's paradigm instead.[37] In rhetoric, on the other hand, the question might be whether to raise taxes or not, or how health care should bē arranged. However, Kuhn also finds that the solution is a social process in many respects similar to voting in rhetorical contexts. The scientists are directed in their decisions by their own taste, group pressures, and their confidence in different paradigms. The reasons for the inadequacy of rational argument are similar in both cases.

5. The postpositivist vacuum and Aristotelian epistemology

In spite of the abundance of Aristotelian insights in the philosophy of science I shall confine myself to two further epistemological remarks. Their main purpose is to show that, in spite of its profundity, the contribution of the positivism critique has at certain points been mainly negative, and an important part of its positive content still remains to be developed. I shall try to argue in the following that we can find valuable material in Aristotle to fill the gaps created by the positivism critique.[38]

The positivist philosophy of science involves an epistemological *foundationalism* in which knowledge is founded on perception, and the other forms of knowledge are justified by observations.[39] The notion of theory in harmony with empiricist foundationalism understands the role of theory *as the systematization of observations*.[40] Empiricist foundationalism, however, suffered a severe blow from the *postulate of theoryladenness of observations* which dominated discussion in the 60s and 70s.[41] Since observations can be shown to be faulty by other observations and theoretical considerations, they cannot form a reliable basis for knowledge and, on the whole, there is no clear-cut epistemological difference between observational and theoretical knowledge. The reliability of perception and theory are rather circularly linked to each other.

Renouncing foundationalism did not, however, lead to an explicit rejection of the conception of theory as a means of systematizing observations. It is in principle still possible to stick to the idea that observations are in some sense primary with respect to theoretical knowledge, in particular, if the theories presupposed by the test-observations can be tested with the help of other observations not presupposing the theory to be tested. This need not be the typical situation, however. Notwithstanding, the scientific realists, for instance, attempt the resolve the question of the existence of theoretical entities by means of the power of theoretical notions to systematize observations.[42] Such an approach retains the empiristic distinction between theory and observations, as if *observational entity* were a notion less theoretical than *theoretical entity*, and as if theory-observation dichotomy were the decisive epistemological distinction in human sciences. I shall suggest that a more sophisticated division of sources of knowledge which does better justice to a vast number of social scientific studies than theory-observation dualism is needed. Since I will not be able to argue for it in any greater detail here, I shall merely indicate that such a change is no minor matter, but involves a fundamental shift in epistemological paradigms.

Even though the researcher may fail to recognize it, invaluable sources of knowledge in several types of social studies are the *commonsense* as well as the *expert* knowledge of both the researcher him/herself and those of his/her objects. These have several functions during the research process. Commonsense and expert knowledge form, firstly, *the object of systematization*, or part of it, secondly, *the means for it* and, thirdly, *its criteria*. This implies that a theory has an essentially different character than is presumed by the positivists. In such an approach, theory is no longer a mere *ordering framework* for observations, but rather a vehicle for *conceptualizing reality*.[43] Since theorization in the latter sense is a basic human activity, not an exclusive right of the scientist, a wise investigator draws upon the experience and theoretization of his/her subjects consciously, though not uncritically.[44]

Let us suppose that the researcher wants to work out the future prospects for trade in consumer products in the early years of Finnish European union membership.[45] For this purpose, it will be natural to interview experts such as the directors of wholesale conglomerates. However expert in the retail trade the researcher may be, he will still get considerably more than empirical observations from his/her interview subjects. If the questions are put judiciously, the answers will contain both things to be classified as observations, and the knowledge systematizing them. This is confirmed by the fact that research advances best when the first interviews open up new perspectives on deepening knowledge through further interviews.

No single interview subject is likely to offer a ready model for systematizing the whole mass of data – the researcher will have to have some competence too. But even if the expert reports were used to serve the conventional realistic viewpoint, in other words to provide information about the external reality, they would by no means only result in a list of loose observations or facts. The experts would not be experts had they nothing to say about the way in which things are interlinked, both through causal relations and semantic structures. These reserves of knowledge should be critically exploited by the researcher in accord with his/her skill, theoretical frame of reference, and the point of view selected. The body-blow directed against positivism by the notion of the theory-ladenness of observations thus turns into a *resource* for directing the systematization of the material.

UNIVERSITY
OF BRISTOL
LIBRARY

The various sources of knowledge, the expert and commonsense knowledge of the research objects and of the investigator him/herself, knowledge from previous studies, and new observations, may all function as criteria of knowledge claims as well. This is acceptable neither to the empiricist who remains prisoner of the theory-observations dichotomy, nor to the scientific realist who uses the *inference to the best explanation* as the criterion of truth while identifying explanation with *explanation of observable phenomena*.[46] The empiricist tendencies run counter to the actual practice of human research which abounds with explanations of various kinds. It is, for instance, fairly common to search support for one's theory by showing the restrictions of previous theories, or argue for the validity of one's description by locating one's sources in their context, such as, assessing the limits of the experiences of the interviewees or of the interview situation itself. Such argument forms exceed the empiricist conception of knowledge, since observations no longer form the sole touchstone of knowledge claims. What is needed is an epistemology which allows for an appropriate extension of the notion of explanatory power and thus takes seriously the implications of the theory-ladenness of observations.

In as much as we are interested in the historical roots and justification of this kind of epistemology it would be wise to turn to Aristotle. This should be no surprise in itself, since his epistemological and methodological comments are based on keen insights into his own studies beside the other contemporary forms of communication. One evident point of reference in this respect is the notion of *endoxa* mentioned above, since *endoxa* for Aristotle are exactly the sort of combination of sources which forms the point of departure for many a social scientist today. *Endoxa* contain both commonsense opinions and knowledge obtained by research, all in a pretty controversial jumble, which a realistically minded researcher should make some sense of. It contains the conceptions of the general public and competing schools, as well as the expert views. Accordingly, in several social studies the fundamentals of research and argument are to be classified in the way we meet with in the Aristotelian conception of *endoxa*.

As to their reliability, it is essential that none of the components of *endoxa* be *a priori* superior to the others. All of them contain theoretical elements, and theory can test expert knowledge, just as the expert can challenge theory, which involves arguing in a circular manner. In consequence theory should not be understood in the positivist manner as systematization of observations, but rather as the conceptual framework for grasping reality. Does Aristotle offer his helping hand at this point? A thorough answer would require a detailed investigation of the practices of present-day social research and their comparison with those of Aristotle. Here I can at best offer some guidelines for such an investigation.

To start with we need some further specifications of the stages of the research process called *saving the phenomena*. As the title indicates, the assumptions of argument are not confined to *endoxa*. They also include another, partly overlapping set of beliefs, called *appearances (phainomena)* by Aristotle. The field of appearances is more extensive than that of *endoxa*, since it covers new observations made by the investigator him/herself, which have not yet reached other researchers or the wider audience. The set of appearances is more restricted on the other hand, because the paradoxical philosophical speculations may not be counted among them.[47]

The various stages on the way to saving the appearances is clearly outlined in the famous methodological passage from the beginning of book VII of the *Nicomachean Ethics* where the theme is weakness of will:

> We must, as in all other cases, set the appearances before us and, after first discussing the difficulties, go on to prove, if possible, the truth of all the reputable opinions about these affections or, failing this, of the greater number and the most authoritative; for if we both resolve the difficulties and leave the reputable opinions undisturbed, we shall have proved the case sufficiently. (1145b2–7)

The investigator has, accordingly, first to gather the appearances relevant to the question under investigation, then seek the problems included in them. After this his/her task is to produce solutions such that at least most or the most important of the prevalent views can be shown to be true or partly true. If successful, this kind of research program leads to a unified theory of the field, constructing a synthesis of the mutually contradictory and complementary given conceptions. This also imposes its own requirements on the justification of the results produced: it is not sufficient to justify them on the basis of observations merely, but one should also rely on the prevalent views. One has thus to appeal to the various sources of knowledge, and it is equally important to specify what is correct and what is false in the given views.[48]

We can hardly say that the Aristotelian model of saving the appearances is valid in today's social research as such. Aristotle was a Kuhnian in the sense that for him a mature science is characterized by one paradigm. The modern researcher on human affairs has to be more modest, since most areas are characterized by a host of paradigms, implying that instead of one only, one has to choose between different theoretical frameworks. A researcher may at best aim at saving the relevant appearances from his/her own theoretical standpoint which results to a *partial* saving of the appearances. In empirical research, one typical situation seems to be that previous theories offer conceptual and methodological principles which are applied to new sub-areas to yield results concerning each and, perhaps, to improve the theoretical results at a more general level. Or, as in our example of investigating the prospects for trade in consumer products, the result has rather the form of *systematic description* of the interview material than abstract theoretical knowledge claims. We may speak about saving the appearances also in this case, even though the range of appearances may be more limited than in connection with theory-generation.

What is the exact nature of argumentation within modern projects of saving the appearances? I am convinced that an inquiry into this problem-area would not only have philosophical interest, but would be most useful to the human and social scientist as well. The numerous methodological ideas developed by Aristotle in his own investigations and comments on his method of saving the appearances would provide that kind of investigation with an abundant source of insights to draw from.

Helsinki School of Economics

NOTES

1 As a vindication of the liveliness of Aristotle's social philosophy one may mention the vivid discussion about virtue ethics raised by Alasdair MacIntyre (1981). Similar traces of Aristotelian social thinking may be found in Amartya Sen's criticism of the utilitarian standpoint of economics. He has suggested that the concept of capability should substitute that of utility, a notion which goes back to Aristotle, as has been shown by Martha Nussbaum (1993). Sen (1993) includes as extensive bibliography on the topic. See Seyla Benhabib (1989) for a detailed presentation of the recent interest in Aristotle's ethical and political writings.

2 Aristotle, *Rhetoric* I 2, 1356a1–20.

3 The Aristotelian notions of *topos, logos*, *ethos* and *pathos* has been successfully applied to studying the rhetoric of science by Ricca Edmondson (1984) and Lawrence J. Prelli (1994).

4 Aristotle, *Rhetoric* I 3, 1358b8–20.

5 Kakkuri-Knuuttila (1989).

6 The concept of *topos* has been studied by Wilhelmus A. de Pater (1965) and (1968), Niels Jørgen Green-Pedersen (1984), Richard Sorabji (1972) and Eleonora Stump (1978). The definition above is argued for by Kakkuri-Knuuttila (1993, pp. 33–9).

7 Aristotle, *Rhetoric* I 2, 1358a2–35.

8 Aristotle, *Topics* II 10, 115a7–11.

9 Aristotle, *Rhetoric* II 18, 1393a1–4.

10 Aristotle, *Rhetoric* I 5, 1360b11–13 for the first two, 4, 1360a15–17 for the following, and 7, 1364a24 for the last one. The first two examples resemble the reasoning in practical syllogism, and their position in the theory of rhetoric shows that the so-called practical syllogism is only one model of reflexion and explanation among others, cf. Georg Henrik von Wright (1971).

11 As Otto Bird (1961) and (1962) has remarked.

12 J. Swales (1990) and Ann-Charlotte Lindeberg (1994) discuss the structure of the introduction to a scientific article, which may be taken as a starting-point for an explication of introduction–*topoi*.

13 Heikki Ylikangas (1988), see Kakkuri-Knuuttila (1992). I am only going to consider one type of historical research, which Ylikangas seems to represent. However, I am not claiming that Ylikangas would be the same in the rest of his production.

14 Ylikangas (1988, pp. 21–9).

15 Ylikangas (1988, p. 20), Kakkuri-Knuuttila (1992, p. 28). Edmondson (1984, ch. 5) has discussed a sociological study, and worked out two of the *topoi* used. One she calls *the model of the ordinary person*, and the other *the model of the reasonable person*. Historical explanation which describes the action of a particular individual could be titled as *the model of a particular person*. See also Stephen P. Turner (1980, p. 79).

16 Ylikangas (1988, pp. 30–2).

17 Research in narrative historiography has not dealt with cases where the plot itself has an argumentative structure. For instance, Hayden White (1987, p. 43) states that the "arguments are more properly considered as a commentary on, rather than part of, the narrative." This may often be true, though not always, as shown by our case study.

18 Michael Scriven (1962, p. 176) emphasizes in his criticism of the Hempel–Oppenheim model that the mere *description* of individual incidents sometimes works as an *explanation*. This depends on the *context*. For a brief account of the Scriven criticism, see Matti Sintonen (1984, pp. 8–11).

19 By the enthymematic structure I mean an argument in which not all premises or inference steps have been expressed. This represents in my view the Aristotelian *strategic means* of making an *enthymema*, or a rhetorical syllogism convincing, and forms thus no part of its definition, contrary to the common view. On the *enthymema*, see Aristotle's *Rhetoric* I 2, 1356a15–18; 1357a16–21; II 22, 1395b25–27 and III 18, 1415a18–19; cf. Burnyeat, forthcoming. Cf. Jon Elster (1993, pp. 4–5 and 449) who requires the explication of causal mechanisms, usually only referred to by the historians.

20 Carl G. Hempel and P. Oppenheim (1948) and Hempel (1965). Hempel's demand for general laws in historical explanation is associated with the nomothetic tradition in social science, represented by John Stuart Mill (1965). Mill assumes mathematical physics with its causal laws to be the ideal of social science, which helps to explain complex phenomena caused by several factors.

21 Hempel (1965, pp. 236–8, 240). At the end of his article (p. 75), Ylikangas also discusses the possibility of generalisation, stating that a historian will reach *a more profound understanding* if he/she finds other, similar cases. It might be replied to this that the object of research changes with the attempt at generalisation, and the deeper understanding will no longer apply to the same thing. Different forms of generalisation are discussed, for instance, by Pertti Alasuutari (1995) and Robert K. Yin (1989).

22 Other examples of the *topos* rules used by Ylikangas:

If a text expresses a systematic, corrective solution for an evil he is familiar with, it is the author's intention to correct the wrong (Ylikangas 1988, pp. 44–61).

If a writer considers a certain person as the instigator of a certain wrong, it is the writer's intention to criticize this person (Ylikangas 1988, pp. 42–9).

23 Scriven (1962, pp. 198–9) also establishes in his criticism of Hempel–Oppenheim that an explanation need not have the general law explicitly mentioned.

24 Contrary to what William Dray (1957, pp. 25, 33 ff.) suggests, the covering-law model does not fail for the reason that the law needs to be specified until it covers the one and only instance to be explained. The uniqueness of historical explanations results rather from the particular combination of principles and events the former are applied to.

Ylikangas (1988, pp. 40–1) makes a brief comment on methodical questions, where he criticizes explanations in the history of ideas which merely point out the associations of the text contents with some earlier text. An explanation of this kind is content with one explicative model of the following type:

If a text T includes similarities with an older text T′, then text T′ is the reason of the origin of text T (Kakkuri-Knuuttila 1992, p. 33).

25 Aristotle, *Rhetorics* II 25, 1402a29–1403a15; 26 1403a24–33. Aristotle points out that a new conception is supported by refuting the competing views in scientific argument as well (*Eudemian Ethics* I 3, 1215a5–7; *De Caelo* I 10, 279b6–7; II 1, 283b30–284a2; II 9, 291a6–7).

26 Hempel (1965, pp. 236–8 and 243).

27 The Kakkuri-Knuuttila typescript discusses the Aristotelian forms of communicative operations further. A similar Aristotelian approach shows in innumerable studies by Jaakko Hintikka, which, in spite of their formal-logical emphasis, aim at finding the underlying forms of rationality instead of giving normative instructions.

28 Scriven (1962, pp. 205–8) criticizes the Hempel–Oppelheim explanation model for not taking into account that the explanation is always given to a particular audience, and can thus function as an explanation only if properly related to the public's fore-knowledge. This notion has been developed by Sintonen (1984, p. 92) on the basis of the Hintikka's more general concept of conclusive answers; see Hintikka (1976, p. 41 ff.), also Kakkuri-Knuuttila and Martin Kusch (1991, p. 175).

29 One critical approach to historical and sociological research is offered by the model of Aristotle's ethics, which will be presented in section 3. Another type of research with a critical attitude to common sense thinking is based on insight into the *historical nature* of given ways of thought and action. This means that modes of thinking and acting are created by human beings, contrary to regularities in nature, and thus transitory, and subject to change. This is one of Karl Marx's fundamental insights, and it is repeated in various forms by the constructivist school, for instance.

30 The latter is also discussed by Hempel (1965, pp. 242–3).

31 Kuhn (1970, pp. 199–200).

32 Aristotle, *Topics* I 1, 100b21–23.

33 Aristotle, *Topics* I 10, 104a33–34.

34 Aristotle, *Nicomachean Ethics* VII 1, 1145b2–7; *Metaphysics* III 1, 995a24–29. On saving the phenomena, see Geoffrey E.R. Lloyd (1968), G.E.L. Owen (1968), Jonathan Barnes (1980), Martha Nussbaum (1986, pp. 240–63), Robert Bolton (1990), and Kakkuri-Knuuttila (1993, pp. 165–73).

35 Kakkuri-Knuuttila (1993, pp. 139–64); four other criteria for solving disputes, presented as interpretations of the Aristotelian dialectic, are discussed on pp. 104–38.

36 Aristotle, *Rhetoric* I 2, 1357a4–7.

37 I am not committed to Kuhn's idea that normal science is based on one paradigm, and that the comparison between different paradigms aims at the adoption of one paradigm and abandonment of the other. Most fields in human science are multiparadigmatic, which implies that the Kuhnian crisis is a normal situation. Thus it seems obvious that the differing paradigms are mutually complementary. This again demands the ability to judge the appropriateness of different paradigms for the actual research problem, which in Kuhn's view demands a philosophical approach. Kakkuri-Knuuttila, forthcoming.

38 These ideas are developed in more detail in Kakkuri-Knuuttila, typescript.

39 On foundationalism see William P. Alston (1989, pp. 11–2), Jonathan Dancy (1985, pp. 53–4), and Keith Lehrer (1990, p. 13).

40 Andrew Sayer (1993, p. 55).

41 Pierre Duhem (1954), N.R. Hanson (1958) and I. Scheffler (1967, ch. 2).

42 J.L. Aronson (1984, pp. 135–6).

43 Sayer (1993, pp. 50–1).

44 Sayer (1993, p. 116) points out without making any further epistemological inferences that the actors already have some knowledge of the causal mechanisms.

45 I owe this example to my colleague Niilo Home (1995).

46 Peter Lipton (1991).

47 Robert Bolton (1987, pp. 121–30) and (1991, p. 11).

48 For a review of Aristotle's forms of justification see Kakkuri-Knuuttila (1993, pp. 260–8).

REFERENCES

Alasuutari, Pertti (1995). *Researching Culture: Qualitative Method and Cultural Studies*. London: Sage.

Alston, William P. (1989). *Epistemic Justification: Essays in the Theory of Knowledge*. Ithaca/London: Cornell University Press.

Aristotle (1984). *The Complete Works of Aristotle: The Revised Oxford Translation I and II,* Jonathan Barnes (Ed.). Princeton: Princeton University Press.

Aronson, J.L. (1984). *A Realist Philosophy of Science*. London/Basingstoke: Macmillan.

Barnes, Jonathan (1980). Aristotle and the Methods of Ethics. *Revue Internationale de Philosophie* **133–4**, 490–511.

Benhabib, Seyla (1989). In the Shadow of Aristotle and Hegel. *Philosophical Forum* **21**, 1–31.

Bird, Otto (1961). The Re-discovery of the Topics. *Mind* **70**, 534–9.

Bird, Otto (1962). The Tradition of the Logical Topics: Aristotle to Ockham. *Journal of the History of Ideas* **23**, 307–23.

Bolton, Robert (1987). Definition and scientific method in Aristotle's *Posterior Analytics* and *Generation of Animals*. In: *Philosophical Issues in Aristotle's Biology*. Cambridge: Cambridge University Press, pp. 120–66.

Bolton, Robert (1990). The Epistemological Basis of Aristotelian Dialectic. In: Daniel Devereux and Pierre Pellegrin (Eds.) (1990), pp. 185–236.

Bolton, Robert (1991). Aristotle's Method in Natural Science: *Physics* I. In: Lindsay Judson (Ed.) (1991), pp. 1–29.

Burnyeat, M.F. (typescript). Enthymeme: Aristotle on the Logic of Persuasion.

Chalmers, A.F. (1986). *What is this Thing called Science*. Philadelphia: Open University Press/Milton Keynes.

Clay, Marjorie (Ed.) (1989). *Knowledge and Skepticism*. Boulder: Westview Press.

Dancy, Jonathan (1985). *An Introduction to Contemporary Epistemology*. Buckinghamshire: Blackwell.

De Pater, Wilhelmus A. (1965). *Les Topiques d'Aristote et la dialectique Platonicienne: La methodologie de le définition*. (Thomistische Studien 10) Fribourg Suisse: St. Paul.

De Pater, Wilhelmus A. (1968). La Fonction du lieu et de l'instrument dans les *Topiques*. In: G.E.L. Owen (Ed.) (1968), pp. 164–88.

Devereux, Daniel and Pierre Pellegrin (Eds.) (1990). *Biologie, Logique et Metaphysique chez Aristote*. Paris: Éditions du CNRS.

Dray, William (1957). *Laws and Explanation in History*. Oxford: Oxford University Press.

Duhem, Pierre (1954). *The Aim and Structure of Physical Theory*. Princeton: Princeton University Press. (1914)

Edmondson, Ricca (1984). *Rhetoric in Sociology*. London/Bassingstoke: Macmillan.

Elster, Jon (1993). *Political Psychology*. Cambridge: Cambridge University Press.

Feigl, Herbert, Michael Scriven and Grover Maxwell (Eds.) (1958). *Concepts, Theories, and the Mind–Body Problem, Minnesota Studies in the Philosophy of Science II*. Minneapolis: University of Minnesota Press.

Feigl, Herbert and Grover Maxwell (Eds.) (1962). *Scientific Explanation, Space, and Time, Minnesota Studies in the Philosophy of Science, Vol. III*. Minneapolis: University of Minnesota Press.

Gotthelf, Allan and James G. Lennox (Eds.) (1987). *Philosophical Issues in Aristotle's Biology*. Cambridge: Cambridge University Press.

Green-Pedersen, Niels Jørgen (1984). *The Tradition of the Topics in the Middle Ages: The Commentaries on Aristotle's and Boethius' 'Topics'*. (Philosophia) München/Wien.

Hanson, N.R. (1958). *Patterns of Discovery: An Inquiry into the Conceptual Foundations of Science*. Cambridge: Cambridge University Press.

Hempel, C.G. (1965a). *Aspects of Scientific Explanation and Other Essays in the Philosophy of Science*. New York: The Free Press.

Hempel, C.G. (1965). The Function of General Laws in History. In: C.G. Hempel (1965a), pp. 231–43.

Hempel, C.G. and Oppenheim, P. (1948). Studies in the Logic of Explanation. *Philosophy of Science* **15**, 135–75.

Hintikka, Jaakko (1976). *The Semantics of Questions and the Questions of Semantics: Case Studies in the Interrelations of Logic, Semantics, and Syntax, (Acta Philosophica Fennica Vol. XXVIII)*. Amsterdam: North Holland Publishing Company.

Hintikka, Jaakko (1989). Knowledge Representation and the Interrogative Model of Inquiry. In: Marjorie Clay (Ed.) (1989), pp. 155–83.

Home, Niilo (1995). *Kauppa 2000 – EU-jäsenyyden vaikutus päivittäistavarakaupan rakenteeseen ja toimintaan* (Commerce in 2000 – The influence of EU-membership to the structure and function of commerce of daily goods – in Finnish). Helsinki: Helsinki School of Economics D-223.

Judson, Lindsay (Ed.) (1991). *Aristotle's 'Physics': A Collection of Essays*. Oxford: Clarendon Press.

Kakkuri-Knuuttila, Marja-Liisa (1989). The *Topos*-Structure of Deliberative Newspaper Editorials. Jussi Niemi (Ed.) (1989)

Kakkuri-Knuuttila, Marja-Liisa (1992). *PAM – Pienten askelten selitysmalli yhteiskuntatieteissä* (Stepwise Model of Explanation for Human Sciences, in Finnish). Helsinki: Helsingin kauppakorkeakoulun julkaisuja D-166 (Reports from the Helsinki School of Economics D-166).

Kakkuri-Knuuttila, Marja-Liisa (1993). *Dialectic and Inquiry in Aristotle*. Helsinki: Helsinki School of Economics.

Kakkuri-Knuuttila, Marja-Liisa (forthcoming a). Miksi tarvitaan tieteellisiä kliseitä? (Why do we need clichés in science? – in Finnish)

Kakkuri-Knuuttila, Marja-Liisa (forthcoming b). Mitä tieteenfilosofit sanovat argumentaatiosta? (What do the philosophers of science say about argumentation – in Finnish)

Kakkuri-Knuuttila, Marja-Liisa (typescript). Aristotelian Epistemology for Post-positivists.

Kakkuri-Knuuttila, Marja-Liisa and Martin Kusch (1991). LSP-Research, Philosophy of Science, and the Question-Theoretical Approach – some Tentative Suggestions. In: Hartmut Schröder (Ed.) (1991), pp. 167–98.

Kitcher, Philip and Wesley C. Salmon (Eds.) (1989). *Scientific Explanation, Minnesota Studies in the Philosophy of Science*, Vol. XIII. Minneapolis: University of Minnesota Press.

Kuhn, Thomas S. (1970). *The Structure of Scientific Revolutions*. (International Encyclopedia of Unified Science, Vol. 2, Number 2). Chicago: The University of Chicago Press. (1962) (2. enlarged edition)

Lehrer, Keith (1974). *Knowledge*. Oxford: Clarendon Press.

Lindeberg, Ann-Charlotte (1994). Rhetorical Conventions in Scholarly Articles in Economics and Business Sciences: A Study of Introductions with a Special Reference to Knowledge Claims. F. Danes (Ed.), *Writing vs Speaking: Language, Text, Discourse, Communication, (Proceedings of the Conference held at the Czech Language Institute of the Academy of Sciences of the Czech Republic, Prague, October 14–16, 1992)*. Tübingen: Gunter Narr, 317–24.

Lipton, Peter (1991). *Inference to the Best Explanation*. London/New York: Routledge.

Lloyd, G.E.R. (1968). *Aristotle: The Growth and Structure of his Thought*. Cambridge: Cambridge University Press.

MacIntyre, Alasdair (1981). *After Virtue: A Study in Moral Theory*. Notre Dame: University of Notre Dame Press.

Mill, John Stuart (1965). *On the Logic of the Moral Sciences (A System of Logic, Book VI)*. Indianapolis: Bobbs-Merrill. (1843)

Niemi, Jussi (Ed.) (1989). *Papers from the Eleventh Scandinavian Conference of Linguistics, Vol. 2*. (Studies in Languages 134). Joensuu.

Nussbaum, Martha Craven (1986). *The Fragility of Goodness: Luck and Ethics in Greek Tragedy and Philosophy*. Cambridge: Cambridge University Press.

Nussbaum, Martha C. (1993). Non-Relative Virtues: An Aristotelian Approach. In: Martha C. Nussbaum and Amartya Sen (Eds.) (1993), pp. 242–69.

Nussbaum, Martha C. and Amartya Sen (Eds.) (1993). *The Quality of Life*. (Wider Studies in Development Economics). Oxford: Clarendon Press.

Owen, G.E.L. (Ed.) (1968). *Aristotle on Dialectic: The Topics*. (Proceedings of the Third Symposium Aristotelicum). Oxford: Clarendon Press.

Owen, G.E.L. (1986a). *Logic, Science and Dialectic. Collected Papers in Greek Philosophy*. Martha Craven Nussbaum (Ed.). London: Duckworth.

Owen, G.E.L. (1986). Tithenai ta Phainomena. In: G.E.L. Owen (1986a), pp. 239–51. (1961)

Polanyi, Michael (1969). *Knowledge and Being*. Marjorie Grene (Ed.). Chicago: The University of Chicago Press.

Pollock, John L. (1986). *Contemporary Theories of Knowledge*. Totowa, New Jersey: Rowman & Littlefield.

Prelli, Lawrence J. (1994). *A Rhetoric of Science: Inventing Scientific Discourse*. (Studies in Rhetoric-Communication). Carroll C. Arnold (Ed.). University of South Carolina Press. (1987)

Sayer, Andrew (1993). *Method in Social Science: A Realist Approach*. London/New York: Routledge. (1984) (2. edition)

Scheffler, I. (1967). *Science and Subjectivity*. Indianapolis: The Bobbs-Merrill Co.

Scriven, Michael (1958). Definitions, Explanations, and Theories. In: Herbert Feigl, Michael Scriven and Grover Maxwell (Eds.) (1958), pp. 99–195.

Scriven, Michael (1962). Explanations, Predictions and Laws. In: Herbert Feigl and Grover Maxwell (Eds.) (1962), pp. 170–230.

Sen, Amartya (1993). Capability and Well-Being. In: Martha C. Nussbaum and Amartya Sen (Eds.) (1993), pp. 30–53.

Sintonen, Matti (1984). *The Pragmatics of Scientific Explanation. Acta Philosophica Fennica* vol. **37**.

Sorabji, Richard (1972). *Aristotle on Memory*. London: Duckworth.

Stump, Eleonore (1978). *Boethius's De topicis differentiis*. Translated, with notes and essays on the text. Ithaca/London: Cornell University Press.

Swales, J. (1990). *Genre Analysis: English in Academic and Research Settings*. Cambridge: Cambridge University Press.

Turner, Stephen P. (1980). *Sociological Explanation as Translation*. Cambridge: Cambridge University Press.

White, Hayden (1987). *The Content of the Form: Narrative Discourse and Historical Representation*. Baltimore/London: The Johns Hopkins University Press.

von Wright, Georg Henrik (1971). *Explanation and Understanding*. London: Routledge & Kegan Paul.

Yin, Robert K. (1989). *Case Study Research: Design and Methods*. (Applied Social Research Method Series Volume 5). Newbury Park: Sage.

Ylikangas, Heikki (1988). Tuomari Olaus Petrin Vaa'assa (The Judge assessed by Olaus Petri, in Finnish). In: *Valta ja väkivalta keski- ja uudenajan taitteen Suomessa* (Power and Violence at the Turn of the Medieval and Modern Times in Finland). Porvoo/Helsinki/Juva: WSOY.

Poznań Studies in the Philosophy of the Sciences and the Humanities
1997, Vol. 51, pp. 41–60

Martin Kusch

THEORIES OF QUESTIONS IN GERMAN-SPEAKING PHILOSOPHY AROUND THE TURN OF THE CENTURY

1. Introduction

Since about the mid-seventies two of Jaakko Hintikka's major philosophical concerns have been the study of the semantics of questions, and the development of an interrogative model of inquiry (e.g. Hintikka 1974, 1976, 1981, 1984, 1987). His work in these interrelated areas has been used by others in various endeavours, e.g. in rational reconstructions of the arguments of Newton's *Optics* or Darwin's *Origin of Species*, or in the investigation of the process of text interpretation (e.g. Garrison 1988; Sintonen 1990; Kusch & Schröder 1989; Kakkuri-Knuuttila & Kusch 1991). Up to this point, little work has been done, however, on the *history* of philosophical question theory. This is all the more surprising given that Hintikka's systematic work, say on dialogue-games or modalities, has often led to important historical studies, either by Hintikka himself, or else by his many students. This paper constitutes a modest attempt to partially remedy this situation. It seeks to save from oblivion one noteworthy episode in the history of philosophical question theory.

Between, roughly, 1880 and the First World War, German-speaking philosophers displayed a lively interest in the nature of questions. The list of contributors to question theory in this period – Hermann Cohen, Benno Erdmann, Edmund Husserl, Theodor Lipps, Alexius Meinong, Paul Natorp, Heinrich Rickert, and Christoph Sigwart, amongst others – reads like a *Who's Who?* of German-speaking philosophy at the turn of the century. All the more surprising then that, with the exception of an important paper by Schuhmann and Smith on Husserl and Daubert (Schuhmann and Smith 1987), historians of philosophy have paid little attention to this literature. Within the limits of a journal article, I can only provide something of a bird's-eye-view of the various contributions. I shall summarise the main ideas and disputes, and conclude by making some suggestions on why philosophers at the time regarded the logic and psychology of questions as an important issue.

2. *Questions and Judgements*

One of the key problems for any theory of questions is to decide whether questions can somehow be interpreted as variants of statements or requests. In the literature considered here, this issue was addressed by considering how questions related to "judgements" (*Urteile*). In his influential *Logik* of 1880, Hermann Lotze had argued that "being a question" was one of three types of *quality* that a judgement could have – the others being affirmation and negation. Lotze presented this triad as an improvement on Kant's three-partite distinction between affirmative, negative, and infinite judgements. Affirmative and negative judgements were constructed by Lotze as combinations of two thoughts: "the thought which is still open to question"

[1] (that) S is P

and the "supplementary thoughts" of "validity" (*Gültigkeit*) and "reality" (*Wirklichkeit*) on the one hand, and their denials on the other hand:

[2] It is valid [that S is P]
[3] It is not valid [that S is P]

Lotze suggested that the content of [1] was most naturally expressed as a question (1880, p. 61). What guided him to this idea must have been the – not self-evident – assumptions: first, that [1] is roughly equivalent to [4]

[4] It is as yet unknown whether it is (in)valid [that S is P];

second, that questions contain neither affirmation nor negation; and third, that questions characterise propositional contents as indeterminate with respect to validity and reality.

A number of philosophers held similar views. Hermann Cohen must have thought that the equation of questions with judgements needed no further argument. Cohen simply stated for a fact that a question "is a kind of judgement" (Cohen 1902, p. 69). Edmund Husserl's position was close to Bernard Bolzano's theory. Bolzano proposed that questions asserted the questioner's desire to know, and that questions could be true or false: a question was true if it truthfully expressed the questioner's desire to know, and false if it did not (Bolzano 1837, p. 88). For Husserl question-sentences got their meaning from inner perceptions of one's mental states of desiring knowledge, and thus were but disguised variations of statements or assertions (Husserl 1901, §§67–70). And according to Friedrich Jodl questions were "twofold judgements" (*Doppelurteile*): they combined presentations – this being the defining feature of judgements – and they characterised the questioner's state of mind as one involved in inquiry (1903, p. 197).

Other philosophers disagreed with the idea that *all* questions are judgements of sorts. This opposite conception could take two forms. One could replace the "all" with a "some", or else deny that any questions are judgements. Although the evidence of their own writings is somewhat inconclusive on this point, Karl Groos and Christoph Sigwart were taken to task for allegedly claiming that one of the two major kinds of questions were judgements. Alexius Meinong attributed this view to Groos for the case of propositional questions, while Josef Kreibig criticised Sigwart for reducing wh-questions to judgements.

Meinong felt justified in his attribution because Groos called propositional questions "questions with a judgement-core", and because Groos spoke of such questions as expressing "conjectures" (*Vermuthungen*) (Groos 1901, p. 149). In other words, for Groos all propositional questions of the form [5] had the meaning of [6]

[5] Is S P?
[6] I *suspect/conjecture* that S is P.

Meinong argued against this interpretation of propositional questions in the following way: "... the questioner as such does not form a judgement with respect to the issue he is asking about; his question aims to enable him to form a judgement on that very issue with respect to which the questioner at the time of his question is *unable* to judge". Although Meinong insisted that questions never "expressed" judgements, he did allow that they had judgements "as their essential presuppositions" (Meinong 1902, p. 122). In a reply published in the same year as Meinong's criticism, Groos conceded Meinong's point but went on to propose a new analysis of propositional questions that also contradicted Meinong's views (Groos 1902, p. 363). (I shall return to this issue later.)

Turning from Groos and propositional questions to Sigwart and wh-questions, Sigwart wrote that wh-questions "contained" judgements, e.g. the question [7] contained the judgement [8]

[7] "Which plants have sense organs?"
[8] "There are plants with sense organs" (Sigwart 1893, p. 303).

To read Sigwart's "contain" as an "is" was perhaps an uncharitable substitution by his critic, Josef Kreibig. Kreibig emphasised that the act of judging [8] is prior to the question [7], and thus not identical with it: "To ask a question is not an act of knowing; it presupposes an act of knowing as a psychological *prius* ..." (1914/15, p. 181; for a similar, earlier view, see Wahle 1890, p. 319). To treat wh-questions as judgements was also unacceptable for Kreibig because it created a strange opposition between wh-questions and propositional questions. In Kreibig's opinion it was "impossible" to treat propositional questions as judgements of sorts (1914/15, p. 181).

As the last few paragraphs already suggested, Kreibig and Meinong were united in their insistence that questions could not be construed as judgements. They also agreed that Lotze had been wrong in claiming that questions contained neither affirmation nor negation. In their opinion the opposition between sentences like [9] and [10]

[9] Will you come tomorrow?
[10] Will you not come tomorrow?

showed that Lotze had been wrong on this issue (Meinong 1902, p. 122; Kreibig 1914/15, p. 153).

Authors who took a similar stand against the Lotze–Bolzano view of questions were Wilhelm Jerusalem ("the question is a formulated wondering, it is no judgement, but the desire ... to form or complete a judgement"; 1895, p. 172), Theodor Lipps (a question is "an unfinished ... not a genuine judgement"; 1897, p. 31), Ed. Martinak (questions are "incomplete judgements"; 1906, p. 334); and Benno Erdmann (questions differ from judgements in that they lack "objective certainty"; 1907, p. 391). Erdmann's further suggestion to call both questions and judgements "assertions" (*Aussagen*, 1907, p. 1) was, however, criticised by Kreibig as an unnecessary revision of ordinary language (1914/15, p. 196).

Of writers who treated questions as falling under the genus of requests and desires, only Meinong made an attempt to characterise the *differentia specifica* of questions. Questions differed from other requests insofar as the object of the request was not directly expressed. The difference between [11] and [12] brings this out clearly:

[11] I desire/request that you come tomorrow

[12] When will you come?

In [11] the object of the request, i.e. your coming tomorrow, is, as it were, visible on the surface structure of the sentence. In [121, however, the object of the request, i.e. the questioner's *knowing* the time of your coming, is not so apparent.

Meinong (1902, p. 124) also found that whereas the object of other desires can usually be expressed as a that-clause, the object of a desire to know must be expressed as a whether-clause; cf. [11] and [13]:

[13] I desire to know whether you will come.

3. Taxonomies of Questions

Turn-of-the-century German philosophers paid much attention to the distinction between "wh-questions" and "propositional questions", although they used different terms for the two elements of the dichotomy. The various attempts to draw this distinction in precise terms neatly exemplified the conceptual tools of different philosophical schools at the time.

One early solution was put forward by Groos (1901), a philosopher with a strong leaning towards experimental psychology. Groos sought to separate the two types of questions as different stages in the psychological process of knowledge-acquisition (*Erkenntnis*). The first stage was a state of wondering. The second phase was "the desire for a *specific kind* of logical relation ... without the concrete solution, the specific contentful relation, having as yet emerged." And, finally, the third stage was "the first, still uncertain, emerging of a solution in the form of a conjecture (*Vermuthung*)" (1901, p. 148). Wh-questions – called "empty questions" by Groos – belonged to the second phase,' whereas propositional questions – "conjecture-questions", or "questions with a judgement-core" – expressed the third stage (1901, p. 149).

Groos' proposal was soon rejected by Meinong (1902). In order to understand his position, we need to note his three-partite distinction between judgements, presentations (*Vorstellungen*), and assumptions (*Annahme*). Judgements fulfilled the criteria of first, being believed and known, and second, of being either affirming or denying. Presentations met neither condition. Assumptions fell between the two categories: they shared with judgements the feature of being either positive or negative, but they behaved like presentations in not being believed or known. Finally, what was assumed in assumptions, and believed and known in judgements were, according to Meinong, *"Objektive"*, i.e. states of affairs. Linguistically they were expressed by that-clauses.

On this basis, Meinong (1902, p. 125), and following him his student Ed. Martinak (1906, p. 333), suggested the following characterisation of the propositional question, or the "decision question" in their terminology. A propositional question consisted of two parts: the assumption of a complete *Objektiv*, and the request to the respondent to take a definite stand with respect to this same *Objektiv*. Thus the propositional questions [14] and [15] were composed not only of the assumptions [16] and [17] (respectively) but also of the requests to the addressee to replace the assumption with a positive or negative judgement (and with respect to the same *Objektiv*) [18] and [19], respectively:

[14] Will he come tomorrow?
[15] Will he not come tomorrow?
[16] I assume (but do not believe or know) that you will come tomorrow.
[17] I assume (but do not believe or know) that you will not come tomorrow.
[18] Tell me whether you will come tomorrow.
[19] Tell me whether you will not come tomorrow.

The analysis for wh-questions, i.e. "supplement-questions", went differently. In uttering a supplement-question one did not form an assumption concerning a complete *Objektiv*. Instead, one formed a judgement concerning an incomplete *Objektiv*. Thus the psychological process underlying the supplement question [20] consisted of the judgement [21] and the request to the hearer to tell the questioner who the X in question was (Meinong 1902, p. 125; Martinak 1906, pp. 333–4):

[20] Who is coming?
[21] I judge [believe & know] that X is coming.

Given this theory, it no longer comes as a surprise that Meinong rejected Groos' equation of decision-questions with conjectures. Conjectures were judgements rather than *Annahmen*, and although decision-questions were based on *Annahmen*, they could not possibly be equated with judgements (1902, p. 122).

In his reply to Meinong's criticism, Groos agreed that questions could not be reduced to judgements, but insisted that "within" a decision-question one could always find "a problematic judgement". In other words, every question of the (logical) form [22] *contained* a judgement of the form

[22] Is it the case that p?
[23] It is possible that p (Groos 1902, p. 363).

Groos now conceded having been wrong to suggest earlier that all propositional questions could adequately be characterised as conjectures. For instance, the sentence

[24] "Did the stone drop onto the street or did it remain on the church-roof?"

did not express a conjecture, although each of the two clauses "contained" a problematic judgement (1902, p. 364).

Despite these concessions, Groos rejected Meinongian *Annahmen* which he – mistakenly – saw as no more as a synonym for his own "conjectures" (1902, p. 363).

Kreibig linked the distinction between the two types of questions to syntactical features, and spoke of *"W-Fragen"* and *"Ob-Fragen"*, respectively. This referred to the fact that supplement-questions contained interrogative pronouns or adverbs starting with a 'w', whereas decision-questions could be rephrased into the form

[25] Ich frage, *ob* (nicht) p (1914/15, pp. 176–77).

(In passing we might note that this surface criterion works better in German than in English, since although questions with "whether" (*ob*) are wh-questions in terms of their surface structure, they are of course, logically speaking, propositional (nexus-) questions.)

Kreibig's view on the logical difference between decision-questions and supplement-questions differed from both Groos' and Meinong's construals. Kreibig did not accept *Annahmen* as an irreducible category, calling them "presentations of judgements" (*Vorstellungen von Urteilen*) (1914/15, p. 180). They figure in his view as implications of questions about irreal objects. For instance, when one asked a question like

[26] "Is the existence of a *perpetuum mobile* dependent upon the non-existence of friction?"

one did not commit oneself to the judgement that a *perpetuum mobile* exists. All one committed oneself to was the existence of the presentation of such judgement.

In line with this more austere ontology, Kreibig drew the distinction

UNIVERSITY
OF BRISTOL
LIBRARY

between decision-questions and supplement-questions in terms of the completeness or incompleteness of the "matter of the answer-judgement" (*Materie des Antworturteils*). The matter of a judgement contrasted here with the act of judging, and it consisted of the "logical subject, the logical predicate and the logical copula." All of these were in turn presentations (1914/15, pp. 156, 177). We also need to note the meaning of Kreibig's technical term "anticipation of a question": it was "the set of presentations and judgements which the question-thought prescribes for the content of the aimed-for judgement or conclusion." Now, in the case of a decision-question, the question contained "as anticipation the whole matter of the answer-judgement except for affirmation or denial." In the case of supplement-questions, the question presented "as anticipation only a part of the matter of the answer-judgement while including affirmation or denial" (1914/15, pp. 177–78).

Finally, it also deserves to be mentioned that both Kreibig and Martinak conceived of a bridge between decision-questions and supplement-questions. This bridge was formed by the idea that supplement-questions like [27] could be rephrased as disjunctive decision-questions [28]

[27] Who is coming?
[28] Is John coming, or is Peter coming, or is David coming, or is Colin coming or ...? (Kreibig 1914/15, p. 184; Martinak 1906, p. 336).

Although the dichotomy of propositional and wh-questions was the most often treated distinction, some authors also introduced and discussed other taxonomies for questions.

For instance, Wilhelm Jerusalem (1895, pp. 176–80) proposed a classification of questions in terms of what we today would call "pragmatic" criteria. He first divided judgements into four groups, and then asked to which types of questions these judgements typically gave rise. Thus "judgements about perceptions" [29] gave rise to "questions about elements of judgements" [30], "judgements about remembered events" [31] triggered "truth-questions" [32], "judgements about expectations" [33] led to "justification-questions" [34] and "causal questions" [35], and "judgements about concepts" [36] were followed by "clarification-questions" [37]:

[29] I see a fire.	[30] Where do you see it?
[31] I saw a fire yesterday.	[32] Did you really?
[33] I think she will come tomorrow.	[34] What made you think that?
	[35] Why would she?
[36] A human being is a rational animal.	[37] What is a 'rational animal'?

Kreibig (1914/15, p. 173) felt that earlier treatments had overlooked what he regarded as the most important distinction: the opposition "judgement-questions" versus "conclusion-questions." Judgement-questions concerned "the objective existence of states of affairs", e.g. [38], whereas "conclusion-questions" aimed for an evaluation of what follows from given premises, e.g. [39]:

[38] Who was here?
[39] Does $a > c$ follow from $a > b$ and $b > c$?

Although Kreibig insisted that this distinction was crucial for both psychological and logical reasons, he nowhere argued his case in detail.

Martinak distinguished between "normal" and "unnatural" questions. The first category referred to cases where the questioners did not already know the answer, and where they really wanted to get an answer from their interlocutor. "Unnatural questions" covered the area of "examination questions" (where the questioner already knows the answer), and "oratory questions" (rhetorical questions), amongst others (Martinak 1906, p. 336).

In this classificatory scheme the status of "suggestive questions" (*Suggestionsfragen*) remained somewhat unclear, and this despite the fact that suggestive questions received considerable attention from several writers. Kreibig (1914/15, p. 175) defined "suggestion" as "the producing of an experience by triggering the idea (*Vorstellung*) of this experience." Accordingly he thought that suggestive questions were those questions "which facilitate the forming of a judgement or a conclusion of a given content by transmitting the idea of an answer with that very same content." Realising that – in their pragmatic context – most questions are at least weakly suggestive by this criterion, Kreibig restricted "real suggestion" to those cases "where the anticipation tries to cancel out the impartiality of the answer by means of tricks or surprises" (1914/15, p. 176).

Martinak submitted that supplementary questions were less suggestive than decision question, and that of the two basic forms of decision questions – positive and negative – the negative form [40] suggested a positive answer more strongly than the positive form [41] proffered a negative answer (1906, p. 335):

[40] Isn't it the case that p?
[41] Is it the case that p?

In order to pinpoint further types of suggestive questions, Martinak identified cases where the suggestion was due to "the body of the question", i.è. the meaning of the linguistic structure, and cases where it was due to "the tone of the question, i.e. the intonation. Thus a stress on the word 'you' in [14] suggested a negative answer, whereas the body of the question – i.e. the question read with normal intonation – did not (1906, p. 335).

Martinak also noticed the importance of the concept of "presupposition" in analysing suggestive questions. He rightly noticed that [42] is a presupposition of [43] and that [44] is a presupposition of [45]:

[42]	"You have done it."	[43]	"Why did you do it?"
[44]	"There are buttons on his coat."	[45]	"How many buttons are there on his coat?"

Martinak defined the "presupposition of a question" as "all that which the person asked may read off the body (not the tone) of the question; i.e. all that which he can, may, and must treat as knowledge held by the questioner, as facts of which the questioner is aware" (1906, p. 335). Martinak held that only supplement-questions had presuppositions in this sense. His surprising view in this matter derived from his adoption of Meinong's analysis of decision-questions: only judgements could have presuppositions, *Annahmen* could not. Decision-questions were based on *Annahmen*, and, *ergo*, they had no presuppositions (1906, pp. 335–36).

4. Kreibig's "Beiträge"

With the exception of an unpublished manuscript by Johannes Daubert (discussed at length in Schuhmann and Smith 1987), the most detailed account of the "psychology and logic" of questions was Josef Kreibig's "Beiträge zur Psychologie und Logik der Frage" (1914/15). Several of his central ideas deserve closer scrutiny.

The first respect in which Kreibig broke new ground was in his insistence that logical and psychological studies of questions ought to be kept separate. He gave the following psychological and logical definitions of questions:

> From the perspective of psychology, asking a question is a *volition* such that this volition aims at reaching a judgement or conclusion about a partially specified subject matter. ... For logic the question is a *sentence*. This sentence expresses that the will aims for one of the following two: either a valid assertion about a specified state of affairs or else the valid sequence of the parts of an inference (1914/15, p. 155).

To put it another way, looked at from the psychological perspective, a question is a mental "act", whereas for the logician a question is a linguistic structure. Kreibig's distinction here was in line with his view of judgements. The psychologist treated judgements as "psychological acts", whereas the logician dealt with a judgement "as a sentence by means of which a specific state of affairs is expressed as existing objectively" (1914/15, p. 155).

Kreibig paid more attention to the psychology of questions than to their logic. He divided the process of asking a question into three stages, and emphasised that from the psychological viewpoint the process of asking a question had both an emotional and an intellectual side.

The first stage was a *"Bewußtseinslage"* called *"nativum"*. *"Bewußtseinslage"* was a technical term in German psychology of the time. It referred to a mental state that could not be reduced to sensations, presentations (*Vorstellungen*), and feelings. In the case under discussion, the *Bewußtseinslage* of the *nativum*, the mental state in question was due to an inhibition or interruption of the normal flow of presentations and judgements. It was the state of mind in which a question, as it were, arose – a state of mind that caused tension and discomfort. During the second phase of asking a question, the *nativum* gave rise to a volition, to a *"Wissenwollen"*, a wanting-to-know. This volition, called *"rogativum"* by Kreibig, was triggered by the emotional discomfort and directed at re-establishing the normal, uninterrupted flow of presentations and judgements. Finally, during the third stage, the *"responsivum"* cancelled the volition and exorcised the earlier discomfort felt. The *responsivum* was the mental experience of being provided with, or of having arrived at, a satisfactory answer (1914/15, p. 163).

Although Kreibig was less detailed in his remarks on the logical aspects of questions, he did introduce an important distinction between the "anticipation", *"quaesitum"* and "supplement" of a question. As already seen, the "anticipation" of a question consisted of "the set of presentations and judgements which the question-thought prescribes for the content of the aimed-for judgement or conclusion." The *"quaesitum"* was the "logical subject" of the question, i.e. it was that part of the anticipation which needed to be filled by possible answers. Finally, those elements of possible answers that could take the place of the *quaesitum* were "supplements" of the question (1914/15, p. 164). It seems that given a question like [42], Kreibig took [43] to be the "anticipation", "x" to be the *"quaesitum"*, and something like [44] to be possible supplements (1914/15, p. 164):

[42] Who lives here?

[43] x lives here
[44] {x | x is such entity as could live here}

Kreibig proved himself to be innovative also in his attempt to analyse the "sufficiency" and "insufficiency" of answers. According to Kreibig, such in/sufficiency needed to be analysed into three dimensions: the psychological, logical, and epistemological dimension. An answer was *psychologically sufficient* if and only if it calmed down the questioner's will to know. It was *logically sufficient* if and only if the answer provided a supplement of the form prescribed by the quaesitum. And finally, an answer was *epistemologically sufficient* if and only if the answer was true or probable.

An answer could be sufficient in one or two of these three respects, without being sufficient in the remaining dimension(s). Let the question be, for instance, [45], and let [46] to [51] be suggested answers to [45]:

[45] Who is the author of *Knowledge and Belief*?

[46] Yes.
[47] Ludwig Wittgenstein.
[48] Jaakko Hintikka.
[49] Some Finn.
[50] Jaakko Hintikka, my supervisor.
[51] I don't know.

Obviously, any answer could be *psychologically sufficient* for a given questioner only if she were in such a mental state as to be satisfied with the answer. Thus ultimately, even the answer [46], although neither logically nor epistemologically sufficient, could be psychologically sufficient for a given questioner. [47] is an answer that Kreibig would have regarded as logically sufficient although epistemologically insufficient. It would be logically sufficient insofar as the set of possible answers – fixed by Kreibig's *quaesitum* – is the set of human individuals, and Ludwig Wittgenstein is a member of that set. [51], on the other hand, would be logically insufficient but epistemologically sufficient, only if it correctly expressed the knowledge of the respondent. Kreibig called such answers "inappropriate". [49] would also be logically insufficient although true: Kreibig spoke of an "incomplete" answer in this case. Finally the difference between [48] and [50] can highlight Kreibig's distinction between "economical" and "abundant" answers: the first provides just enough information for the questioner to be able to identify the individual (or time, place, ...) asked for, whereas the second type of answer provides more than one description or name of the individual (or time, place ...). (1914/15, p. 191).

5. *Questions and Knowledge*

Given German-speaking philosophers' interest in the logic and psychology of questions, it is hardly surprising that they used question-theoretical terms in their attempts to understand how humans come to understand and know. This move was popular with Neokantian philosophers in particular. As early as 1865, Otto Liebmann was calling the question "the source of all knowledge", and seeking to derive the Kantian categories of substance and causality from the children's questions "What is this?" and "Why is this?". He also submitted that intellectual growth could best be conceptualised "as a step-by-step exchange of interrelated questions and answers", and pointed out "that it is *the* peculiarity of the theoretical intellect that whenever it has studied and grasped some field of knowledge through an alternating questioning and answering, it never concludes with an answer *but always on a question*" (1865, pp. 53–60).

Hermann Cohen, too, called the question "the beginning of knowledge" (1902, p. 69), and Paul Natorp taught that the question was "the decisive moment in the thought process itself; in all genuine thinking the question is that moment in virtue of which thinking is a *procedere*: the primordial moment of *direction* towards determinate knowledge" (1910, p. 42).

Heinrich Rickert (1904) used question-theoretical ideas in defence of his basic epistemological claim, i.e. that the "epistemological subject" is a subject that affirms and denies. The starting point of the argument was the claim that a judgement could not be reduced to one or several presentations. Rickert sought to secure this point by suggesting that the best way to identify the essence of a judgement was to see it as an answer to a (decision-)question. He granted immediately that this analysis might not be psychologically adequate. But this psychological inadequacy of the solution was "not of great importance for the theory of science. Logically the problem precedes the solution to the problem, i.e. wherever knowledge is sought for its own sake, it must be sought as an answer to a question" (1904, p. 95). Next Rickert argued that a decision-question and its answer-judgement had the same content of presentations. They differed from one another only insofar as the answer-judgement contained an additional, non-presentational, and thus non-psychological element: the element of affirmation or denial. The logical structure of a judgement for Rickert was thus something like [52] and [53] (cf. Wittgenstein 1953, §22):

[52] Is it the case that p? Yes.
[53] Is it the case that p? No.

For Rickert it was the 'Yes' or 'No' which marked the essence of a judgement, and which distinguished the judgement from the question. And insofar as knowledge-acquisition consisted of forming judgements, he could also claim "that to come to know is – as far as its logical value is concerned – *affirming and denying*, or: the theoretical subject must be understood as an affirming or denying subject" (1904, p. 103).

A. Stadler (1908), another Neokantian, also felt that focusing on questions was of crucial importance for epistemology. For him questions were basic conditions of the possibility of experience. Only through appreciating the "transcendental character" of questions was it possible to convincingly tease out the "transcendental apriori" from the "physiological apriori" – a key problem for Neokantian philosophy. Two insights were the starting point of epistemology for Stadler: one, that theoretical reason (*Vernunft*) wanted to acquire knowledge, and two, that reason was bound to deem such acquisition possible. It was the task of epistemology to clarify what was presupposed by, and involved in, this want and in this inevitable assumption. And here epistemology could do no better than focus on the

> most primordial function of understanding: the question. ... All understanding in science, in the child, and indeed everywhere, starts from questions ... Therefore it must be possible to find all categories and basic concepts of knowledge via an analysis of the question .... The question 'which questions are to count as the primordial facts of knowledge-acquisition?' is immediately answered by remembering our theoretical will: we want to know 'what something is', and 'why it is' (1908, p. 246).

Stadler granted that asking questions was also linked to human physiology, "for if we were organised differently [physiologically differently], we might well be unable to ask questions". This physiological *factum* was transformed into a transcendental apriori, however, since the question "as the postulate of all knowledge is the condition of the possibility of all possible knowledge" (1908, p. 247).

The project of using question-theoretical terms for understanding knowledge-acquisition was not restricted to Neokantians. Karl Groos (1901&1902) also used an investigation into questions to epistemological ends, although his was an empirical-psychological investigation into what kinds of questions the enquiring human mind actually formed. Groos agreed with Rickert that to understand knowledge-acquisition was to understand the forming of judgements. Therefore Groos began his inquiry by distinguishing between two dimensions for classifying judgements. Judgements varied in the two dimensions of "new" vs. "repeated", and "natural" vs. "artificial." A new judgement was one that arose from solving a problem for oneself and for the first time; a repeated judgement was a problem-solution that one happened to remember; a natural judgement was one that occurred spontaneously on the basis of one's present experiences; and an artificial judgement was due to a philosopher's attempt to trigger a judgement in himself for the sole purpose of observing it introspectively. Groos believed that epistemology and psychology should busy themselves with an experimental study of the genesis of new and natural judgements. And this, Groos proposed, could best be done by bringing experimental subjects into situations in which they were encouraged to ask questions. After all, "questions contain great numbers of burgeoning acts of knowing, and at least some of the latter have the character of new [and natural] judgements" (1901, p. 146). Accordingly, Groos confronted his students with short narratives and encouraged them to enquire further about the content of these stories. He then analysed their questions according to their "relation of thought" (*Denkbeziehung*). His results showed that about 9% of the questions were directed at spatial features, 7% at temporal relations, 2% at quantities, 15% at teleological relations, and 31% at causal relations. 54% of the questions were propositional, and 46% wh-questions (1901, pp. 151–62). In a further study (1902), Groos found that the number of questions about causality increased with maturity, and suggested that his experiments be done in order to find differences amongst the Germans and the British (1902, p. 368).

Finally, Groos also submitted that one could distinguish between two types of knowledge-acquisition: a "passive waiting for knowledge", and "an active search for the truth." These two types found their respective expression in categorical syllogisms and wh-questions, on the one hand, and mixed-hypothetical inferences and propositional questions, on the other hand. Groos' example was that of a city-child who wakes up in the countryside and notices from its window the contours of a big bird. Choosing the path of "passive waiting", the child will ask "What kind of bird is this?", and wait. If the bird eventually shouts 'cock-a-doodle-do', the child will conclude that the bird is a rooster. The child's reasoning here has the logical form of a categorical syllogism:

[54] A bird that shouts cock-a-doodle-do is a rooster.
This bird shouts cock-a-doodle-do.
____
This bird is a rooster.

The child can also choose a different strategy of knowledge-acquisition, however. Rather than wait for further evidence, it may go out to find it by itself. Based on its earlier experience the child might first form the decision-question: "Is this possibly a rooster?" Subsequently, the child might start to approach the bird, and focus its attention on those characteristics of birds – size, form and length of feathers, etc. – that will enable it to decide its guiding question. It might check whether any of these conditions are met, and then, given the right kind of perceptual evidence, conclude that the bird in question is indeed a rooster. Groos put much emphasis on the observation that strictly speaking the reasoning of the child here is not logically valid, i.e. that it is an instance of the fallacy of affirming the consequent:

[55] If the bird is a rooster then it has feathers of type X.
The bird has feathers of type X.
――――――――――――――――――――
The bird is a rooster.

Groos concluded: "Nothing is more characteristic of the difference between logic and psychology than the fact that the form of inference which is the most important inference of all ... is not permissible on purely logical grounds" (1902, p. 370).

6. Summary and Conclusion

Above I have summarised the main contributions to philosophical question theory in German-speaking philosophy around the turn of the century. The more controversial issues were the relation between questions and judgements, and the proper way to distinguish between propositional and wh-questions. A number of further taxonomies of questions were also suggested. Moreover, some authors dealt with the intricate problems of answerhood, presuppositions, and suggestive questions. And several authors used question-theoretical terms in order to conceptualise the process of coming-to-know.

The corpus of texts considered above falls squarely between the years 1880 and 1915. With the rare exceptions of Bolzano (1837) and Liebmann (1865) I have not found theories of questions or interrogative models of enquiry before that date (unless of course one goes as far back as Kant). As concerns the *terminus ad quem*: although the post-war period saw the occasional philosophical treatment of questions, the topic was now much more peripheral (see Tumlirz 1919, Ingarden 1925, Löw 1928 – Ingarden's paper alone was original in its treatment of questions). The reason for this lay in the decline of Neokantian philosophy, and the general shift towards *Lebensphilosophie*. This shift occurred also within phenomenology where the stars of the 1920s were Scheler and Heidegger rather than Husserl.

Given the philosophical climate of the 1920s, a treatise on the logic of question and answer would have been labelled "dry", "pedantic", and, *"lebens-"* and *"weltfremd."* (For a detailed characterisation of the philosophical climate in Weimar Germany, see Kusch 1995, Chapter Eight.)

But why were philosophers interested in questions in the pre-war period? While there is no straightforward and simple answer to this question, a number of factors can be adduced here.

First of all, Neokantian philosophy was the dominating philosophy of the period, and although the Neokantians often strayed far afield from Kant's own writings, they were at least "raised" on them. Kant's major critical writings, of course, abound with explicit references to "questions that trouble human reason", "questions that never end", and "questions that human reason is unable to answer." Kant's own writings suggested that the process and the limits of human inquiry could be couched in question-theoretical terms. His Neokantian followers took their cue from him.

Second, most of the philosophical logic of the day was – despite Frege's and Schröder's heroic efforts – non-mathematical and informal. Even while remaining by and large within the classical scheme of a treatment of concepts, judgements, and syllogisms, logicians often paid some attention to the variety of linguistic expressions. Thus they frequently ended up making at least some passing comments on questions and commands.

Third, this last-mentioned tendency was strengthened by the fact that for many philosophers of the time the borderlines between logic, epistemology, linguistics, and psychology were blurred or even non-existent (see, e.g. Kusch 1995). Indeed psychology was still regarded by many as *a* – if not *the* – central part of philosophy. Since psychology studied all phenomena of the mind, it could hardly escape the attention of good introspectionists that questions (or acts of questioning) figure prominently in human thinking. Elaborating further on this part of human reasoning, the philosopher/psychologist would subsequently draw on all kinds of theoretical resources – including logic, grammar, and epistemology. Eventually some philosophers would of course emphasise a sharp division between logic and psychology, but that was well after questions had already found their way into the philosophical literature. As we have seen, question theory could be used both by Neokantian defenders of a sharp divide between logic and psychology, and by writers like Kreibig, Meinong, and Groos who were much less concerned with drawing boundaries.

Fourth and finally, it should be mentioned that by the early twentieth century questions had become a major issue for both theoretical and applied psychology. Applied psychologists showed a great interest in forensic psychology in general, and interviewing techniques and suggestive questions in particular. This line of research originated with Alfred Binet in France (Binet 1892, 1907), but was soon picked up by German philosopher/psychologists (see Kusch, under submission). Theoretical psychologists, on the other hand, hotly debated the work of the "Würzburg School" in the psychology of thought. The Würzburgers used a novel experimental technique in which questions to the experimental subject were crucial. It is ample evidence of a link between these psychological concerns and the work summarised above that Kreibig used the technical term *Bewußtseinslage* throughout his paper. After all, it was precisely the existence of *Bewußtseinslagen* that the Würburgers claimed to have proven by means of their interrogative experimental technique (see Kusch, under submission).

Science Studies Unit
University of Edinburgh
Dept. of Sociology
21 Buccleuch Place
Edinburgh EH8 9LN

REFERENCES

Binet, A. (1892). *Les altérations de la personnalité*. Paris: Alcan.

Binet, A. (1907). *La suggestibilité*. Paris: Alcan.

Bolzano, B. (1837). *Wissenschaftslehre*. Vol. I. Sulzbach: Seidel.

Cohen, H. (1902). *Die Logik der reinen Erkenntnis*. Berlin: Cassirer.

Erdmann, B. (1907). *Logische Elementarlehre*. Second edition. Halle: Niemeyer.

Garrison, J.W. (1988). Hintikka, Laudan and Newton: An Interrogative Model of Scientific Inquiry. *Synthese* **74**, 145–71.

Groos, K. (1901&1902). Experimentelle Beiträge zur Psychologie des Erkennens. *Zeitschrift für Psychologie und Physiologie der Sinnesorgane* **26**, 145–67, and **29**, 358–71.

Hintikka, J. (1974). Questions about Questions. In: M.K. Munitz and P.K. Unger (Eds.), *Semantics and Philosophy*. New York: NYU Press, pp. 103–58.

Hintikka, J. (1976). *The Semantics of Questions and Questions of Semantics*. (*Acta Philosophica Fennica*, vol. **28**, No. 4). Amsterdam: North-Holland Publishing Company.

Hintikka, J. (1981). The Logic of Information-Seeking Dialogues: A Model. In: W. Becker and W.K. Essler (Eds.), *Konzepte der Dialektik*. Frankfurt am Main: Klostermann, pp. 212–31.

Hintikka, J. (1984). The Logic of Science As a Model-Oriented Logic. In: P.D. Asquith and P. Kitcher (Eds.), *PSA 1984*. Vol. I. East Lansing, Michigan: The Philosophy of Science Association, pp. 177–85.

Hintikka, J. (1987). The Interrogative Approach to Inquiry and Probabilistic Inference. *Erkenntnis* **26**, 429–42.

Husserl, E. (1901). *Logische Untersuchungen*. Vol. II. Halle: Niemeyer.

Ingarden, R. (1925). Essentiale Fragen. *Jahrbuch für Philosophie und philosophische Forschung* **7**, 125–304.

Jerusalem, W. (1895). *Die Urtheilsfunction*. Wien: Braumüller.

Jodl, F. (1903). *Lehrbuch der Psychologie*. Vol. II. Second Edition. Stuttgart: Cotta.

Kakkuri-Knuuttila, M.-L. and M. Kusch (1991). LSP-Research, Philosophy of Science, and the Question-Theoretical Approach – Some Tentative Suggestions. In: H. Schröder (Ed.), *Subject-oriented Texts: Languages for Special Purposes and Text Theory*. Berlin: de Gruyter, pp. 167–98.

Kreibig, J. (1914/15). Beiträge zur Psychologie und Logik der Frage. *Archiv für die gesamte Psychologie* **23**, 152–212.

Kusch, M. (1995). *Psychologism: A Case Study in the Sociology of Philosophical Knowledge*. London: Routledge.

Kusch, M. (under submission). Recluse, Interlocutor, Interrogator: Natural and Social Order in Turn-of-the-Century Psychological Research Schools.

Kusch, M. and H. Schröder (1989). The Question-Theoretical Approach in Hermeneutics and LSP-Research. In: C. Lauren and M. Nordman (Eds.), *Special Language: From Human Thinking to Thinking Machines*. Philadelphia: Multilingual Matters, pp. 53–71.

Liebmann, O. (1865). *Kant und die Epigonen: Eine kritische Abhandlung*. Stuttgart: Cotta.

Lipps, T. (1893). *Grundzüge der Logik*. Hamburg and Leipzig: Voss.

Lotze, H. (1880). *Logik: Drei Bücher*. Edited and selected by G. Misch. Leipzig: Meiner, 1912. (The original was first published in 1880.)

Löw, F. (1928). Logik der Frage. *Archiv für die gesamte Psychologie* **66**, 357–436.

Martinak, E. (1905). Das Wesen der Frage. Eine psychologisch-logische Untersuchung. *Atti del V Congresso Internazionale di Psicologia*. Roma: Forzani, pp. 332–36.

Meinong, A. (1902). Ueber Annahmen. *Zeitschrift für Psychologie und Physiologie der Sinnesorgane*, Ergänzungsband 2. Leipzig: Barth. Leipzig 1902.

Natorp, P. (1910). *Die logischen Grundlagen der exakten Wissenschaften*. Leipzig: Teubner.

Rickert, H. (1904). *Der Gegenstand der Erkenntnis*. Second edition. Tübingen: Mohr.

Schuhmann, K. and B. Smith (1987). Questions: An Essay in Daubertian Phenomenology. *Philosophy and Phenomenological Research* **47**, 353–83.

Sigwart, C. (1893). *Logik*. Vol. II. Second edition. Freiburg i. Br.: Mohr.

Sintonen, M. (1990). The Interrogative Model of Inquiry in Evolutionary Studies. In: L. Haaparanta, M. Kusch, and I. Niiniluoto (Eds.), *Language, Knowledge, and Intentionality: Perspectives on the Philosophy of Jaakko Hintikka*. (*Acta Philosophica Fennica*, vol. **49**). Helsinki: Akateeminen Kirjakauppa, pp. 473–487.

Stadler, A. (1908). Die Frage als Prinzip des Erkennens und die Einleitung der Kritik der reinen Vernunft. *Kantstudien* **13**, 238–48.

Tumlirz, O. (1919). *Das Wesen der Frage: Beiträge zu ihrer Psychologie, Gegenstandstheorie und Pädagogik*. Prague: Haase.

Wahle, R. (1890). Zur Psychologie der Frage. *Zeitschrift für Psychologie und Physiologie der Sinnesorgane* **1**, 310–20.

Wittgenstein, L. (1953). *Philosophical Investigations*. Transl. by E. Anscombe. Oxford: Blackwell.

Poznań Studies in the Philosophy of the Sciences and the Humanities
1997, Vol. 51, pp. 61–84

Nils-Eric Sahlin

'HE IS NO GOOD FOR MY WORK'*
On the Philosophical Relations between Ramsey and Wittgenstein

Was it Frank Ramsey that gave us the later Wittgenstein?

In his *Logic, Philosophy, and Language*, Georg Henrik von Wright remarks that: 'Wittgenstein's later thinking is, as far as I can see, completely independent of a philosophical tradition and without literary sources'[1]. von Wright is far from being the only one entertaining this view. But Wittgenstein's later thinking is, of course, not the genesis of a new philosophy. What is surprising, however, is that although many interesting links and connections have been pointed out, the most obvious of them all seems to have been almost totally overlooked or neglected by most Wittgensteinscholars. Thus, since it is far better to be thought-provoking than provocative let me straightaway rephrase the opening question and instead ask to what extent Frank Ramsey may have influenced the thinking of the later Wittgenstein.

Having written *The Philosophy of F. P. Ramsey*, I was left with the impression that Wittgenstein was far more influenced by Ramsey than Ramsey was influenced by Wittgenstein.[2] It is well known that Wittgenstein got a solid grip of the very young Frank Ramsey's philosophical panorama. But, what is less well known is that Ramsey came to change his view of philosophy and in doing so became a major force in Wittgenstein's upheaval of the Tractarian view. And that from 1929 and on one can clearly spot a remoulded Ramsey in the work of Wittgenstein.

There are no incontrovertible arguments in favour of this thesis. However, there are enough circumstantial pieces of evidence to support the hypothesis and combining these concurring pieces gives it not too low a probability. I will present my arguments in three parts. First, a few biographical glimpses have to be given. They will give us a picture of what we know about Ramsey's and Wittgenstein's philosophical discussions, when and where they met, and what was probably on their agenda. Second, something has to be said about Ramsey's British pragmatism. The core of Ramsey's philosophical and scientific work consists of no more than, say, 15 papers. But they are more than 15 brilliant essays on 15 disparate subjects. As I read them I found that almost all of them contain the same view of philosophy – a method of analysis – merging a sound portion of realism, or to use G. E. Moore's words 'the soundest of common sense', with Ramsey's kind of pragmatist philosophy. For my purpose it is of crucial importance to show that Ramsey came to entertain this view of philosophy in the mid 20's. Third, I will give a few examples of where this British pragmatism can be spotted in Wittgenstein's work after 1930.

1. Biographical glimpses

Frank Plumpton Ramsey was born on the 22nd of February 1903 and died at the age of 26 on the 19th of January 1930. He became a Fellow of King's in 1924 at the age of 21. Two years later he was made a lecturer in mathematics at Cambridge.

With the *Tractatus*, Wittgenstein made sure that the professional philosophers were fully occupied, and, for better *and* worse, he forced Oxbridge into a special mode of philosophical thought. Perhaps Wittgenstein is the philosopher that, together with Russell, had the greatest importance for Ramsey's early philosophical development. Not that Ramsey blindly accepted their views. Ramsey was, for example, far too clear-sighted not to perceive the defects of *Tractatus* (and *Principia Mathematica*) at an early stage.

In 1922 Wittgenstein's *Logisch-Philosophische Abhandlung* of 1921 was published in a bi-lingual version, this time under the title *Tractatus Logico-Philosophicus*, a title which Moore is said to have suggested. Wittgenstein's aphoristic style caused many, including Moore, to doubt whether it was possible to translate the book at all. Ramsey was extremely interested in Wittgenstein's text and C. K. Ogden suggested to him that he should try to translate it. At this time, Ramsey was 18 years old. During the winter of 1921–1922 Ramsey went now and again to Mrs. Pate's typing office and dictated straight off a translation of the *Tractatus*. Having Ramsey translate the book proved to be a stroke of genius.

But it was not until the summer of 1923 that Ramsey met Wittgenstein. Ogden had written to Wittgenstein in Puchberg explaining that Ramsey would like to meet him. Ramsey got an invitation from Wittgenstein and went to visit him. This gave him the opportunity of discussing the difficulties he had with the *Tractatus*. It is well-known that Ramsey's and Wittgenstein's conversations resulted in both the English translation and the German original being altered, alterations that were made in the second edition of the *Tractatus*.

Ramsey's review of *Tractatus*, published in *Mind* (October 1923), although written before they met gives us some ideas of what they actually discussed at their meetings. But, this is what happened on the philosophical level; there is, however, also a psychological dimension. Ramsey admired Wittgenstein a lot. *Tractatus* was the hub around which Ramsey's philosophical mind was spinning. It was no doubt a devoted and enthusiastic student that arrived in Puchberg to learn from his master. In an often quoted letter to his mother, written on September 20, 1923, Ramsey says:

> He is great. I used to think Moore a great man but beside W!

In the same letter he also tells us something about how their discussions went on:

> He is prepared to give 4 to 5 hours a day to explaining his book. I have had two days and got through 7 (+ identical forward references) out of 80 pages. And when the book is done I shall try to pump him for ideas for its further development which I shall attempt.

and a few lines later on continues:

> He has already answered my chief difficulty which I have puzzled over for a year and given up in despair myself and decided he had not seen.

That Ramsey forced Wittgenstein to defend his position is clear from the following passage:

> It's terrible when he says "Is that clear" and I say "no" and he says "Damn it's *horrid* to go through that again". Sometimes he says I can't see that now we must leave it. He often forgot the meaning of what he wrote within 5 min[ute]s, and then remembered it later. Some of his sentences are intentionally ambiguous having an ordinary meaning and a more difficult meaning which he also believes.

In 1923 Ramsey, having translated the *Tractatus*, went to Puchberg hoping to 'pump' Wittgenstein for ideas for its further development, something that he himself wanted to attempt. I do not think that Ramsey got any new ideas or insights from his discussions with Wittgenstein, at least none that he later came to develop. A vigilant reading of the letter to his mother suggests that Wittgenstein did not, contrary to what one might believe, act as a teacher or advisor, hinting at new possibilities, generously giving away fruitful ideas. Instead it seems as if Wittgenstein was under attack and had to stand up for his ideas. In translating *Tractatus* Ramsey had already spotted the weaknesses of the building. But his aim was not to tear it down, to build something new; rather he wanted to spend time repairing and developing the construction and improving its foundation.

Ramsey went back to Vienna in March 1924 to see Wittgenstein again. But he does not seem to have had much philosophical enjoyment from the conversations with him on the few occasions that they met. In May, having seen Wittgenstein one more time, he writes:

> ... he is no good for my work.[3]

In less than a year Ramsey had turned around completely. He no longer wished to work with Wittgenstein on the foundations of mathematics. He had realized that Wittgenstein was no good for his work. We can only guess what reasons he had for this change of mind. However, as early as 1923 he must have sensed that Wittgenstein had nothing new to offer him. That summer Wittgenstein also explained to Ramsey that: 'he himself will do nothing more'. Thus, if Ramsey had what he thought promising ideas that he wanted to develop, that he saw things that Wittgenstein hadn't seen or didn't agree on, why should he not pursue them on his own? We should be grateful that Ramsey realized this and decided to cultivate his own ideas. What came out of this decision was a couple of first-rate papers, for example, 'Universals', 'The Foundations of Mathematics' (both 1925), and 'Mathematical logic' (1926).

But there is one more thing of importance in this context. In 1922 Ramsey wrote a review of Keynes' *A Treatise on Probability*. And he criticised this theory so effectively that Keynes himself abandoned it. Ramsey also read C. S. Peirce on probability and was introduced to American pragmatism.[4] Keynes, whose theory he thought so flawed, and Peirce, whose work I do not think he knew very well (from what we can tell he just read bits and pieces of Peirce's writings), came to influence him considerably. Dissatisfied with their views on probability he developed a radically new theory of probability. However, in doing so he also laid the foundation for the British pragmatism that he then came to entertain and develop till his death.

After the summer of 1924 Ramsey and Wittgenstein seem to have lost touch, at least when it comes to philosophy. They met again in 1925 at

Keynes' place and had a heated spat, but unfortunately not on philosophy. There is also a letter from 1927. Apart from these two sporadic encounters, there are no indications of any other correspondence whatsoever or philosophical trysts. However, when Wittgenstein moved to Cambridge in January 1929, he and Ramsey resumed their acquaintance. It is known that they met frequently and spent considerable time discussing philosophy, but exactly what was on the agenda is mostly entombed in oblivion.[5]

In the introduction to *Philosophical Investigations* Wittgenstein announces the significance which Ramsey's 'certain and forcible' criticism had for his philosophical reorientation:

> For since beginning to occupy myself with philosophy again, sixteen years ago, I have been forced to recognize grave mistakes in what I wrote in that first book. I was helped to realize these mistakes – to a degree which I myself am hardly able to estimate – by the criticism which my ideas encountered from Frank Ramsey, with whom I discussed them in innumerable conversations during the last two years of his life.

What Wittgenstein alludes to must be the 1923 discussions as well as the discussions they had in 1929. But the Ramsey that Wittgenstein met in 1929 was surely not the same young mathematician as he had to defend his work to in the summer of 1923. With 'Truth and probability' (1926) Ramsey had begun to develop his pragmatist view of philosophy. And in 1929 he had a unified and quite complete theory of belief, truth, knowledge and probability. In fact he had been working on a book manuscript, now published as *On Truth*, dealing with these topics. This manuscript dates from the period 1927–29.[6] It is essentially elements of this British pragmatism of Ramsey's that we can find traces of in the work of the later Wittgenstein.

2. Ramsey's British Pragmatism

Ramsey concludes 'Facts and propositions' (1927) by saying:

> The essence of pragmatism I take to be this, that the meaning of a sentence is to be defined by reference to the actions to which asserting it would lead, or, more vaguely still, by its possible causes and effects.[7]

In 'Facts and propositions' Ramsey uses his pragmatist philosophy to outline a theory of truth. Ramsey's theory has been misunderstood in later philosophical literature. The reason for this is, I take it, that no one has clearly comprehended the intimate connection between his theories of truth, partial belief (the subjective theory of probability) and knowledge.

In his paper 'Truth and Probability', written in 1926, Ramsey laid the foundations of the modern theory of subjective probability. He showed how people's beliefs and desires can be measured by use of a traditional betting method. What we want to do is to measure a person's belief by proposing a bet, and 'see what are the lowest odds which he will accept'[8]. Ramsey took this method to be 'fundamentally sound', but saw that it suffered from 'being insufficiently general, and from being necessarily inexact. ... partly because of the diminishing marginal utility of money, partly because the person may have a special eagerness or reluctance to bet, ...'[9]. To avoid these difficulties he laid the foundations of the modern theory of utility. He then went on to show that if people in their behaviour *follow a set of rules*, the measure of our 'degrees of belief' will satisfy the laws of probability.

Ramsey was the first one to prove the celebrated Dutch book theorem; he had a proof of the value of collecting evidence; he took higher order probabilities seriously; and he had the notion important for Bayesian statistics of 'exchangeability'. In 'Truth and probability' he also laid the foundations of modern decision theory. Finally, it is important to emphasize that Ramsey was far from being the narrow-minded subjectivist/Bayesian that others have often presented him as. He did not, for example, believe that 'probabilities do not exist', meaning objective probabilities; rather he saw that some types of probability are a matter for physics, not for psychology and not for logic.

In this context it is especially fruitful to look upon 'Truth and probability' as a theory of rule-following. It tells that we can describe a person's actions in terms of rule-following. It, for example, tells us that if our distribution of degrees of belief follows the rules of probability theory, a book cannot be made against us.

But let us now return to Ramsey's theory of truth. In 'Facts and propositions' Ramsey argues that 'if we have analysed judgment we have solved the problem of truth'.[10] To carry out such an analysis successfully one has to say what the content of a belief is without falling into a regress by appealing to the meaning of sentences understood as truth conditions.

There is an important paragraph in the paper where Ramsey clearly indicates how such an analysis can be carried out:

> . . . it is, for instance, possible to say that a chicken believes a certain sort of caterpillar to be poisonous, and mean by that merely that it abstains from eating such caterpillars on account of unpleasant experiences connected with them. . . . An exact analysis of this relation would be very difficult, but it might well be held that in regard to this kind of belief the pragmatist view was correct, i.e. that the relation between the chicken's behaviour and the objective factors was that the actions were

> such as to be useful if, and only if, the caterpillars were actually poisonous. Thus any set of actions for whose utility *p* is a necessary and sufficient condition might be called a belief that *p*, and so would be true if *p*, i.e. if they are useful'.[11]

In a note Ramsey adds: 'It is useful to believe *aRb* would mean that it is useful to do things which are useful if, and only if, *aRb*; which is evidently equivalent to *aRb*'.

This pragmatic theory of truth is something rather different than the redundancy theory of truth credited to Ramsey. If propositions are the carriers of truth value, then to say that '"it is true that Caesar was murdered" means no more than that Caesar was murdered'. But, Ramsey does not find this a very interesting analysis of truth. Far more challenging is to say what it means to have a true belief and to do this without appealing to the meaning of sentences. To succeed in this a pragmatic analysis seems to be the correct way to go.

Ramsey's theory of truth, like his theory of probability, tells us something about rule-following. The chicken in Ramsey's example can be seen as having a decision problem, it has a choice between the two actions: (i) eat the caterpillar; (ii) refrain from eating the caterpillar. If the chicken chooses to eat the caterpillar, this choice will lead to one of two consequences, depending on whether the caterpillar is poisonous or edible. If the caterpillar is poisonous, the chicken gets an upset stomach; if it is edible, the chicken gets a good lunch. If, on the other hand, the chicken refrains from eating the caterpillar, this means that it has either avoided an upset stomach or missed its lunch.

This is a well-defined decision problem and we can therefore use Ramsey's theories of subjective probability, utility and decision to solve it. 'Truth and probability' tells us that if a chicken does not know whether the caterpillar is poisonous or not, he should 'act in the way [he] think[s] most likely to realize the objects of [his] desires'[12]; i.e. maximize his subjective expected utility. However, a truth-problem is not one of degrees of belief, but of full belief. We want to make clear what is meant by saying that the chicken believes fully, i.e. believes, that the caterpillar is poisonous. What it means is that the chicken refrains from eating the caterpillar: an action that is useful if and only if the caterpillar is poisonous (and the chicken wants to avoid an upset stomach).

This is the gist of Ramsey's theory of truth. It is an obvious example of a pragmatic theory of truth, but also of a type of rule-following epistemology. Having a true belief is having a more or less complicated rule, which if put to use, always leads to success.

In 'Knowledge', written in 1929, Ramsey uses his pragmatic theory to give an analysis of what it means to have knowledge. 'I have', he says, 'always said that a belief was knowledge if it was (i) true, (ii) certain, (iii) obtained by a reliable process.'[13] On the surface this definition of knowledge looks very much the same as the traditional, true-justified-belief theory, but working out the details of this theory one discovers that it diverges significantly from that account of knowledge. Of special interest is his third condition.

Ramsey requires that *X*'s belief (that *p*) has been obtained by a reliable process. It is not sufficient that *X* has evidence for believing that *p*; the way in which we acquire our beliefs should be reliable. The reliability condition thus tells us that the provenance of knowledge is of decisive importance. To have full belief is not enough, not even if the belief is supported by heaps of evidence. But as important as the provenance of the beliefs we call knowledge is their future use. A belief, being a map by which we steer, being a rule to follow, must guide our future actions. A full belief, obtained by a reliable method, is definitely not knowledge if it leads us on the wrong track; to be knowledge it must help us to avoid errors. Thus, knowledge is simply not true justified belief but rather: *A belief is knowledge if it is obtained by a reliable process and if it always leads to success.*[14] Knowledge is simply a special type of rule-following activity.

3. Ramsey and Wittgenstein

It should then be obvious in what ways Ramsey must have influenced Wittgenstein. In the introduction to *Philosophical Investigations* Wittgenstein himself mentions how Ramsey forced him to 'recognize the grave mistakes' he made in *Tractatus*, but also the 'innumerable conversations' they had in 1929. One thing that must have been on the agenda during these conversations is the pragmatism that Ramsey had begun to develop already in 1926 with 'Truth and probability'. An essential element of this pragmatist philosophy is, as we have seen, various aspects of rule-following. Rational decision making, acting on true beliefs, and knowledge are all examples of rule-following activities.

It is not my intention to compare the writings of Ramsey and Wittgenstein word by word or even sentence by sentence. I find such a task utterly uninteresting and more often than not it leads nowhere. I will also try to avoid being caught in a tedious and unproductive exegesis. There are many readings of the later Wittgenstein's work, but it seems to me quite clear that they have no bearing on my work.

I have chosen to compare Ramsey's and the later Wittgenstein's writings on the following points: (a) rule-following; (b) general propositions and theories; and (c) induction and probability.

3.1. Rule-following

Colin McGinn opens his book *Wittgenstein on Meaning* by saying: 'However much interpreters of Wittgenstein's philosophy may disagree, there is one point in which no dispute is to be expected: that Wittgenstein held that the proper way of understanding and resolving philosophical problems lies in arriving at a correct conception of language'[15]. If you are looking for a Ramsey in the later Wittgenstein's writings, do not expect to find the ideas presented as they are in his own work. Rather what you have to look for is a Ramsey ground through the mind of a person with a diametrically different philosophical temperament who was primarily concerned with the philosophy of language; with language as a method of analysis.[16]

For Ramsey the meaning of a sentence is to be defined by reference to the actions to which asserting it would lead. If this is your view of meaning, you have to study rule-following: so did Ramsey, and so did Wittgenstein. Wittgenstein, however, did not aim at a definition, but to him there is no meaning without rule-following and no rule-following without meaning.[17]

Ramsey teaches us how people's beliefs and desires can be measured by use of a traditional betting method. He also shows that if people in their behaviour *follow a set of rules*, the measure of our 'degrees of belief' will satisfy the laws of probability. His primary concern is to give a *description* of sound decision making, to depict what it is to maximize expected utility. McGinn tells us that one thing Wittgenstein aimed at was to give a 'characterisation [a description] of what it is actually like to follow a rule, to remind us of the "phenomenology" of ... using a word'[18] Wittgenstein's entire project is more sweeping, but the core of it is essentially Ramsey's decision theoretical project applied to language. Note, however, that 'giving a description' for Ramsey is to 'solve a problem', for Wittgenstein to 'dissolve a problem'.

A Wittgenstein familiar with Ramsey's descriptive view of decision making, ought to ask Ramsey a number of thought-provoking questions about sound decision making. Is it, for example, possible to obey the postulated rules of rationality 'privately'? And, '[i]s what we call ["maximizing expected utility"] something that it would be possible for only *one* man to do, and to do only *once* in his life?' (*Philosophical Investigations*, 199).

What is decision making but a practice? Most of us believe ourselves to master the techniques of rational decision making; that, for example, we are not violating the rules of probability calculus. But, thinking that one is a sound decision maker, obeying the rule of maximizing expected utility, does not entail that one is; 'to *think* that one is obeying a rule is not to obey a rule' (*Philosophical Investigations*, 202). The only way to find out seems to be to conduct an experiment. Beliefs and desires have to be separated and measured. Measurement, however, demands stability, our preference rankings should not be drifting. We must also be able to identify one and the same outcome or consequence in different preference rankings and from one time to another. One might now argue that this can be accomplished only from a third-person perspective; that the rationality of a decision maker is something that has to be experimentally verified.[19] One might argue that it is the surroundings that make a decision rational or irrational. Exactly what constitutes this outer frame of reference is an interesting question, but not one that has to be dealt with in this context. One answer, however, would be to say that what is needed is more than one decision maker who speaks the language of decision. (In *Remarks on the Foundations of Mathematics*[20] Wittgenstein, talks about this problem: 'What sort of public must there be if a game is to exist, if a game can be invented?' and 'What, in a complicated surrounding, we call "following a rule" we should certainly not call that if it stood in isolation'.)

What I have designed is a 'private language argument' in disguise. The argument is presented as it might occur to you after you have read some descriptive decision theory. I do not say that it is a very good argument. What I do imply, however, is that someone who is primarily concerned with language may well have found Ramsey's theory of decision thought-provoking and transmuted it into a theory of language.

In 'Truth and probability' Ramsey discusses how his theory relies on mathematical expectation. He argues that 'mathematical expectations in terms of money are not proper guides to conduct', but that the 'dissatisfaction often felt with this idea is due mainly to the inaccurate measurement of goods'.[21] Ramsey solves this difficulty by deriving measures of desire rather than measures of goods. However, we might want to ask him what the arguments are for maximizing expected utility in a unique situation. The standard justification for using mathematical expectation as a decision maxim is the average utility in the long run. But, if only one man plays a game only once in his life, a long-run justification is hardly admissible.

Assume, for example, that you arrive at a remote spot where they sell Coca-Cola for 5 cents a bottle. The price of a bottle has risen to 6 cents recently but the machine only takes nickels. If every sixth bottle in the machine is empty in a random way, the expected price per bottle will be six cents. For the owner of the machine and those who often use it this will be quite satisfactory, but for you, the visitor passing through, the system is of doubtful value.[22]

Reading Wittgenstein on rule-following reminds one of some of the key elements in Ramsey's theories of belief, truth and probability. It is like coming back to a city destroyed by war and rebuilt again. We recognize the major roads, a few buildings, the rest is untrodden ground. I might be wrong about this, but in Wittgenstein's later writings I see a theory of rule-following taking off from a Ramseyian analysis of sound decision making, from a theory of decision making as rational rule-following.[23]

3.2. General propositions

Ramsey argued that the logical form of a belief determined its causal properties. A belief is 'a map of neighbouring space by which we steer'.[24] The difference between the belief 'not-*p*' and the belief '*p*' lies in their causal properties. Thus disbelieving '*p*' and believing its negation have the same causal properties. They express, as Ramsey puts it, really the same attitude: 'It seems to me that the equivalence between believing "not-*p*" and disbelieving "*p*" is to be defined in terms of causation, the two occurrences having in common many of their causes and many of their effects'.[25] One of the advantages that Ramsey found in this theory is how it avoids the ontological proliferation of Russell's theory; negative facts, for example, are not needed.[26]

A causal property theory of this kind also has to handle more complex beliefs. What precise differences are there between the various logical forms of a belief and its causes and effects. Disjunctive beliefs engender no problems. To 'believe *p* or *q* is to express agreement with the possibilities *p* true and *q* true, *p* false and *q* true, *p* true and *q* false, and disagreement with the remaining possibility *p* false and *q* false'.[27] However, quantification introduces a set of problems which are not that easily handled.

In 'Facts and propositions' Ramsey follows Johnson and Wittgenstein and sees general propositions as the logical products and the logical sums of atomic propositions. 'For all x, ϕx' is to be interpreted as: a is ϕ, b is ϕ, c is ϕ ... and 'There is an x such that ϕx' consequently is equivalent to the logical sum of the values of 'ϕx'. If all propositions are truth functions of elementary propositons, traditional quantification leads to truth functions of an infinite number of arguments. With this analysis the causal property theory is easily extended to cover also the case of general propositions: 'Thus general propositions, just like molecular ones, express agreement and disagreement with the truth-possibilities of atomic propositions, but they do this in a different and more complicated way. Feeling belief towards "For all x, ϕx" has certain causal properties which we call its expressing agreement only with the possibility that all the values of ϕx are true'.[28]

Two years later Ramsey no longer found this a tenable analysis. In 'General propositions and causality', written in the summer of 1929, he has four arguments against analysing 'For all x, ϕx' as a conjunction. First, 'For all x, ϕx' cannot be written out as a conjunction. Second, it is never used as a conjunction. The statements are different as a basis for action. Third, 'For all x, ϕx' exceeds by far what we know or have knowledge of. What we know are, at most, a few instances of this generalization.

> A belief of the primary sort is a map of neighbouring space by which we steer. It remains such a map however much we complicate it or fill in details. But if we professedly extend it to infinity, it is no longer a map; we cannot take it in or steer by it. Our journey is over before we need its remoter parts.[29]

Fourth and finally, he argues that what we can be certain about is the particular case, or a finite set of particular cases. Of an infinite set of particular cases we could not be certain at all. Thus, 'For all x, ϕx' expresses, as Ramsey puts it, an inference we are at any time prepared to make, not a belief of the primary sort.

But, if general propositions are not conjunctions and thus not propositions, and assuming that general facts do not exist, how then are we to look upon sentences of this type? What status do they have; in what way can they be right or wrong? Unsurprisingly, Ramsey gives a pragmatic answer to this question. That general propositions are neither true nor false, that they carry no truth value, does not imply that they are meaningless. This type of sentence is the very foundation of the expectations that direct our actions. If I accept that for all x, ϕx, this means that when I have an x, I act as if it is ϕ. As Ramsey puts it, a general proposition *is not a judgment but a rule for judging*: it cannot be negated but it can be disagreed with.

In his *Wittgenstein*, Georg Henrik von Wright tells us that in one of the first converstations he and Wittgenstein had in 1939, Wittgenstein said that 'the biggest mistake he made in the *Tractatus* was that he had identified general propositions with infinite conjunctions and disjunctions of singular propositions.'[30]

It is well-known that the later Wittgenstein took a different view of quantification, a view which in fact is very similar to the one adopted by Ramsey in 'General propositions and causality'.[31] In *Philosophical Grammar*, for example, Wittgenstein argues that '[a]n hypothesis is a law for forming propositions'. 'You could also say', he continues, 'an hypothesis is a law for forming expectations'.[32] Ramsey could have written these words. What they show is that Wittgenstein came to foster the view that propositions containing quantifiers are not 'genuine' propositions.[33] Wittgenstein uses the word Gesetz, i.e. law, and one could then succintly state the later Wittgenstein's position using Ramsey's words: general propositions are not judgments, but *laws* for forming judgments, they are expectations-forming: 'If I meet a ϕ, I shall regard it as a ψ'[34].

We cannot say if Wittgenstein got these ideas from Ramsey, or via some other source. Not too bold a conjecture is, however, that the discussions they had in 1929 must have made Wittgenstein see his earlier view of quantification as the biggest mistake he made in *Tractatus*.

Ramsey got his view of general propositions from reading Weyl. In 'The foundations of mathematics', first read in November 1925, there are references to Hermann Weyl's *Das Kontinuum*, and a reference to Weyl's paper 'Über die neue Grundlagenkrise der Mathematik', a paper published in *Mathematische Zeitschrift* in 1921. The F. P. Ramsey Collection[35] contains a short transcript (6 handwritten pages) in German of bits of this (40-page) paper. The transcript is not dated, but it must have been written before 1925. Something that caught Ramsey's attention is what Weyl has to say about quantification:

> *Ein Existentialsatz* – etwa "es gibt eine gerade Zahl" – *ist überhaupt kein Urteil im eigentlichen Sinne, das einen Sachverhalt behauptet*; Existential-Sachverhalte sind eine leere Erfindung der Logiker. "2 ist eine gerade Zahl": das ist ein wirkliches, einem Sachverhalt Ausdruck gebendes Urteil; "es gibt eine gerade Zahl" ist nur ein aus diesem Urteil gewonnenes *Urteilsabstrakt*.[36]

Ramsey then jumps a few lines in Weyl's text and continues his annotations:

> Ebensowenig ist das generelle "jede Zahl hat die Eigenschaft E" – z. B. "für jede Zahl m ist $m + 1 = 1 + m$" – ein wirkliches Urteil, sondern eine generelle *Anweisung auf Urteile*.[37]

At the time when Ramsey made these annotations he was not the least impressed with or convinced by Weyl's view of mathematics.[38] But obviously he came to change his mind, and in the process maybe Wittgenstein's too.

3.3. Theories

Ramsey saw that the best way to understand how the theoretical entities of a theory function is to picture them as existentially bound variables. If the entities of our theory are α, β and γ, the 'best way to write our theory' according to Ramsey is: '($\exists\alpha$, β, γ): dictionary . axioms'.[39] This is the theory's 'Ramsey sentence'. The existentially bound variables are the carriers of ontological commitment; if the Ramsey sentence is true, they tell us what there is.[40]

Ramsey's view of theories has many advantages. First, Ramsey sentences help us understand the dynamics of scientific theories and scientific growth. Second, they explain the phenomenon of 'incommensurability'.[41] We note, for example, that no proposition of a theory 'can be understood apart from the whole theory to which it belongs. If a man says "Zeus hurls thunderbolts", that is not nonsense because Zeus does not appear in my theory, and is not definable in terms of my theory. I have to consider it as part of a theory and attend to its consequences, e.g. that sacrifices will bring the thunderbolts to an end'.[42] Thus, the 'adherents of two such theories could quite well dispute, although neither affirmed anything the other denied'.[43]

What Ramsey says about scientific theories should be compared to what Wittgenstein says about language. In *The Blue Book* he argues: 'The sign (the sentence) gets its significance from the system of signs, from the language to which it belongs. Roughly: understanding a sentence means understanding a language'.[44]

In this respect, Wittgenstein's view merely extends Ramsey's view of scientific theories to the whole of language.[45]

3.4. Induction and probability

What has been called Wittgenstein's theory of probability is, as a theory of probability, somewhat incomplete.[46] The classical definition of probability says that probability is the ratio of the number of favourable cases to the number of all equipossible cases. Transformed into the world of *Tractatus* 'the number of all equipossible cases' is the number of truth-grounds of a given proposition *q*; and 'the number of all favourable cases' is the number of truth-grounds of a proposition, *p*, which also are truth-grounds of *q*. Assume that these numbers are *m* and *k*, respectively. Then the conditional probability, the probability of *p* given *q*, is k/m, 'the degree of *probability* that the proposition ['*q*'] gives to the proposition ['*p*']' (5.15).

von Wright[47] reminds us that when Wittgenstein wrote *Tractatus* he obviously believed that logically independent propositions give one another the probability 1/2. The classical view of probability needs a set of exclusive and exhaustive alternatives. Wittgenstein must have thought that his system of propositions was sufficiently rigorous to allow a straightforward transformation of the classical definition. But it all hangs on the notion of independence. If two propositions are logically independent, if they have no truth-arguments in common (5.152), (i.e. if they have no elementary propositions in common), they do not necessarily give one another the probability 1/2. von Wright tells us that Wittgenstein must have realized his mistake and in the second edition we are told that two *elementary propositions* give one another the probability 1/2.[48] What we then have is a set of exclusive, exhaustive and equally probable elementary propositions, i.e. a hidden principle of indifference.[49]

For Wittgenstein probability is a logical relation between propositions. In *A Treatise on Probability* Keynes advocates a similar idea. Keynes began his work on probability in 1906 and it was almost completed in 1911; however, it was not published until 1921. In the preface Keynes says that he has 'been much influenced by W. E. Johnson, G. E. Moore, and Bertrand Russell, that is to say by Cambridge'. It is known that Johnson already around 1907 entertained ideas similar to Keynes', but they were not published until after his death.[50] Thus, when Wittgenstein came to Cambridge in the autumn of 1911, probability theory must have been on the philosophical agenda.[51] It seems therefore safe to say that Wittgenstein was as much influenced *by Cambridge* as Keynes. However, as a logical theory of probability Wittgenstein's attempt in *Tractatus* is but a pale and incomplete counterpart of the theories of Keynes and Johnson.

At the bottom of Keynes' theory we have a primitive logical (probability) relation. When this relation is measurable it tells us how strong an inference is from one proposition to another. The conclusive inferences of deductive logic are in Keynes' theory replaced by objective inconclusive inferences. 'Objective' means that for any two propositions one and only one probability relation holds. If this relation has the degree k/m, it is irrational to have any other degree of belief in the derived conclusion.

Ramsey's key argument against Keynes is that this probability relation does not seem to exist.[52] Ramsey says that he cannot perceive it and that he shrewdly suspects that no one else can. What is, he asks, the probability relation between 'That is red' and 'That is blue'? Is it 1, 1/4, 1/9, 1/25, 1/64, 1/169, or? One could, of course, give up the idea of a primitive logical probability relation and instead make probability depend on ordinary logical relations. But can a logical relation justify a unique degree of belief? What logical relations justify what degrees of belief?[53]

In a paper written in the autumn of 1922 and read at one of the meetings of the Apostles (20 October 1923) Ramsey briefly touches upon Wittgenstein's theory of probability. In the first part of the paper he attacks Keynes' view of induction. He then gives a simple example of the Tractarian view of probability. If p and q are elementary propositions, the probability of p, given *p* or *q*, is 2/3. If the disjunction is to be true, the propositions cannot both be false. Thus it is easily seen that *p* is true in 2 out of 3 cases.

Ramsey finds two objections to this theory. First, it is of almost no practical use. How do we know 'the logical forms of the complicated relations of every day life'?[54] Second, it definitely does not justify induction. The theory does not tell us how to make an inference from one set of facts to another distinct set of facts. And he might have added, third, it does not allow us to learn by experience. Following Wittgenstein, the probability of an elementary proposition, given any conjunction of elementary propositions (from which it is not entailed), is 1/2.

Ramsey's own theory of probability avoids these difficulties. The basic idea is that probability is to be interpreted as degree of belief, i.e. to give the notion of probability a subjective interpretation. This requires the measurement of partial beliefs. He therefore showed how people's beliefs and desires can be measured by use of a betting method and also gave a joint axiomatization of probability and utility. Given some intuitive rules of rational behaviour, a system of preferences among options, he could prove that the measure of our 'degrees of belief' satisfies the laws of probability.

A classical blemish that Ramsey succeeds in getting rid of is the principle of indifference. Keynes thought it possible to base it on purely logical conditions, but did not succeed in doing so. A careless reading of *Tractatus* seems to suggest that Wittgenstein could do without the principle. But this is not true, he simply hid it in the fabric of elementary propositions. Ramsey's probability theory also gives a justification for the axioms of the calculus; Keynes' theory has to assume the existence of a probability relation. In Ramsey's theory probable knowledge is effectively accommodated.[55] In Keynes' theory my rational degree of belief is given by the probability relation (between the hypothesis and what is known for certain, the evidence) and in Wittgenstein's theory by some logical relation (between sets of elementary propositions).

von Wright tells us that there are two poles in Wittgenstein's thinking about probability. The one pole is the logical theory of probability as it is sketched in *Tractatus*. The other pole is the epistemological view of probability as it is briefly outlined in *Philosophical Remarks* and *Philosophical Grammar*. von Wright emphasizes that Wittgenstein's later theory of probability is linked to the notions of imperfect knowledge and incomplete descriptions. But so is Ramsey's theory! A subjective theory of probability handles probable knowledge and incomplete descriptions: that is the whole point. A subjectivist does not need any logical relations to guide his or her probability assessments.

We want our subjective probabilities to stem from as complete and accurate knowledge as possible. They should be well calibrated. The best way to calibrate them is to take account of well-established frequencies and objective probabilities. To bet 1 to 1 on the toss of an American penny is not to be well calibrated.[56] Similarly it seems rather stupid not to follow the probabilities given by accepted physical theories. The same ideas I think we can find in the later Wittgenstein's writings on probability. To give an account of the relationship between frequencies and probabilities, you have to expand the bulk of knowledge by various hypotheses.[57]

Wittgenstein says that the logic of probability is only concerned with the state of expectation in the sense in which logic is concerned with thinking. This could well be Wittgenstein's understanding of what Ramsey was doing.

Wittgenstein's upheaval of the Tractarian view also forced him to reconsider his view on probabilities. von Wright says that the bridge between the two poles in Wittgenstein's thinking on probability is the idea of a probability which is relative to the bulk of our knowledge. One could say that some of the reasons for, and the drawings and material for this bridge come from Ramsey.

In *Tractatus* Wittgenstein argues that induction consists in accepting as true the simplest law that harmonizes with our experience (6.363). This procedure, however, has no logical justification, only a psychological one (6.3631). Ramsey, however, thought it would 'be a pity, out of deference to authority, to give up trying to say anything useful about induction'.[58]

Ramsey concludes the paper on induction that he read to the Apostles in 1923 by saying that: 'a type of inference is reasonable or unreasonable according to the relative frequencies with which it leads to truth and falsehood. Induction is reasonable because it produces predictions which are generally verified, not because of any logical relation between its premiss and conclusion. On this view we should establish by induction that induction was reasonable, . . .'[59] In 'Truth and probability' we find the same idea again, but this time more fully developed. He says: 'We are all convinced by inductive arguments, and our conviction is reasonable because the world is so constituted that inductive arguments lead on the whole to true opinions. We are not, therefore, able to help trusting induction, nor if we could help it do we see any reason why we should, because we believe it to be a reliable process'.[60] That is, our conviction is justified because the world houses reliable processes; inductive arguments on the whole lead to success.

Hume's problem is a problem of justification or validity. The premises of an inductive argument do not logically entail its conclusion. But what is it that has to be certified? The truth of the belief? Of course not! General beliefs carry no truth value; they 'are not judgments but rules for judging "If I meet a ϕ, I shall regard it as a ψ"'.[61] What has to be certified is the effectiveness of our general inferential habits or beliefs. The only way in which this can be done adequately is by assuming the existence of underlying reliable processes.[62]

Instead of accepting the Tractarian view Ramsey showed why some type of rule-following, some beliefs, are better habits qua basis for action than others. It is not because they are backed up by more evidence; that they have proved successful in the past. It is because there are underlying reliable processes or mechanisms accounting for our habits. Our habit of acting as if all men are mortal is successful simply because there is an underlying biological mechanism which more or less rapidly breaks down our minds and bodies. We do not need to assume that we can account for the underlying mechanisms or the reliable processes. No one has a clue to the enigma of aging. But this fact does not make our habit less successful; it is successful because there is an underlying mechanism.

Ramsey's and the later Wittgenstein's views on induction merit discussion. In *Philosophical Investigation* Wittgenstein rejects the traditional demand for a justification of our expectations.[63] Induction tells us that if we drop a book it will fall to the flor. According to Wittgenstein 'we don't need any grounds for *this* certainty . . .'. Because '[w]hat could justify the certainty *better* than success?' (324) Wittgenstein also tells us that '[j]ustification by experience comes to an end'. 'If it did not it would not be justification'.

To me some of Wittgenstein's renowned remarks are but echoes of Ramsey. Induction does not need the type of justification that we have traditionally been looking for. Wittgenstein is not talking about underlying mechanisms. He saw that it is not the truth of our rule-following habits that has to be certified, what has to be certified is the effectiveness of our rule-following habits, and nothing can do this better than success.[64]

4. Conclusions

Jaakko Hintikka and Mathieu Marion have showed us that the break in Wittgenstein's philosophy occurred in October 1929.[65] Wittgenstein did not live in a vacuum. To me he is very much a Cambridge product. The early Wittgenstein was influenced by Russell, Moore, Keynes, Johnson and other Cambridge philosophers: the later Wittgenstein by Ramsey.

Cambridge devoted much of the thirties to digesting Wittgenstein. D. H. Mellor has argued that, 'the whole philosophical community, would have been at least as well employed trying to digest Ramsey; and indeed that, but for Ramsey's early death, Wittgenstein's own work, on which Ramsey had a strong influence, would have developed more profitably than it did and been digested less uncritically than it was'.[66] Mellor is no doubt right about this and I hope that what has been said above at least hints at the influence Ramsey had on Wittgenstein. Ramsey is not just another key, but the key to unlocking the interpretation of Wittgenstein.[67] I am convinced that Wittgenstein research would gain a lot from looking more closely at this philosophical tryst.

Gothenburg University and
Lund University
Sweden

NOTES

* Thanks are expressed to Ingar Brinck, Bengt Hansson, Sören Halldén, Jaakko Hintikka, Mathieu Marion, Brian McGuinness, D. H. Mellor, Stig Nystrand, Margaret Paul and Fredrik Stjernberg for their many constructive comments and suggestions.

A condensed version of the second half of this paper, 'On the philosophical relations between Ramsey and Wittgenstein' was published in the *Proceedings of the 17th International Wittgenstein Symposium: The British Tradition in the 20th Century Philosophy*, ed. by J. Hintikka and K. Puhl, Hölder-Pischler-Tempsky, Wien 1995.

1 Doxa 1993, first published 1957. My translation from Swedish.

2 Cambridge University Press, Cambridge 1990, p. 227.

3 I am grateful to Mrs Margaret Paul for showing me this interesting remark of her brother's. She will discuss it further in the biography she is writing about Frank Ramsey.

At one point (in 1929) Ramsey had obviously told Wittgenstein straight out that he didn't like his method of arguing: 'I don't like your method of arguing'. Ramsey's statement must have left Wittgenstein in disquietude; if not, why should he bother to tell Moore about it? See G. E. Moore, 'Wittgenstein's lectures in 1930–31', in *Philosophical Papers*, Allen and Unwin, London 1959, p. 254.

4 In 'Facts and propositions' (1927) he says: 'My pragmatism is derived from Russell; and is, of course, very vague and undeveloped' (p. 51). But in 1927 he had read Peirce. Thus, one reading of this line is that the type of pragmatism he wanted to develop was not the sort he found in Peirce.

All references to *Philosophical Papers*, ed. by D. H. Mellor, Cambridge University Press, Cambridge 1990. See also Mellor's 'Ramsey, Frank Plumpton', in *The Encyclopedia of Philosophy*, 1994, for a concise and excellent presentation of Ramsey's philosophical work.

5 A few things are mentioned in Moore's *Philosophical Papers*, *op. cit.*. We also have the letter that Ramsey sent to Trinity College in 1929 (on Moore's request) in support of Wittgenstein (*op. cit.*, pp. 253–54). The recently published *Ludwig Wittgenstein: Wiener Ausgabe*, ed. by M. Nedo, Springer-Verlag, Wien 1994, reveals that one thing they talked about was general propositions and infinity.

6 See 'Editor's introduction', *On Truth*, ed. by N. Rescher and U. Majer, Kluwer, Dordrecht 1991.

7 P. 51.

8 P. 68.

9 P. 68.

10 P. 39.

11 P. 40.

12 P. 69.

13 P. 110.

14 See N.-E. Sahlin, 'Obtained by a reliable process and always leading to success', *Theoria* **57**, 1991, pp. 132–49.

15 Basil Blackwell, Oxford 1989, p. xi.

16 It is astonishing how different they were. Philosophy for Ramsey was a pleasure, a joyful activity. For Wittgenstein it seem to have been tormenting, almost causing him physical pain. Ramsey's thinking was constructive; it aimed at solving problems, producing theories. Wittgenstein's thinking was destructive; it aimed at showing what cannot be done, what is impossible. Ramsey's work is well thought-out, all arguments are secured. Wittgenstein's work is open to a sea of interpretation, full of loose ends.

17 'Language, . . ., relates to a *way* of living. In order to describe the phenomenon of language, one must describe a practice, not something that happens once, *no matter of what kind*.' (*Remarks on the Foundations of Mathematics*, 3rd ed., Basil Blackwell, Oxford 1978, p. 335.)

18 *Op. cit.*, p. 20.

19 Cf. McGinn's succinct discussion of the reading-example, *op. cit.*, p. 47.

20 Basil Blackwell, Oxford 1978, pp. 334–5.

21 P. 79.

22 See I. Hacking, *The Emergence of Probability*, Cambridge University Press, Cambridge 1975, p. 94.

23 Wittgenstein says: 'Following a rule is a human activity', *RFM*, 3rd ed., p. 331. I have not, for reasons that are too obvious to be mentioned, touched on all aspects of the so-called "rule-following argument". However, I believe that also its "inductive" facets can be viewed from my perspective.

The Wittgenstein scholar might argue that if Wittgenstein was influenced by Ramsey on rule-following, there should not be a time lag between cause and effect. I do not think that this is too good an argument. Not all seeds start to grow at once: it depends on how germinative they are, but also on the quality of the soil.

24 P. 146.

25 P. 44.

26 In *The Philosophy of F. P. Ramsey*, Chapters 4 and 8, I point out that Ramsey's contribution to this problem is highly dependent on his theory of universals. In 'Universals' he forcefully argued that 'the whole theory of particulars and universals is due to

mistaking for a fundamental characteristic of reality what is merely a characteristic of language' (p. 13). There is no intrinsic difference between universals and particulars. There are no negative or complex properties. Negative facts, like complex properties, are superstition and lead to a distorted view of reality. Ramsey found Wittgenstein's theory in *Tractatus*, 'that neither is there a copula, nor one specially connected constituent, but that, . . ., the objects hang one in another like the links of a chain' (p. 17) dogmatic, for it does not 'really explain any difference in the mode of functioning of subject and predicate' (p. 17). But it was the insufficient theories of Wittgenstein, Johnson and Russell, that inspired Ramsey to work on this clasical problem. See also my 'Obtained by a reliable process and always leading to success', *Theoria* **57**, 1991, pp. 132–49, and 'Ramsey, Frank Plumpton', in *A Companion to Metaphysics*, ed. by J. Kim and E. Sosa, Basil Blackwell, 1994, pp. 429–30.

27 Pp. 45–6.

28 P. 49.

29 P. 146.

30 P. 151, note 28, Basil Blackwell, Oxford 1982.

31 Compare Ramsey's paper with what Moore writes in 'Wittgenstein's lectures in 1930–33', *op. cit.*, pp. 297–8:

> In order to make clear exactly where the mistake lay, he [Wittgenstein] first said that in the case of such a universal proposition as 'Everybody in this room has a hat' (which I will call 'A'), he had known and actually said in the *Tractatus*, that, even if Smith, Jones and Robinson are the only people in the room, the logical product 'Smith has a hat, Jones has a hat and Robinson has a hat' cannot possibly be identical with A, because in order to get a proposition which entails A, you obviously have to add 'and Smith, Jones and Robinson are the only people in the room'. But he went on to say that if we are talking of 'individuals' in Russell's sense . . ., the case is different, because, in that case, there is no proposition analogous to 'Smith, Jones and Robinson are the only people in the room'. The class of things in question, if we are talking of 'individuals', is, he said, in this case, determined not by a proposition but by our 'dictionary': it is 'defined by grammar'. . . . He went on to say that one great mistake which he made in the *Tractatus* was that of supposing that in the case of *all* classes 'defined by grammer', general propositions were identical either with the logical products or with logical sums. . . . He said that, when he wrote the *Tractatus*, he had supposed that *all* such general propositions were 'truth-functions'; but he said now that in supposing this he was committing a fallacy, which is common in the case of Mathematics, e.g. the fallacy of supposing that 1+1+1 . . . is a sum, whereas it is only a *limit*, . . . He said he had been misled by the fact that $(x).fx$ can be replace by $fa\ .\ fb\ .\ fc\ .\ .\ .$, having failed to see that the latter expression is not always a logical product: that it is only a logical product if the dots are what he called 'the dots of laziness', . . .

What Wittgenstein says 1930–33 is not too far away from what Ramsey wrote in 1929. In 'General propositions causality' Ramsey is objecting to the infinite character of the logical products and sums. He is also talking about a form of laziness, that 'which we cannot express for lack of symbolic power' (p. 146); and points out the relation to previous mistakes in Mathematics, 'Now this we never do with these variable hypotheticals except in mathematics in which it is now recognized as fallacious' (p. 147). My guess is that it was Ramsey (the mathematician) who taught Wittgenstein that he in fact had committed a fallacy. In his lectures Wittgenstein also uses the word 'dictionary' a word frequently used by Ramsey in 'Theories' (written in 1929) a paper in which he is discussing very similar problems.

For a detailed discussion of Wittgenstein's view of general propositions see Merrill B. Hintikka and Jaakko Hintikka, *Investigating Wittgenstein*, Basil Blackwell, Oxford 1986.

32 P. 219, Basil Blackwell, Oxford 1974.

33 See von Wright, *op. cit.*, p. 151.

34 In 'General propositions and causality', p. 149.

35 The Archives of Scientific Philosophy in the Twentieth Century, Hillman Library, Pittsburgh University. Document number 007-04-01.

36 P. 54 in Weyl's paper.

37 P. 55 in Weyl's paper.

38 In 'Mathematical logic' Ramsey discusses Weyl's position: 'Weyl says that an existential proposition is not a judgment, but an abstract of a judgment, and that a general proposition is a sort of cheque which can be cashed for a real judgement when an instance of it occurs. ... [I]f, as Weyl says, an existential proposition is a paper attesting the existence of a treasure of knowledge but not saying where it is, I cannot see how we explain the utility of such a paper, except by presupposing its recipient capable of the existential knowledge that there is a treasure somewhere' (pp. 233–4).

39 P. 131.

40 It might be interesting and productive to compare Ramsey and W. V. Quine on this point (R. C. Jeffrey has hinted that this is an interesting relation). Cf. Quine's famous slogan 'to be is to be the value of a variable'. See *From a Logical Point of View*, Harper Torchbooks, New York 1953. A more general comparison would also be interesting. Cf., for example, Ramsey and Quine on (i) logic (realism), (ii) truth (disquotational); (iii) scientific theories and (iv) pragmatism (American vs. British).

41 See *The Philosophy of F. P. Ramsey*, pp. 146–9.

42 Pp. 137–8.

43 P. 133.

44 *Blue and Brown Books*, Basil Blackwell, Oxford 1958, p. 5. See also 'Some remarks on logical form', in *Ludwig Wittgenstein: Philosophical Occasions 1912–1951*, ed. by J. C. Klagge and A. Nordmann, Hackett, Indianapolis 1993, pp. 29–35.

45 Ramsey's 'Theories' seems to have influenced Wittgenstein also in other ways. See Mathieu Marion, 'Wittgenstein in transition: Phenomenology, hypothesis, and assertions', unpublished paper.

46 For a discussion of this 'theory' see M. Black, *A Companion to Wittgenstein's "Tractatus"*, Cambridge University Press, Cambridge 1964, and G. H. von Wright, 'Wittgenstein on probability', *Wittgenstein*, Basil Blackwell, Oxford 1982, pp. 137–62.

47 *Op. cit*, p. 141.

48 It is most likely that it was Ramsey who taught him this in Puchberg. According to C. Lewy, 'A note on the text of the *Tractatus*', *Mind* **76**, 1967, pp. 417–23, the relevant correction was made in the German text and is in Wittgenstein's handwriting. In a letter to Wittgenstein, dated 15 of September 1924, Ramsey talks about '... a lot of

corrections we made to the translation' (see Lewy's paper; *Ludwig Wittgenstein: Letters to C. K. Ogden*, ed. by G. H. von Wright, Basil Blackwell, Oxford 1973, p. 86; and von Wright, *op. cit.*, pp. 108–9).

49 An interesting hypothesis is that we here have the seed of Wittgenstein's abandonment of the completeness idea. However, it would take us too far away to discuss this surmise and I thus have to leave it for the time being.

50 See S. L. Zabell, 'Ramsey, truth, and probability', *Theoria* **57**, 1991, pp. 211–38.

51 Some parts of Keynes' theory are discussed by Russell in his *The Problem of Philosophy*, Oxford University Press, Oxford 1986 (1912).

52 See 'Mr Keynes on probability', *The Cambridge Magazine* **11**, no. 1, 1912–21, pp. 3–5 (reprinted in *The British Journal for the Philosophy of Science* **40**, 1989); 'Truth and probability', 56–62; and some of the notes in *Notes on Philosophy, Probability and Mathematics*, ed. by M. C. Galavotti, Bibliopolis, Napoli 1991.

53 'Truth and probability', p. 61.

54 *Notes on Philosophy, Probability and Mathematics*, p. 300.

55 See S. L. Zabell, 'Ramsey, truth, and probability', *Theoria* **57**, 1991, pp. 211–38.

56 Don Saari has told me that throwing an American penny results in heads about 29% of the time. This frequency is supported by experiments carried out by his students every year since 1972.

57 See von Wright, *op. cit.*, p. 159.

58 P. 87.

59 *Notes on Philosophy, Probability and Mathematics*, p. 301.

60 P. 93.

61 P. 149.

62 See D. H. Mellor, *The Warrant of Induction*, Cambridge University Press, Cambridge 1988. Mellor gives detailed arguments for reliable processes as a way of warranting induction; and also N.-E. Sahlin, 'Obtained by a reliable process and always leading to success', *op. cit.*

63 See McGinn, *op. cit.*, pp. 24–5.

64 In the *Wiener Ausgabe*, Band 2, *op.cit.*, there are a number of short remarks on probability. Wittgenstein, for example, says that:

> Wahrscheinlichkeit und Galtonsche Photographie.
> Die Galtonsche Photographie is das Bild einer Wahrscheinlichkeit.
> Die Fragen über die Wahrscheinlichkeit hängen auf irgend eine Weise mit denen über die "Unbestimmtheit" der Sinnesdaten zusammen.
> . . .
> Das Gesetz der Wahrscheinlichkeit ist das Naturgesetz was man sieht wenn man blinzelt. (pp. 101–2)

Wittgenstein relates a probability to an indistinct or foggy picture. This is odd because in one respect a probability statement is as exact or determined as anything else. On the other hand we might well percieve a state of uncertainty as inprecise and undetermined; as a picture of many states of the world.

Wittgenstein is also, like Ramsey, talking about "expectations".

Alle /"begründete"/ Erwartung ist Erwartung daß eine bis jetz beobachtete Regel weiter gelten wird.

Die Regel /aber/ muß beobachtet worden sein und kann nicht selbst wieder /nur/ erwartet werden.

Die Theorie der Wahrscheinlichkeit hat es nur in so fern mit /dem Zustand/ der Erwartung zu tun wie etwa die Logik mit dem Denken.

Die Wahrscheinlichkeit hat es vielmehr mit der Form und einem (gewissen) Standard der Erwartung zu tun.

Es handelt sich um die Erwartung daß die zukünftige Erfahrung einem Gesetz entsprechen wird, dem die bisherige Erfahrung entsprochen hat. (p. 129).

To my ears this sounds a lot more like "Truth and probability" than *Tractatus*. I also believe that it supports what has been said above about rule-following.

See also *Philosophical Grammar* and *Philosophical Remarks*, *op.cit.*

65 Merrill B. Hintikka and Jaakko Hintikka, *Investigating Wittgenstein*, Basil Blackwell, Oxford 1986, and Mathieu Marion, 'Wittgenstein in transition: Phenomenology, hypothesis, and assertions', unpublished paper.

66 See the introduction to Ramsey's *Philosophical Papers*, p. xvii.

67 In his 'Ludwig's apple tree: On the philosophical relations between Wittgenstein and the Vienna Circle', *Scientific Philosophy: Origins and Developments*, ed. by F. Stadler, Kluwer, Dordrecht 1993, pp. 27–46, Jaakko Hintikka argues that Wittgenstein's philosophical views around 1932 were close to the views of some of the members of the Vienna Circle (e.g. Carnap). And that they interpreted the *Tractatus* more faithfully than 'Wittgenstein's own alleged followers'. This is probably correct, but what Hintikka does not point out is that there is support for his thesis in Ramsey's work. Ramsey was, I think, a trustworthy (faithful, whatever that means) interpreter of *Tractatus*, and his reading of the book can preferably be compared to the interpretation by members of the Vienna Circle. But, saying this, it is also important to emphasize that Ramsey was *not* a positivist or logical positivist (see *The Philosophy of F. P. Ramsey*, Chapter 5, *op. cit.*).

II

FORMAL TOOLS: INDUCTION, OBSERVATION AND IDENTIFIABILITY

Poznań Studies in the Philosophy of the Sciences and the Humanities
1997, Vol. 51, pp. 87–99

Theo A.F. Kuipers

THE CARNAP–HINTIKKA PROGRAMME IN INDUCTIVE LOGIC[1]

1. Qualitative and quantitative confirmation theory

According to the hypothetico-deductive method a theory is tested by examining its implications. The result of an individual test of a general hypothesis stated in observation terms, and hence of a general test implication of a theory, can be positive, negative or neutral. If it is neutral the test was not well devised, if it is negative, the hypothesis, and hence the theory, has been falsified. *Qualitative* confirmation theory primarily aims at further explicating the intuitive notions of neutral and positive test results. Some paradoxical features discovered by Hempel (1965) and some queer predicates defined by Goodman (1955) show that this is not an easy task.

Assuming that a black raven confirms the hypothesis "All ravens are black" and that confirmation is not affected by logically equivalent reformulations, Hempel argued that not only a non-black non-raven but, even more counterintuitively, also a black non-raven confirms it. Goodman argued that not all predicates guarantee the "projectibility" of a universal hypothesis from observed to non observed cases. For example, if "grue" means "green, if examined before t" and "blue, if not examined before t", a green emerald discovered before t would not only confirm "All emeralds are green" but also "All emeralds are grue", and hence, assuming that consequences are also confirmed, even "All emeralds not examined before t are blue". Whereas Goodman succeeded in formulating criteria for acceptable predicates, in terms of their relative "entrenchment" in previously successfully projected generalizations, up to now nobody has given a generally accepted solution to Hempel's qualitative riddles.[2]

So far for the explication of non-negative test results. Negative results are indirectly explicated by explicating non-negative results. However, further analysis of negative results, including falsification, is on its turn interesting and important for qualitative confirmation theory, as has become evident from the Duhem–Quine problem.

Below we treat *quantitative*, more specifically, probabilistic confirmation theory, which aims at explicating the idea of confirmation as increasing probability due to new evidence. Carnap (1950) introduced this perspective and pointed confirmation theory toward the search for a suitable notion of logical or inductive probability. Generally speaking, such probabilities combine indifference with inductive properties. For this reason, probabilistic confirmation theory in this style is also called inductive probability theory or, simply, inductive logic.

In Section 2 Carnap's continuum of inductive methods, and its generalization by Stegmüller, will be presented. In Section 3 Roberto Festa's proposal is sketched for estimating the (generalized) optimum method. Section 4 deals with the approaches of the present author and Brian Skyrms, and hints upon one by Maria di Maio, of designing Carnapian-like systems that take considerations of analogy by similarity and proximity into account.

Section 5 presents Hintikka's first approach to universal hypotheses, a plausible generalization, Hintikka's second approach, together with Ilkka Niiniluoto, and their mutual relations. The section also hints upon the extension of such systems to polyadic predicates by Raimo Tuomela and to infinitely many (monadic) predicates, e.g. in terms of partitions by Sandy Zabell.

Finally, Section 6 indicates some of the main directions of application of the Carnap–Hintikka programme.

The Carnap–Hintikka programme may be seen as a special version of the Bayesian approach to quantitative confirmation and inductive inference. See Howson and Urbach (1989) for a handbook on the Bayesian approach of confirmation of deterministic and statistical hypotheses, paying some attention to the Carnap–Hintikka version and to the classical statistical approach. There are also non-Bayesian approaches, from Popper's degrees of corroboration to L.J. Cohen's "baconian probabilities" (Cohen 1977/1991) and the degrees of confidence of orthodox statistics. Cohen and Hesse (1980) present several quantitative and qualitative perspectives.

2. *The continuum of inductive systems*[3]

Mainly by his *The Continuum of Inductive Methods* (1952) Carnap started a fruitful research programme centering around the famous λ-continuum. The probability systems in this programme can be described in terms of individuals and observation predicates or in terms of trials and observable outcomes. The latter way of presentation will be used here in an informal way. Moreover, we will presuppose an objective probability process, although the systems to be presented can be applied in other situations as well.

Consider a hidden wheel of fortune. You are only told, truthfully, that it has precisely four coloured segments, BLUE, GREEN, RED, and YELLOW, without further information about the relative size of the segments. So you do not know the objective probabilities. What you subsequently learn are only the outcomes of successive trials. Given the sequence of outcomes e_n of the first n trials, your task is to assign reasonable probabilities, $p(R/e_n)$, to the hypothesis that the next trial will result in, for example, RED.

There are several ways of introducing the λ-continuum, but the basic idea behind it is that it reflects gradually learning from experience. According to Carnap's favourite approach $p(R/e_n)$ should depend only on n and the number of occurrences of RED thus far, n_R. Hence, it is indifferent to the colour distribution among the other $n-n_R$ trials. This is called the principle of *restricted relevance*. More specifically, Carnap wanted $p(R/e_n)$ to be a special weighted mean of the observed relative frequency n_R/n and the (reasonable) initial probability 1/4. This turns out to leave room for a continuum of (C-)systems, the λ-continuum, $0<\lambda<\infty$:

$$p(R/e_n) = (n_R+\lambda/4)/(n+\lambda) = n/(n+\lambda) \,.\, (n_R/n) + \lambda/(n+\lambda) \,.\, (1/4)$$

Note that the weights $n/(n+\lambda)$ and $\lambda/(n+\lambda)$ add up to 1 such that for increasing n the former gradually increases from 0 to 1 and hence the latter from 1 to 0. Moreover, the larger the value of λ the slower the first increases, at the expense of the second, i.e., the slower one is willing to learn from experience.

C-systems have several attractive indifference, confirmation and convergence properties (for a systematic treatment see Kuipers (1978)). The most important are these:

- *initial indifference*, $p(R/e_0) = p(R)=1/4$,
- *order indifference* or *exchangeability*: the resulting prior probability, $p(e_n)$, for e_n, does not depend on the order of the results of the trials, i.e., $p(e_n) = p(e_n{}^*)$ for any permutation $e_n{}^*$ of e_n; this property implies
- *conditional order indifference* or *conditional exchangeability*, i.e., $p(RB/e_n) = p(BR/e_n)$, where the first probability concerns the hypothesis that, given e_n, first RED and then BLUE will occur, and the second the reverse order;

- *instantial confirmation*: if e_n is followed by RED this is favourable for RED, i.e., $p(R/e_nR) > p(R/e_n)$,
- *instantial convergence*: $p(R/e_n)$ approaches n_R/n for increasing n.

C-systems satisfy moreover:

- *universal-instance confirmation*, i.e., suppose for example that RED and BLUE are the only colours that have occurred according to e_n, then $p(R \vee B/e_n) < p(R \vee B/e_nR \vee B)$. The name "universal-instance confirmation" is plausible in view of the fact that the first probability concerns the hypothesis that, given e_n, the next result will also be RED or BLUE, i.e., will be in agreement with the universal hypothesis that all future results are RED or BLUE, and the second probability concerns the same hypothesis for the next trial, but now not only given e_n, but also that it was followed by RED or BLUE, i.e., a result in agreement with the universal hypothesis,
- *universal-instance convergence*, i.e., $p(R \vee B/e_n)$ approaches 1 for increasing n if only RED and BLUE remain occurring according to e_n.

Though C-systems have the properties of universal-instance confirmation and convergence, some direct form of confirmation of, and convergence to, universal hypotheses is excluded. The reason is that C-systems in fact assign zero prior probability to all universal generalizations, for instance to the hypothesis that all results will be RED. Of course, this is desirable in the described situation, but if you were only told that there are *at most* four coloured segments, you would like to leave room for this possibility.

Stegmüller (1973) generalized C-systems to (GC-)systems, in which the uniform initial probability 1/4 is replaced by arbitrary non-zero values, e.g. $p(R)$, leading to the special values:

$$p(R/e_n) = (n_R + p(R).\lambda)/(n+\lambda)$$

Apart from the intended weakening of restricted relevance all properties of C-systems continue to apply.

Johnson (1932) anticipated C-systems already in 1932, including their axiomatic foundation. He introduced C-systems as a natural representation of symmetric so-called Dirichlet distributions, which are well-known in Bayesian statistics. Zabell (1982) rediscovered this and showed that non-symmetric versions are GC-systems. For a detailed analysis of the relation between GC-systems and Dirichlet distributions, see Festa (1993). Skyrms

(1993a) points out that the analogue of C- and GC-systems for a continuum of possible values can be found in the statistical literature.

3. Optimum inductive systems

Carnap (1952) proved that for certain kinds of objective probability processes, such as a wheel of fortune, there is an optimal value of λ, depending on the objective probabilities, in the sense that the average mistake may be expected to be lowest for this value. Surprisingly, this optimal value is independent of n. In fact he proved the existence of such an optimal value for any multinomial process, i.e., sequences of independent experiments with constant objective probabilities for a fixed finite number of outcomes. Of course, in actual research, where we do not know the objective probabilities, this optimal value cannot be calculated. Carnap did not raise the question of a reasonable estimate of the optimal value for a specific objective process, for he saw the problem of selecting a value of λ primarily as a choice that each scientist had to make in general. However, the question of a reasonable estimate has attracted the attention of later researchers.

In reaction to some "hyperempiricist" estimations of the optimal value of λ Kuipers (1986), solely based on the available evidence, Festa (1993) proposed basing the estimate on "contextual" knowledge of similar processes in nature. For example, in ecology one may know the relative frequencies of certain species in different habitats before one starts the investigation of a new habitat. Festa first generalizes the proof of the existence of an optimal value of λ to the classes of GC-systems with the same initial probabilities. Then he formulates for such a class two solutions of the estimation problem, which turn out to be essentially equivalent. Moreover, one of these solutions relates the research area of inductive logic to that of Popperian truth approximation: the optimum solution may be expected to be the most efficient way of approaching the objective or true probabilities.

Unfortunately, wheels of fortune do not constitute a technological (let alone a biological) kind, for which you can use information about previously investigated instances. But if you had knowledge of a random sample of all existing wheels of fortune, Festa's approach would work on the average for a new, randomly drawn, one.

4. Inductive analogy by similarity and proximity

Carnap struggled with the question of how to include analogy considerations into inductive probabilities. His important distinction between two kinds of analogy – by similarity and by proximity – was posthumously published Carnap (1980).

Suppose you find GREEN more similar to BLUE than to RED. Carnap's intuition of *analogy by similarity* is that, for instance, the posterior probability of GREEN should be higher if, other things being equal, some occurrences of RED are replaced by occurrences of BLUE. The background intuition, not valid for artificial wheels of fortune, is that similarity in one respect (e.g. colours) will go together with similarities in other respects (e.g. objective probabilities). However, explicating what precise principle of analogy one wants to fulfil, has turned out to be very difficult.

Niiniluoto (1988) discusses, among others, proposals for inductive analogy by similarity by D. Costantini, T. Kuipers, I. Niiniluoto and W. Spohn. He concludes that my approach (Kuipers 1984, 1988) is the most elegant one, for which reason it may not be too immodest to select this approach for further attention. It goes in terms of *virtual trials*, which is a way of determining how analogy influence is to be distributed after n trials. To present the main idea, it is convenient to start from another heuristic for introducing C-systems. A special value $p(R/e_n)$ can be seen as the ratio of n_R *real* plus $\lambda/4$ *virtual* occurrences of RED and n real plus λ virtual trials. Similarly, we may introduce virtual occurrences and trials, accounting for analogy influences. The number of *virtual analogy occurrences* obtained by RED depends on the particular number of occurrences of the other colours and the distances of them to RED, expressing their dissimilarity. To put it in a nutshell for the case indicated above, BLUE gets more analogy credits than RED from an occurrence of GREEN. The (total) number of *virtual analogy trials* determines how much analogy influence in total is to be distributed after n trials.

It is interesting to note that Carnap's favourite way of presenting C-systems can now also be extended to the special values accounting for analogy. They can be rewritten as the weighted mean of the observed relative frequency, the initial probability and an analogy factor.

The resulting systems of virtual analogy have the same confirmation and convergence properties as C-systems (see Section 2). Moreover, they satisfy a principle of analogy that is in general more plausible than Carnap's principle indicated above. They only satisfy Carnap's principle under special conditions. Roughly, the new principle says that replacement of a previous occurrence of BLUE by GREEN makes less difference for the probability that the next trial will be BLUE, than replacement of a previous occurrence of RED by GREEN makes for the probability that the next trial will be RED.

Unfortunately, systems of virtual analogy are not exchangeable: the order of results makes a difference, e.g. the initial and posterior probability of the sequence RB differs in general from that of BR. There are, however, exchangeable systems with some kind of analogy by similarity. For example, Skyrms (1993b) has sketched a Bayesian approach that uses information about possible gravitational bias of a symmetric wheel of fortune to determine the influence of analogy. To be more precise, Skyrms's basic idea, the full elaboration of which still has to start, is the following. Let (only in this paragraph) the wheel of fortune be symmetric, i.e., let the four segments be of equal size, with the possibility that it is gravitationally biased in favour of one colour, the most strongly at the expense of the opposite colour and moderately at the expense of the remaining two. Analogy influences may now be objectively based on the resulting four possibilities and implemented in a Bayesian conditionalized probability system along the following lines. Assuming that we know the relative positions of the coloured segments, but do not know in favour of which colour the possible bias is, suggests four metahypotheses, to be implemented in a Bayesian conditionalized probability system by imposing for each conditional system a GC-system with suitable initial values in accordance with the relevant metahypothesis. For example, the value 0.5 for the favoured colour, 0.1 for the opposite one and 0.2 for the two other colours. Starting with equal probabilities for the metahypotheses, it is evident that the resulting "hyperCarnapian" system realizes influences of analogy by similarity of sorts.

It is also possible to construct a completely different kind of exchangeable systems that have analogy by similarity influences of another sort (Kuipers 1984). They are obtained by assuming the right kind of prior distribution in systems leaving room for universal generalizations, to be dealt with in the next section.

The most recent exchangeable approach is by Di Maio (1996). She weakens restricted relevance by a combination of a plausible suggestion by Carnap (1980) with a newly invented, also plausible, principle in such a way that it leads to exchangeable systems, most of which are sensitive to analogy by similarity. However, the resulting systems still are everything but transparant.

Future research has to make clear whether there is any relation between the three indicated exchangeable approaches.

Carnap also set the stage for *analogy by proximity*: differences may arise due to the number of trials between different occurrences of the same outcome. This may concern occurrences in the past, i.e., in the evidence, or in the future, i.e., in the hypothesis. Both are studied in Kuipers (1988). For past influences it is again possible to use the idea of virtual occurrences. For future proximity influences a general approach is still missing, but the problem also redirects the attention to the question of how to approach a so-called Markov chain. Piecemeal approaching it by using C-systems as component systems is the plausible idea, which has been elaborated by Skyrms (1991a) in Bayesian terms and derived by Zabell (1995) by an extension of Johnson's (1932) axiomatic way of deriving C-systems.

5. Universal generalizations[4]

Carnap was well aware that C-systems do not leave room for non-analytic universal generalizations. In fact, all systems presented thus far have this problem. Although tolerant of other views, Carnap himself was inclined to downplay, for this and other reasons, the theoretical importance of universal statements. However, as mentioned before, if you are only told that the wheel of fortune has at most four coloured segments, you might want to leave room for statements claiming for example that all results will be RED.

Hintikka (1966) took up the problem of universal statements. His basic idea for such a system, here called H-system, is a Bayesian conditionalized probability system of the following kind: (1) a prior distribution of non-zero probabilities for (the outcome spaces of) mixed (existential and universal) statements telling that all and only the colours so-and-so will occur in the long run, (2) for each of these so-called constituents an appropriate C-system as conditional system, and (3) Bayes's theorem for conditionalizing posterior probabilities is applied. Hintikka, more specifically, proposed a two-dimensional system, the so-called α-λ-continuum, by taking a uniform value of λ for all conditional systems and the prior distribution proportional to the probability that α virtual individuals or trials are, according to the C-system for all possible colours, compatible with the relevant constituent. Here α is a finite number to be chosen by the researcher. The larger it is chosen, the lower the probability of all constituents based on fewer outcomes than possible.

However, as shown in Kuipers (1978), H-systems in general have already, besides the instantial, and universal-instance, confirmation and convergence properties, the desired property of universal confirmation, i.e., the probability of a not yet falsified universal statement increases with another instance of it. For example, assuming that only RED occurred according to e_n, $p(H_R/e_n)$ is smaller than $p(H_R/e_nR)$, where H_R indicates the constituent that only RED will occur in the long run. H-systems have, moreover, universal convergence, i.e., the probability of the strongest not yet falsified universal statement converges to 1. For example, $p(H_R/e_n)$ approaches 1 for increasing n as long as only RED continues to occur. For the special case of H-systems belonging to Hintikka's α-λ-continuum holds that, for increasing parameter α, universal confirmation is smaller and universal convergence slower.

An easily definable and transparant special class of H-systems, SH-systems, has become particularly interesting from a finitistic point of view, viz. the subclass of H-systems based on an arbitrary prior distribution and with conditional C-systems with λ proportional to the corresponding number of possible outcomes. Note that, although SH-systems may have Hintikka's originally proposed prior distribution, they nevertheless do not belong to the α-λ-continuum in that case, due to non-uniform character of λ in SH-systems.

SH-systems became interesting due to a *prima facie* completely different type of systems introduced by Hintikka and Niiniluoto (1976/1980). Kemeny (1963) showed already that for three or more outcomes C-systems could be axiomatically introduced by imposing the indifference properties of exchangeability and (positive) restricted relevance. By weakening restricted relevance in some way Stegmüller extended this to the already mentioned GC-systems (see Section 2). Hintikka and Niiniluoto showed that exchangeability and another weakening of restricted relevance, according to which $p(R/e_n)$ may not only depend on n and n_R but also on the number of different colours reported in e_n, leads to a very interesting class of systems. Such NH-systems are determined by "finite" principles and parameters, i.e., principles and parameters related to finite sequences of outcomes.

Now, the theorem can be proved (Kuipers 1978) that the classes of NH- and SH-systems coincide. The consequence is that SH-systems can be axiomatically introduced as NH-systems. After calculating the corresponding prior distribution, the attractive confirmation and convergence properties and many other characteristics can be studied by the transparant SH-representation of these systems.

H-systems, and hence SH/NH-systems, can be generalized by using plausibly related initial probabilities for the conditional systems. Such GH- (GSH/GNH-) systems keep all properties, except some further weakening of restricted relevance for the conditional C-systems and the NH-systems. Tuomela (1966) showed how to extend Hintikka's approach to polyadic predicates or ordered outcomes. It still needs to be investigated whether the same idea can be used to extend H-systems and even GH-systems to include polyadic predicates.

There is a plausible "delabeled" reformulation of H-systems, hence of SH/NH-systems, which can be extended to the very interesting case of an unknown denumerable number of possible outcomes (Kuipers 1978). Presenting the delabeling in terms of exchangeable partitions, Zabell has studied this extension by principles, in strength between those of C- and NH-systems, leading to an interesting class of systems with three parameters (Zabell to appear), anticipated in Zabell (1992).

6. *Applications*

In contrast to Lakatos's claim (Lakatos 1968) the Carnap–Hintikka programme in inductive logic, combines a strong internal dynamics with applications in several directions. Systems of inductive probability were primarily intended for explicating confirmation as increasing probability. One may even define a quantitative degree of confirmation; namely, as the difference or ratio of the posterior and the initial or prior probability. There are other interesting types of applications as well. Carnap (1971) and Stegmüller (1973) stressed that they can be used in decision making, and Skyrms (1991b) applies them in game theory by letting players update their beliefs with C-systems. Costantini et al. (1982) use them in a rational reconstruction of elementary particle statistics. Festa (1993) suggests several areas of empirical science where optimum inductive systems can be used. Finally, for universal hypotheses, Hintikka (1966), Hilpinen (1968)[5] (see also Hintikka and Hilpinen 1966) and Pietarinen (1972)[6] use systems of inductive probability to formulate rules of acceptance, i.e., the task of inductive logic in the classical sense. Niiniluoto (1987) uses such systems to estimate degrees of verisimilitude, hence giving rise to a second fusion of inductive logic and the Popperian truth approximation programme.[7]

Department of Philosophy of Science, Logic and Epistemology
Faculty of Philosophy
University of Groningen
A-weg 30
9718 CW Groningen, Netherlands
fax: 50-3636151
e-mail: T.A.F.Kuipers@PHILOS.RUG.NL

NOTES

1 This article is a revised and extended version of my "Confirmation Theory" which will appear as lemma in the new edition of the *Encyclopedia of Philosophy*, edited by Edward Craig, and published by Routledge, London.

2 Scheffler (1967) still offers a lucid and detailed analysis of qualitative confirmation, including the confirmation paradoxes and Goodman's problem. See Glymour (1980) for a refined version of Hempel's definition of qualitative confirmation, which attracted the most attention.

3 For a technical survey of most of the systems presented in Sections 2 and 5 see Kuipers (1980).

4 See note 3.

5 Hilpinen (1968) gives a detailed analysis of Hintikka's two-dimensional continuum and corresponding rules of acceptance proposed in Hintikka and Hilpinen (1966).

6 Pietarinen (1972) gives an adequate survey of the main ideas and problems in the Carnap–Hintikka programme around 1970.

7 I like to thank Roberto Festa, Michel ter Hark, Alfons Keupink, Erik Krabbe, Jeanne Peijnenburg and Sjoerd Zwart for their comments on the first extended version.

REFERENCES

Carnap, R. (1950). *Logical Foundations of Probability*. Chicago: University of Chicago Press.

Carnap, R. (1952). *The Continuum of Inductive Methods*. Chicago: University of Chicago Press.

Carnap, R. (1971). Inductive Logic and Rational Decisions. In: R. Carnap and R. Jeffrey (Eds.), *Studies in inductive logic and probability. Vol. I.* Berkeley: University of California Press, pp. 5–31.

Carnap, R. (1980). A Basic System of Inductive Logic, Part 2. In: Jeffrey (1980), pp. 7–155.

Cohen, L.J. (1977/1991). *The Probable and the Provable*. Oxford: Clarendon Press, 1977. Second Edition: Aldershof: Gregg Revivals, 1991.

Cohen, L.J. and M. Hesse, (Eds.), (1980). *Applications of Inductive Logic*. Oxford: Clarendon Press.

Costantini, D., M.C. Galavotti and R. Rosa (1982). A Rational Reconstruction of Elementary Particle Statistics. *Scientia* **76** (117), 151–9.

Festa, R. (1993). *Optimum Inductive Methods*. Dordrecht: Kluwer.

Glymour, C. (1980). *Theory and Evidence*. Princeton: Princeton University Press.

Goodman, N. (1955). *Fact, Fiction, and Forecast*. Cambridge: Harvard University Press.

Hempel, C. (1965). *Aspects of scientific explanation*. New York: The Free Press.

Hilpinen, R. (1968). *Rules of Acceptance and Inductive Logic*. Amsterdam: North-Holland.

Hintikka, J. (1966). A Two-dimensional Continuum of Inductive Methods. In Hintikka and Suppes (1966), pp. 113–32.
Hintikka, J. and R. Hilpinen (1966). Knowledge, Acceptance and Inductive Logic. In: Hintikka and Suppes (1966), pp. 96–112.
Hintikka, J. and I. Niiniluoto (1976/1980). An Axiomatic Foundation for the Logic of Inductive Generalization. In: M. Przelecki, K. Szaniawski and R. Wojcicki (Eds.), *Formal Methods in the Methodology of the Empirical Sciences*. Dordrecht: Reidel, 1976, pp. 57–81. Reprinted in: Jeffrey (1980), pp. 157–81.
Hintikka, J. and P. Suppes, (Eds.), (1966). *Aspects of Inductive Logic*. Amsterdam: North-Holland.
Howson, C. and P. Urbach (1989). *Scientific Reasoning. The Bayesian Approach*. La Salle: Open Court.
Jeffrey, R., (Ed.), (1980). *Studies in Inductive Logic and Probability Vol. II*. Berkeley: University of California Press.
Johnson, W. (1932). Probability: the Deductive and Inductive Problems. *Mind* **41** (164), 409–23.
Kemeny, J. (1963). Carnap's Theory of Probability and Induction. In: P.A. Schillp (Ed.), *The Philosophy of Rudolf Carnap*. LaSalle: Open Court, pp. 711–38.
Kuipers, T. (1978). *Studies in Inductive Probability and Rational Expectation*. Dordrecht: Reidel.
Kuipers, T. (1980). A Survey of Inductive Systems. In: Jeffrey (1980), pp. 183–92.
Kuipers, T. (1984). Two Types of Inductive Analogy by Similarity. *Erkenntnis* **21**, 63–87.
Kuipers, T. (1986). Some Estimates of the Optimum Inductive Method. *Erkenntnis* **24**, 37–46.
Kuipers, T. (1988). Inductive Analogy by Similarity and Proximity. In: D.H. Hellman (Ed.), *Analogical Reasoning*. Dordrecht: Kluwer, pp. 299–313.
Lakatos, I. (Ed.) (1968). *The Problem of Inductive Logic*. Amsterdam: North-Holland.
di Maio, M.C. (1996). Predictive Probability and Analogy by Similarity in Inductive Logic. *Erkenntnis* **43**, 369–94.
Niiniluoto, I. (1987). *Truthlikeness*. Dordrecht: Kluwer.
Niiniluoto, I. (1988). Analogy and Similarity in Scientific Reasoning. In: D.H. Hellman (Ed.), *Analogical Reasoning*. Dordrecht: Kluwer, pp. 271–98.
Pietarinen, J. (1972). *Lawlikeness, Analogy and Inductive Logic*. Amsterdam: North-Holland.
Scheffler, I. (1967). *The Anatomy of Inquiry*. New York: Alfred A. Knopf.
Skyrms, B. (1991a). Carnapian Inductive Logic for Markov Chains. *Erkenntnis* **35**, 439–60.
Skyrms, B. (1991b). Inductive Deliberation, Admissable Acts and Perfect Equilibrium. In: M. Bacharach and S. Hurley (Eds.), *Foundations of Decision Theory*. Oxford: Blackwells, pp. 220–41.
Skyrms, B. (1993a). Carnapian Inductive Logic for a Value Continuum. In: H. Wettstein (Ed.), *The Philosophy of Science* (Midwest Studies in Philosophy, Vol. 18). South Bend: University of Notre Dame Press, pp. 78–89.
Skyrms, B. (1993b). Analogy by Similarity in HyperCarnapian Inductive Logic. In: J. Earman (Ed.), *Philosophical Problems of the Internal and External Worlds*.

Essays in the Philosophy of Adolf Grünbaum. Pittsburgh: University of Pittsburgh Press, pp. 273–82.

Stegmüller, W. (1973). *Carnap II: Normative Theorie des Induktiven Räsonierens*. Berlin: Springer.

Tuomela, R. (1966). Inductive Generalization in an Ordered Universe. In: Hintikka and Suppes (1966), pp. 155–74.

Zabell, S. (1982). W.E. Johnson's "Sufficientness" Postulate. *Annals of Statistics* **10** (4), 1091–9.

Zabell, S. (1992). Predicting the Unpredictable. *Synthese* **90**, 205–32.

Zabell, S. (1995). Characterizing Markov Exchangeable Sequences. *Journal of Theoretical Probability* **8**, 175–8.

Zabell, S. (to appear). The Continuum of Inductive Methods Revisited. To appear in: J. Earman and J. Norton (Eds.) (1996), *The Cosmos of Science. Pittsburgh–Konstanz Series in the Philosophy and History of Science*. Pittsburgh/Konstanz: University of Pittsburgh Press/Universitätsverlag Konstanz.

Poznań Studies in the Philosophy of the Sciences and the Humanities
1997, Vol. 51, pp. 101–115

Isaac Levi

CAUTION AND NONMONOTONIC INFERENCE

Reasoning from premises $h_1, h_2, \ldots, h_n$ to conclusion g is reasoning from the single premise $h = h_1 \wedge h_2 \wedge \quad \wedge h_n$. Assume that we can represent an agent's state of full belief (in all relevant respects) by a deductively closed set **K** of sentences in some regimented language **L**. The reasoning from h to g is representable as a transformation $T(\mathbf{K},h)$ of the inquirer's *corpus* to another *potential corpus* **K**. In this discussion, I shall suppose that **K** is consistent. To represent reasoning, a transformation $T(\mathbf{K},h)$ should contain h. The conclusions derivable from h via reasoning are the sentences in $T(\mathbf{K},h)$.

Different kinds of reasoning from premises may be construed as different types of transformations of potential corpora. *Expansion by adding h* is an especially important sort of transformation.

> The *expansion* $\mathbf{K}_h^+$ of **K** by adding h is the smallest potential corpus containing $Cn(\mathbf{K} \cup \{h\})$ – i.e., the logical consequences of the sentences in **K** together with h. If no such potential corpus exists, the transformation goes undefined.

I shall assume that the set of potential corpora meets the following condition:

> *Closure under expansion*: For every consistent potential corpus **K** and every h, there is a smallest potential corpus containing $Cn(\mathbf{K} \cup \{h\})$.

In the applications in which we shall be primarily interested, the set of potential corpora will constitute at least a Boolean algebra and following condition will obtain:

Closure under strict expansion: For every consistent potential corpus **K** and every h, $Cn(\mathbf{K} \cup \{h\})$ is a potential corpus.

Under this condition, $\mathbf{K}_h^+ = Cn(\mathbf{K} \cup \{h\})$.[1]

We can represent reasoning from h to g (relative to **K**) as showing that g is a member of $\mathbf{K}_h^+$ – i.e., that g is a deductive consequence of **K** and h. In this case, the reasoning can be called an *inference* and, indeed, a deductive inference from h to g relative to **K**. When **K** is the set of truths of deductive logic, g is a deductive consequence of h.

Not all reasoning is inference. Sometimes the supposition h of the reasoning is added not to the initial belief state **K** but to the *contraction* $\mathbf{K}_{\sim h}^-$ removing $\sim h$ from **K** in an "optimal" way or $\mathbf{K}_h^-$ removing h from **K** in an optimal way. Whatever an optimal contraction might be, it calls for contravening assumptions in the initial belief state so that the reasoning from h cannot be viewed as an inference from h relative to a background that remains unquestioned.

In this discussion, I shall suppose that an optimal contraction satisfies the postulates for contraction reproduced in Gärdenfors (1988) with the exception of the *Recovery Postulate* that stipulates that $\mathbf{K} \subseteq [\mathbf{K}_h^-]_h^+$. A transformation adding a supposition h to a contraction of **K** is currently called a *revision*.

> An *AGM revision* $\mathbf{K}_h^*$ of **K** by adding h is a transformation of consistent **K** satisfying the eight postulates for revision imposed by Alchourrón, Gärdenfors and Makinson (1985). These postulates insure that h is in the revision, that the revision is consistent, that the revision is identical with the expansion of **K** by adding h when h is consistent with **K** and is a consistent potential expansion otherwise. That is to say, $\mathbf{K}_h^* = \mathbf{K}_h^+$ if h is consistent with **K**. If h is not consistent with (consistent) **K**, $\mathbf{K}_h^*$ is the expansion by adding h of the recommended *contraction* $\mathbf{K}_{\sim h}^-$ removing $\sim h$ from **K**.

Alchourrón, Gärdenfors and Makinson (1985) require recommended contractions to satisfy the Recovery Postulate. As Makinson (1987) has noted, however, the AGM postulates for revision remain intact if AGM contraction is replaced by a standard for recommended contraction that agrees with AGM contraction except for the Recovery Postulate.

Reasoning from h to g (relative to **K**) that is representable by an AGM revision is not an inference because the mode of reasoning allows for *belief contravention* when the premise h is inconsistent with **K**. There are two kinds of settings where h can be added to **K** even though it is inconsistent with it. One is where the inquirer genuinely comes to believe that *h* and must then give up some background beliefs in **K**. One may subsequently make inferences from *h* relative to the contracted corpus; but not from *h* and **K**. Adding *h* inconsistent with **K** can also take place in *suppositional reasoning* where a claim is hypothesized to be true for the sake of the argument even though the supposition is judged false. Here too it would be misleading to speak of inference from *h* relative to **K** because some of the background beliefs have to be removed from **K** for the sake of the argument.

> A *Ramsey Revision* $\mathbf{K}_h^{*r}$ is an AGM revision except in the case where *h* is in **K**. When *h* is in **K**, $\mathbf{K}_h^{*r}$ is the expansion by adding *h* of the optimal contraction of **K** removing *h*.

If the Recovery Postulate holds, Ramsey Revision is identical with AGM Revision. If the Recovery Postulate fails, the postulates for Ramsey Revision are weaker than the postulates for AGM revision. See Levi (1996) for details. Ramsey Revision improves on AGM revision in cases like one where it is known that a coin has been tossed and landed heads and one wants to point out that the coin's being tossed cannot account for its landing heads. The supposition that it was tossed does not allow one to reason to the conclusion that it landed heads even though it is known that it was tossed and did land heads. AGM revision insists, however, on a contrary verdict.

My concern here is not, however, to argue for the superiority of Ramsey Revision over AGM Revision – although I do think it furnishes a better account of suppositional reasoning. In the subsequent discussion, I shall, in spite of my reservations with it, use AGM revision. The point I mean to emphasize is that suppositional reasoning, whether it is based on AGM or Ramsey revision, is not inference because beliefs in the background beliefs relative to which reasoning from *h* takes place are contravened.

Inference from *h* to *g* relative to **K** according to expansion is deductive and not belief contravening. It is also *monotonic* in the sense that if *g* is legitimately obtained from *h* relative to **K**, then for any sentence *f*, *g* is legitimately inferred from $h \wedge f$ relative to **K**. Reasoning according to AGM and Ramsey revision is not deductive but can be belief contravening. It is *nonmonotonic*.

Are there any useful transformations of potential belief states into other such states that license non belief contravening inferences and are non monotonic? I believe there are.

Let $E(\mathbf{K})$ be any transformation of potential states into other potential states such that $\mathbf{K} \subseteq E(\mathbf{K})$ for every $\mathbf{K}$. Unless $E(\mathbf{K})$ is the identity function, it shall be called an *ampliative expansion function* or an *inductive expansion function*. When E is the identity function, it is *explicative*. The idea is that if $\mathbf{K}$ is the inquiring agent's belief state at a given time, $\mathbf{K}$ represents the agent's total information at that time. $E(\mathbf{K})$ represents the results of expanding $\mathbf{K}$ by adding some proposition d. When E is the identity transformation, d is already in $\mathbf{K}$ so that the inference is a mere "explication" of the contents of $\mathbf{K}$. When E is ampliative, d is a proposition obtained by genuine inductive or ampliative *inference* from the total evidence in $\mathbf{K}$.

Once we have the notion of a generalized expansion function E, inductive inference from h to g can be understood to be the result of ordinary expansion $\mathbf{K}_h^+$ of $\mathbf{K}$ by adding h followed by an ampliative expansion $E(\mathbf{K}_h^+)$ of $\mathbf{K}_h^+$. This shall be called an *inductively extended* expansion of $\mathbf{K}$. Reasoning from h to g is in this case not belief contravening. It is ampliative or inductive. And it is nonmonotonic.[2]

To see that ampliative inference must be nonmonotonic, let h be any sentence consistent with $\mathbf{K}$ and let $h \wedge f$ also be consistent with $\mathbf{K}$ but inconsistent with $E(\mathbf{K})$. Since both $E(\mathbf{K}_h^+)$ and $E(\mathbf{K}_{h \wedge f}^+)$ are both consistent, $E(\mathbf{K}_{h \wedge f}^+)$ must be belief contravening with respect to $E(\mathbf{K}_h^+)$. The latter cannot be a subset of the former and, hence, inference from h relative to $\mathbf{K}$ via inductively extended expansion cannot be monotonic.

Notice that although $E(\mathbf{K}_{h \wedge f}^+)$ contravenes $E(\mathbf{K}_h^+)$ and this point is critical to the argument, ampliative expansion and inductively extended expansion do not represent shifts from the second of these belief states to the first. Both are to be understood as transformations of $\mathbf{K}$. In both cases, the shift from $\mathbf{K}$ is not belief contravening and, hence, in both cases the reasoning can be understood to be inference (from h and from $h \wedge f$ respectively) relative to $\mathbf{K}$.

One can, of course, imagine a scenario where an inquirer has shifted from $\mathbf{K}$ to $E(\mathbf{K}_h^+)$ via an inductively extended expansion and then added f inconsistent with the inductively extended expansion. Assuming (purely for the sake of the argument) that the appropriate kind of modification is an AGM revision, the result is $[E(\mathbf{K}_h^+)]_f^*$. It is far from obvious that $[E(\mathbf{K}_h^+)]_f^* = E(\mathbf{K}_{h \wedge f}^+)$. However, for ampliative expansion functions E that are *permutable* in the sense that $E(\mathbf{K}_h^+) = [E(\mathbf{K})]_h^*$ this equation will hold.

If ampliative expansion is, indeed, permutable, *inductively extended AGM revision* $E(\mathbf{K}_h^*)$ can be equated with unextended AGM revision of the ampliative expansion of $\mathbf{K}$ – i.e., $[E(\mathbf{K})]_h^*$. In this sense, the principles of nonmonotonic inductive inference of the non belief contravening sort represented by inductively extended expansion can be reduced to the principles of belief contravening nonmonotonic reasoning. To be sure, the reduction is not complete. In addition to the initial corpus **K**, one must be given the value of *E*(**K**). Then one can determine the ampliative expansions of all consistent expansions of **K**. This means that the set of sentences nonmonotonically implied by sentences in **K** is a proper superset of **K**. $T \mid\!\sim h$ @ **K**, which means that the tautology *T* nonmonotonically implies *h* relative to **K**, does not itself require that *h* be in **K** but only that it be in *E*(**K**). Nonetheless, the nonmonotonic inferential relation relative to **K** generated by inductively extended expansion will satisfy all the requirements for a *rational* consistency preserving inferential relation in the sense of Lehmann and Magidor (1992). These requirements may be derived from an AGM revision function provided that *E*(**K**) replaces **K** in the AGM postulates for revision and the definition of the nonmonotonic inferential relation in terms of AGM revision is appropriately modified so that the definition is in terms of inductively extended AGM revision.

All of these claims clearly depend on the ampliative expansion function being permutable. This is a rather important point not only theoretically but practically; for it appears that the goal of default reasoning is to characterize varieties of ampliative or inductive reasoning. Consequently, if it is not plausible that ampliative inference should conform to the demands of permutable *E*-functions, the status of the rational inferential relations and AGM revisions in applications to such default reasoning is in jeopardy.

I believe that permutability is not an attractive property for ampliative expansion transformations to possess. My reasons for this view derive from an attitude towards expansion that insists that such expansion should be rationalized decision theoretically relative to a certain vision of cognitive goals. Jaakko Hintikka and his distinguished students Juhani Pietarinen, Risto Hilpinen and Ilkka Niiniluoto are important pioneers in developing this perspective. Hence, I am happy to be able to indicate some additional reasons to those already abundantly available for attaching importance to decision theoretic approaches to induction of the sort on which Hintikka and I worked more than thirty years ago.

According to the idea that both Hintikka and I then advocated, ampliative or inductive expansion is to be justified by showing that a given expansion strategy is best among an appropriately specified set of available expansion strategies relative to a given set of cognitive goals, initial corpus or state of full belief and state of credal probability judgment. The criterion used for evaluating the options with respect to how well they promoted the given goals was the injunction that expected utility be maximized. In cognitive decision making, the utility function used to determine expected utility is an expression of the commitments generated by the cognitive goals and the probability is derived from the state of credal probability judgment.

I have argued that the credal state need not be representable by a uniquely permissible probability function. The set of probability functions may be convex. I have also suggested that conflicts between goals may be unresolved and reflected in utility representations where not all utility functions are positive affine transformations of one another. For the purpose of this discussion, I shall suppose that such conflicts have been ironed out and that probabilities and utilities are unique.

I have repeatedly argued that it is or should be a common feature of cognitive goals of specific efforts at inductive inference that they be multidimensional. In inquiry, an investigator should be concerned to "shun error" as William James wrote. But if there were no desideratum that tempted one to risk error, the rational agent would never expand beyond the initial corpus. According to James, the desideratum of seeking truth was taken to afford an opportunity of finding a truth that would be forclosed if one focused on shunning error. Unlike Peirce, who had recognized the twin desiderata of seeking truth and shunning error thirty years before James wrote "The Will to Believe", James did not clearly associate seeking truth (i.e., the opportunity for hitting upon the truth) with the acquisition of belief that carried valuable information. Thanks to the stimulus of exchanges with Hintikka and Pietarinen in the mid 1960's (one of which is published as Hintikka and Pietarinen, 1966), I was prompted to make an amendment to my *Gambling with Truth* that was already in press (and was published in 1967). This correction was published in "Information and Inference" in *Synthese* in the same year as the book. (See Levi 1984, ch. 5.) The amendment makes clear that desiderata additional to the concern to avoid error should all be representable by utility functions that partially order the potential answers to a given question with respect to how much they strengthen the information already available. Moreover, the extension of this partial ordering, when completed, should make reference to the "demands for information" the inquirer endorses in confronting his problem. Potential answers may be compared with respect to simplicity, explanatory power or other such considerations. I have not sought to give a detailed account of these considerations. For me, the more interesting philosophical question is how inquirers who differ in the way they assess informational value can reasonably resolve their differences. This question is, in turn, part of a more general question concerning how conflicts in value may be resolved through inquiry. Though the problem is difficult, I doubt that it is hopeless. I have sketched some remarks on the topic in Levi (1984, ch. 7) originally published in *Acta Fennica Philosophica* in 1979.

In any case, I have taken the position that the common feature of the proximate aims of efforts at inductive expansion is the quest for new error free information. Hintikka and Pietarinen (1966) and Hilpinen (1968) have endorsed views that, like my view in *Gambling with Truth* and my earlier publications on this subject, anticipate this perspective. Niiniluoto (1986) has adopted a different view according to which the proximate aim of inductive expansion is the acquisition of new information that is nearer the truth. Like the position I favor, Niiniluoto's view recognizes a trade-off. For me, the trade-off is between minimizing risk of error and acquiring new valuable information. For Niiniluoto, it is minimizing expected distance from the truth and acquiring new valuable information. It is a common feature of all the proposals considered here that the trade-off is representable as a weighted average of two utility functions and that maximizing the expected value of this weighted average will yield a criterion for evaluating inductive expansions that is *caution (or boldness) dependent*. Understanding caution dependence is critical to the critique of the approaches to non-monotonic inference that are being subjected to critical scrutiny.

I am not concerned here to rehearse the differences between the approaches of Hintikka and Pietarinen (1966), Hilpinen (1968) and Niiniluoto (1986) and views I have advanced in Levi (1967, 1980, ch. 2, 1984, ch. 5, 1986, ch. 3, and 1991, ch. 3). What the members of the Finnish School and I share in common is the insistence that ampliative or inductive expansion should satisfy the following requirements:

(1) The shift should be from one consistent theory or corpus **K** closed under logical consequence to another such consistent theory $E(\mathbf{K})$ that contains it. In particular, attention is focused on such ampliative expansions where $E(\mathbf{K})$ is representable as a corpus $\mathbf{K}^{+}_{d}$ where d is a *strongest sentence accepted via induction* from **K** together with the logical consequences of doing so. Any principle of ampliative or inductive expansion licensing such shifts is *deductively cogent*.

(2) The principles of legitimate ampliative expansion should be relativised to a partition (that I call an *ultimate partition* in Levi 1967 and elsewhere). Partition sensitivity is explicitly conceded by Hilpinen (1968) and is tacit in Niiniluoto (1986). Instead of characterizing the conclusion of ampliative expansion by a strongest sentence d accepted via induction, one can instead formulate inductive expansion rules as criteria for rejecting elements of the ultimate partition. d is then equivalent given **K** to the disjunction of elements of the partition surviving such rejection. Any principle of ampliative or inductive expansion licensing such shifts is *partition sensitive*.

(3) The principles of legitimate inductive expansion are derived from the injunction that the "cognitive options" (representable by the set of elements of the ultimate partition rejected by exercising that option) are to be evalu-

ated with respect to their expected utility relative to **K** where the utility function is epistemic in the sense that the values of "payoffs" are determined by how well the options promote cognitive goals. Expected epistemic utility is, therefore, to be maximized. When there are two or more optimal options, the weakest of these is to be favored by appeal to a secondary criterion.

(4) Although the members of the Finnish School disagree with each other and with me on the how these goals are to be characterized (in terms of the epistemic utility functions), we all agree that the utility functions ought to be representable as weighted averages of two desiderata. Hilpinen and I agree that the two desiderata are avoiding error and promoting informational value. We differ as to how informational value is to be assessed. Niiniluoto thinks that they are minimizing distance from the truth and a desideratum that he takes to be informational value. The important point to keep in mind is that in all cases, there is some latitude in how the weights may be assigned so that the injunction to maximize expected utility may be seen as generating a family of inductive expansion rules parameterized by how these weights are specified. One can order members of such a family by an index of *caution* or of *boldness* so that rules with a higher boldness value tend to reject more elements of the ultimate partition than more cautious rules do. Moreover, there are maximum and minimum levels of boldness and corresponding maximum and minimum levels of caution. Any family of principles of inductive or ampliative expansion with this kind of property is *caution dependent*.

I cannot claim that all principles of inductive expansion derivable by some decision theoretic approach or other have the properties of deductive cogency, partition sensitivity and caution dependence that the proposals of the Finnish School and my proposals share in common. Nor conversely need deductively cogent, partition sensitive and caution dependent families of rules be derived from decision theoretic principles and a conception of the goals of ampliative expansion. It is, however, a common feature of the proposals the Finnish School and I make that the families of rules generated have this property. For the purpose of this discussion, I mean to appeal to these shared agreements together with another assumption to which Hilpinen has objected explicitly. For purposes of illustration, I will use my own proposals; but nothing critical to the conclusions I mean to draw will depend on this.

Given an initial corpus **K**, an ultimate partition U and the other relevant features of the context,[3] $E(\mathbf{K})$ is obtained by rejecting all elements x of U such that $Q(x) < qM(x)$ where q increases with increasing boldness and decreases with increasing caution.

Once the shift is made from **K** to $E(\mathbf{K})$, $E(\mathbf{K})$ is the new corpus representing the new state of full belief. The new truncated ultimate partition consists of all surviving elements of *U*, and the new *Q* and *M* functions are conditionalizations of the initial ones. If inductive expansion at the given value for *q* was legitimate relative to **K**, it is also legitimate relative to the new corpus $E(\mathbf{K})$. Consider another expansion at the same level of boldness so that $E[E(\mathbf{K})]$ is obtained. The process may be iterated until a "fixed point" is reached such that $E^{n+1}(\mathbf{K}) = E^{n}(\mathbf{K})$. The net result is an ampliative expansion of **K** via what I called *bookkeeping* in Levi (1967). The sentences added via bookkeeping to **K** are *stably accepted* according to the usage of Levi (1980, ch. 2.7). One can think of this ampliative expansion of **K** as ampliative expansion via a bookkeeping rule belonging to a family of deductively cogent, caution depent and partition sensitive rules.

Hilpinen observes that the original family of rules I propose violates what he takes to be a condition of adequacy (1968, p. 102).

(CA4) If *h* is acceptable on the basis of *e* whereas *g* is not acceptable on the basis of *e*, *g* is not acceptable on the basis of *e&h* either.

As I understand (CA4), it implies that the criterion of acceptability – i.e., the principle of inductive expansion – should be such that a second application of the principle should not yield a stronger corpus. $E[E(\mathbf{K})] = E(\mathbf{K})$. This alleged condition of adequacy is satisfied by stable acceptability so that Hilpinen who seems to have been committed to (CA4) cannot object to bookkeeping rules because they violate (CA4). His objection is to the rules for inductive expansion derived initially from the injunction to maximize expected utility on the basis of which I obtained these rules. The notion of acceptability characterized by these rules is not stable acceptability and does not satisfy (CA4).

I agree that the notion of stable acceptability not only should but, given the way it is characterized in terms of bookkeeping, must satisfy (CA4). Still, I do not share Hilpinen's view of the merits of (CA4). He seems to think that any criterion of inductive acceptability should equate acceptability with stable acceptability. Since he offered no argument for this view and I am at a loss to supply it for him, I can only conjecture that he has confused stable acceptability with acceptability.

The rules Hilpinen favor are caution dependent, partition sensitive and deductively cogent. They satisfy (CA4). If one reiterates their application via bookkeeping, one gets the same result as initially obtained. So they may be seen as yielding stable acceptance via vacuous bookkeeping. But Hilpinen does not obtain satisfaction of (CA4) purely in terms of decision theoretic considerations and bookkeeping. He uses a very special kind of ultimate partition and credal probability distribution to obtain his results. Different ultimate partitions and different probability measures could lead to violations of (CA4). Nonetheless, one can recover (CA4) by bookkeeping to a fixed point while committing no special hostages to a particular way of selecting an ultimate partition and credal probability distribution. I think this feature of the proposals I favor is a virtue.

In any case, since the property (CA4) is shared by my bookkeeping rules and Hilpinen's rules and since I mean to exploit it in the subsequent discussion, I do not think that our differences regarding bookkeeping have much bearing on the conclusions I mean to draw.

I propose to characterize nonmonotonic inferential relations in terms of inductively extended expansions as follows:

> *Nonmonotonic Implication* (NI):
> $h \mid\!\sim g$ @ $<\mathbf{K},E>$ (*h* nonmonotonically implies *g* relative to **K** and *E* in the inductively extended sense) if and only if *g* is a member of $E(\mathbf{K}_h^+)$ and *E* is an ampliative transformation of potential corpora into potential corpora.

I shall assume that *E* is an ampliative transformation obtained by using a rule from a family of deductively cogent, partition sensitive and caution dependent rules. This family may or may not be the product of using bookkeeping. My interest will sometimes be to contrast the two possibilities. Hence, I will explicitly indicate when I understand the *E* one way or the other.

No matter how *E* is understood, $\mid\!\sim$ @ $<\mathbf{K},E>$ is an *inference relation* in the sense of Lindström (1990) satisfying the following three conditions (I omit the @$<\mathbf{K},E>$ suffix to expressions of the type h $\mid\!\sim$ g):

> (*Closure*) If $h \mid\!\sim g_i$ for all g_i in set of sentences **A** and $\mathbf{A} \mid\!\sim f$, then $h \mid\!\sim f$.
> (*Reflexivity*) $h \mid\!\sim h$
> (*Left Logical Equivalence*) If $\vdash h \equiv g$ and $h \mid\!\sim f$, then $g \mid\!\sim f$.

It is worth emphasizing the point that although the proposals made by the Finnish school and myself are probability based, the inductive expansion rules we favor are deductively cogent. This feature together with (NI) insures satisfaction of the three minimal conditions for an inference relation in Lindström's sense.

There is another condition satisfied by $\mid\!\sim$ defined according to (NI) using an ampliative expansion transformation generated by a deductively cogent, partition sensitive and caution dependent family of inductive expansion rules:

(*Consistency Preservation*) If $h \mid\!\sim \neg \mathbf{T}$ where $\mathbf{T}$ is a logical truth, $h \vdash \neg\mathbf{T}$.

Consistency preserving inference relations cannot be obtained from inductive acceptance rules that take high probability to be sufficient as well as necessary for acceptance. Although the proposals offered by myself and the Finnish School resemble high probability rules in being probability based inductive acceptance rules, they do yield consistency preserving inference relations. This shows that probability based rules need not fail to be deductively cogent.

In general, nonmonotonic implication defined by (NI) does *not* satisfy the following conditions:

(*Cut*) If $h \mid\!\sim g$ and $h \wedge g \mid\!\sim f$, then $h \mid\!\sim f$.
(*Cautious Monotony*) If $h \mid\!\sim g$ and $h \mid\!\sim f$, then $h \wedge g \mid\!\sim f$.

(NI) enables us to derive Cut from Hilpinen's (CA4). The failure of Cut, therefore, can be imputed to the fact that using a family of deductively cogent, caution dependent and partition sensitive inductive expansion rules does not guarantee satisfaction of (CA4). However, as has already been noted, we can always derive a family of deductively cogent, caution dependent and partition sensitive rules that guarantees satisfaction of (CA4) by appealing to bookkeeping and the notion of stable acceptability. Relative to such a family of rules, (NI) yields an inferential relation satisfying Cut.

Cautious Monotony, however, will not, in general, be satisfied even if bookkeeping is satisfied unless an additional condition is satisfied – to wit, that the boldest member of the family of deductively cogent, caution dependent and partition sensitive rules that yields stable acceptability is used to characterize E.

However, when such a *maximum boldness* condition is imposed, E is permutable so that $E(\mathbf{K}^{+}_{h}) = [E(\mathbf{K})^{*}_{h}]$ where * is AGM revision. According to (NI) g is a member of $E(\mathbf{K}^{+}_{h})$ if and only if $h \mid\!\sim g$ @ $<\mathbf{K},E>$. Permutability implies that this nonmonotonic inferential dependence holds if and only if g is a member of $[E(\mathbf{K})^{*}_{h})$. This is the definition of the nonmonotonic implication relation proposed by Gärdenfors and Makinson (1994). The inferential relation thereby secured is a *rational relation* in the sense of Lehmann and Magidor (1992).

Consequently, if one uses (NI) to define inferential relations based on ampliative transformations derived from deductively cogent, caution dependent and partition sensitive criteria for inductive expansion and acceptance, we automatically obtain consistency preserving inference relations in the sense of Lindström. If we require the rules secure stable acceptability, we obtain, in addition, the satisfaction of Cut. If, in addition, we require maximum boldness, we obtain rational inference relations in the sense of Lehmann and Magidor and can use a definition of inferential relations proposed by Gärdenfors and Makinson to derive them from AGM revisions of $E(\mathbf{K})$.

At the outset, I suggested that nonmonotonic reasoning can be of two main varieties: belief contravening reasoning that is not strictly speaking a form of reasoning and ampliative or inductive inference. In my judgment, the notion of Ramsey revision is better suited than the notion of AGM revision as the basis of an account of suppositional belief contravening nonmonotonic reasoning. But for the purposes of this discussion, the differences between the two approaches were neglected and attention focused on AGM revision. Our present concern is with the credentials of AGM revision as a basis for an account of ampliative inference – given that such inference is according to caution, dependent, deductively cogent and partition sensitive families of rules. According to the previous discussion, the fortunes of AGM revision in this context depend upon whether bookkeeping with maximum boldness is acceptable.

Stable acceptance with maximum boldness is inappropriate in most contexts. Exercising maximum boldness with bookkeeping is being bold to the point of rashness. If each element of the ultimate partition U is as informative as every other, bookkeeping with maximum boldness recommends accepting the element of U (or the disjunction of elements of U) carrying maximum probability.

Consider a coin known to be fair that is tossed a million times. An inquirer interested in predicting the relative frequency of heads might plausibly rate each of the million and one exact estimates of the relative frequency consistent with the initial corpus to be equally informative. Should that inquirer predict that it will land heads exactly a half million times? It may be plausible to conclude that it will land heads approximately half a million times; but it seems absurd to predict that it will land heads exactly that number of times. This is the kind of recommendation that bookkeeping with maximum boldness recommends. It is the kind of nonmonotonic inference that is represented by rational and consistency preserving inference relations.

Default reasoning seems intended to be the sort of nonmonotonic reasoning I have called ampliative or inductive where one leaps to conclusions. To infer that Tweety flies from the supposition that Tweety is a bird is not explicative even when one appeals to assumptions in background beliefs that most of us would share. The reasoning is nonmonotonic but it is not belief contravening. The supposition that Tweety is a bird contravenes no background assumption – at least in ordinary contexts. Nor does adding the information that Tweety is a penguin. The supposition that Tweety is a penguin contravenes the conclusion reached from the background assumptions and the supposition that Tweety is a bird; but this sort of contravention is quite distinct from contravention of the background assumptions. Nonmonotonicity is not, in this case, attributable to belief contravening supposition but is due rather to ampliative inference. The results reported imply that the principles of nonmonotonic reasoning appropriate to default reasoning should be very weak indeed. They cannot even satisfy the requirements for "cumulative" reasoning favored by Gabbay (1985) and Makinson (1989) or the cumulative and preferential logics of Kraus, Lehmann and Magidor (1990).

This does not mean that there is nothing to say about induction. To the contrary, the decision theoretic approach to ampliative reasoning favored by Hintikka and the Finnish School as well as myself promises a systematic account of ampliative reasoning far more sensitive to the contextual features relevant to such reasoning than either Bayesian probabilists, contemporary philosophers of science or students of computer science and of artificial intelligence have offered. The conception of the aim of ampliative reasoning as seeking to maximize informational value while minimizing risk of error affords a juster understanding of the relation between risk of error and the quest for simple, explanatory and predictively powerful answers in science than is customary to find in contemporary discussions of realism and instrumentalism. It affords a less simplistic perspective on the character of scientific inquiry than probabilistic Bayesianism and most contemporary approaches to evidential support or confirmation provide. And as I have argued in this essay, it offers an interesting perspective on the relations between ampliative reasoning and belief contravening suppositional or hypothetical reasoning.

Columbia University

NOTES

1 Closure under strict expansion is quite natural when the language **L** is restricted to sentences containing no modal, conditional or epistemic connectives or operators. In such cases, it seems quite reasonable to allow all deductively closed theories in **L** to be potential corpora representing potential belief states. This is by no means obvious when modal or conditional connectives are included. In such cases, closure under expansion does not, in general, imply closure under strict expansion. In Levi (1991, ch. 2), I argue for the view that potential states of full belief should constitute a Boolean algebra at least and this argues for adopting linguistic representations guaranteeing that closure under strict expansion obtains. In (1988, ch. 3.2), Gärdenfors offers a set of postulates for expansion which imply the definition of expansion I use here. He then goes on to claim that expansion so construed can be shown to be equivalent to strict expansion. (Theorem 3.1). The proof of this theorem fails because Gärdenfors fails to show that the consequences of **K** and *h* constitute a "belief set" or potential corpus. He does not explicitly assume this to be so and, indeed, in ch. 7 uses a language containing conditionals where closure under strict expansion becomes problematic.

2 Inductively extended expansion was discussed in connection with Ramsey Tests and nonmonotonic reasoning in Fuhrmann and Levi (1992). The results reported here are new.

3 On my view, these include the credal probability function *Q* defined over the Boolean algebra generated by *U* relative to **K** representing the inquirer's state of credal probability judgment and the informational value determining probability *M* defined over *U* used to evaluate informational value (see Levi 1984, ch. 5 originally published in *Synthese*, 1967).

REFERENCES

Alchourrón, C., Gärdenfors, P. and Makinson, D. (1985). On the Logic of Theory Change: Partial meet functions for contraction and revision. *Journal of Symbolic Logic* **50**, 510–30.

Fuhrmann, A. and Levi, I. (1992). Undercutting and the Ramsey test for conditionals. *Konstanzer Berichte: Logik und Wissenschaftstheorie, #23.* to be published in *Synthese*.

Gabbay, D. (1985). Theoretical foundations for nonmonotonic reasoning in expert systems. In: K.R. Apt (Ed.), *Logic and Models of Concurrent Systems*. Berlin: Springer Verlag.

Gärdenfors, P. (1988). *Knowledge in Flux*, Cambridge, Mass.: MIT Press.

Gärdenfors, P. and Makinson, D. (1994). Nonmonotonic Inferences based on Expectations. *Artificial Intelligence* **65** (2), 197–245.

Hilpinen, R. (1968). *Rules of Acceptance and Inductive Logic*. Amsterdam: North Holland.

Hintikka, J. and Pietarinen, J. (1966). Semantic Information and Inductive Logic. In: J. Hintikka and P. Suppes (Eds.), *Aspects of Inductive Logic*. Amsterdam: North Holland, pp. 96–112.

Kraus, S., Lehmann, D. and Magidor, M. (1990). Nonmonotonic Reasoning, Preferential Models and Cumulative Logics. *Artificial Intelligence* **44**, 167–201.

Lehmann, D. and Magidor, M. (1992). What does a conditional knowledge base entail? *Artificial Intelligence* **55**, 1–60.

Levi, I. (1967a). *Gambling with Truth*. A. Knopf reissiued in paperback in 1973 by MIT Press.

Levi, I. (1980). *The Enterprise of Knowledge*. Cambridge, Mass.: MIT Press.

Levi, I. (1984). *Decisions and Revisions*. Cambridge: Cambridge University Press.

Levi, I. (1986). *Hard Choices*. Cambridge: Cambridge University Press.

Levi, I. (1991). *The Fixation of Belief and Its Undoing*. Cambridge: Cambridge University Press.

Levi, I. (1996). *The the Sake of the Argument*. Cambridge: Cambridge University Press.

Lindström, S. (1990). A Semantic Approach to Nonmonotonic Reasoning: Inference Operations and Choice. Unpublished manuscript.

Makinson, D. (1987). On the Status of Recovery in the Logic of Theory Change. *Journal of Philosophical Logic* **16**, 383–94.

Makinson, D. (1989). General theory of cumulative inference. In: M. Reinfrank, J. de Kleer, M.L. Ginsberg and E. Sandewall (Eds.), *Non-Monotonic Reasoning*. Berlin: Springer Verlag, Lecture Notes on Artificial Intelligence, no. 34 6.

Niiniluoto, I. (1986). Truthlikeness and Bayesian Estimation. *Synthese* **67**, 321–46.

Poznań Studies in the Philosophy of the Sciences and the Humanities
1997, Vol. 51, pp. 117–131

Ilkka Niiniluoto

INDUCTIVE LOGIC, ATOMISM, AND OBSERVATIONAL ERROR

In the mid-1960s Jaakko Hintikka made a major breakthrough in the theory of induction by formulating a system of inductive logic where universal generalizations may receive non-zero probabilities (Hintikka 1966). But two decades later, in the context of his new interrogative approach to inquiry, he starts to wonder "why the concept of induction has sometimes occupied the pride place in the philosophy of science" (Hintikka 1987, p. 438). He goes on to suggest that the traditional problems of inductive generalization and curve-fitting have only a very modest place on the lowest "micro-level" of experimental inquiry (Hintikka 1988, 1992).

In this paper, I defend the Carnap–Hintikka programme of inductive logic[1] against Hintikka's recent doubts. His own system is not tied with the "super-dogma" of the "Atomistic Postulate" (*ibid.*, p. 176), since it can be applied to situations where the evidence contains or consists of a universal theory (see Niiniluoto and Tuomela 1973).

On the other hand, Hintikka's plea for considering methodological situations with different degrees of complexity is well-taken. In this spirit, I sketch below a modification of inductive logic which allows for observational error.

1. Standard Inductive Logic

The standard application of inductive logic concerns inferences about a potentially infinite universe U. This universe is described by a monadic first-order language L: its individual names $a_1, a_2, \ldots$ refer to the elements (objects, events) of the domain U, and the primitive predicates $M_1, M_2, \ldots$ designate the properties of these individuals. Sometimes these predicates are organized into a finite number of families $F_1, F_2, \ldots, F_k$, where each family consists of a finite set of mutually exclusive and jointly exhaustive predicates (e.g., color terms, length expressed by a finite precision). Then the *Q-predicates* of L are defined by k-fold conjunctions which contain one predicate from each family. In the simplest case, there is only one family, and its basic predicates define directly the Q-predicates. In another simple case, each of the k families consists of a predicate and its negation, i.e., $F_i=\{M_i, \sim M_i\}$. Then the Q-predicates reduce to conjunctions

$$(1) \quad (\pm)M_1(x)\&\ldots\&(\pm)M_k(x),$$

where $(\pm)$ is replaced by the negation sign $\sim$ or by nothing. The number of the Q-predicates $Q_1,\ldots,Q_K$ of L is then $K=2^k$.

The point of this construction is that the class $\mathbf{Q}$ of Q-predicates is a complete *classification system* for the individuals in universe U: each element of U belongs to one and only one "cell" Q_i in $\mathbf{Q}$.

In some cases it is natural to assume that $\mathbf{Q}$ has a topological and metric structure. The predicates of L may have unequal "widths" as regions in the space $\mathbf{Q}$. The *width* q_i of a Q-predicate Q_i is the product of the widths of its conjuncts. In the case of dichotomous families, the widths of Q-predicates of form (1) are usually taken to be equal, i.e., $q_i=1/K$ for all $i=1,\ldots,K$.

Further, each family F_i may be associated with a distance function d_i, which expresses the dissimilarity between the predicates in F_i. This generates a Euclidean (or, more generally, Minkowskian) metric d between the Q-predicates of L. In this sense, $\mathbf{Q}$ can be viewed as a k-dimensional conceptual space – a qualitative or discrete counterpart to state spaces defined by continuous quantities (see Niiniluoto 1987, Ch. 3.2).

In the case of dichotomous families, the *distance* $d(Q_i,Q_j)=d_{ij}$ between two Q-predicates Q_i and Q_j of form (1) can be defined as the number of "mismatches" concerning $M_1,\ldots,M_k$, divided by k. Then, $0\leq d_{ij}\leq 1$, and $d_{ij}=0$ iff $i=j$. The degree of *similarity* or *resemblance* between Q_i and Q_j is definable by $r_{ij}=1-d_{ij}$. If two individuals a and b in U satisfy the Q-predicates Q_i and Q_j, respectively, then their similarity (relative to language L) is r_{ij} (see Niiniluoto 1988).

The investigator may now explore the universe U by picking out individuals (without replacement) $a_1,\ldots,a_n$ from U and by observing their properties relative to L. Thus, the *singular evidence* statement e_n describes this finite n-fold sample in terms of the classification system $\mathbf{Q}$:

$$(2) \quad e_n = Q_{j_1}(a_1)\&\ldots\&Q_{j_n}(a_n).$$

We let $n_i\geq 0$ be the number of observed individuals with the Q-predicate Q_i, so that n_i/n is the observed relative frequency of cell Q_i $(i=1,\ldots,K)$.

The task of inductive logic is to define an epistemic probability measure P for the statements of L. This can be done by specifying probabilities of the form

(3) $P(Q_i(a_{n+1})/e_n)$.

In Rudolf Carnap's (1952) λ-continuum, this probability is defined as a weighted average between the observed relative frequency n_i/n and the width of Q_i:

(4) $$\frac{n_i + \lambda/K}{n + \lambda},$$

where $0 < \lambda < \infty$. If the width of Q_i is q_i, (4) is generalized by

(5) $$\frac{n_i + q_i\lambda}{n + \lambda}.$$

If λ is chosen as K, (4) reduces to Carnap's confirmation function c^*, which is a generalization of Laplace's rule of succession.

Carnap's system leads to the conclusion that $P(h/e_n)=0$ for any n, when h is a universal generalization in L. Indeed, the λ-continuum assigns the whole prior probability one to the statement that all cells Q_i in $\mathbf{Q}$ are instantiated in U:

(6) $(\exists x)Q_1(x) \& \ldots \& (\exists x)Q_K(x)$.

Hintikka's (1966) λ-α-system, instead, gives non-zero probabilities to *constituents* C_i of the form

(7) $$\bigwedge_{j \in CT_i} (\exists x)Q_j(x) \& (x)[\bigvee_{j \in CT_i} Q_j(x)]$$

which claim that certain cells $\mathbf{CT}_i \subseteq \mathbf{Q}$ are exemplified and others are empty. We let $w_i = card(\mathbf{CT}_i)$ be the *breath* of the constituent C_i. Then (6) is in fact the maximally broad constituent, to be denoted below by C^K. The number of different constituents in language L is 2^K-1 (assuming that U is non-empty).

Hintikka (1966) relativized probabilities (3) to constituents C_j by the formula

(8) $$P(Q_i(a_{n+1})/e_n \& C_j) = \frac{n_i + \lambda/w_j}{n + \lambda},$$

where $Q_i \in \mathbf{CT}_j$. This implies the fundamental theorem about the asymptotic behavior of posterior probabilities: when c is the number of cells exemplified in evidence e_n, and C_j is a constituent compatible with e_n,

$$(9) \quad \begin{aligned} P(C_j/e_n^c) &\to 1, \text{ if } w_j = c \\ &\to 0, \text{ if } w_j > c \end{aligned}$$

for $n \to \infty$, c fixed. As each quantificational statement of L can be expressed as a disjunction of constituents, (9) guarantees that inductive generalization can be studied in Hintikka's system.

In Hintikka's "generalized combined system", which is a generalization of Carnap's c^*, λ is chosen as equal to w_j, so that (8) reduces to

$$(10) \quad \frac{n_i + 1}{n + w_j}.$$

Hintikka's and Niiniluoto's axiomatic treatment leads to a system where prior probabilities are regulated by $K-1$ free parameters, and in probabilities of the form (8) λ/w_j is a constant (cf. Niiniluoto 1977; Kuipers 1977). Hence, in this *SH*-system (as Kuipers calls it), (8) is reduced to

$$(11) \quad \frac{n_i + \lambda/K}{n + w_j\lambda/K}.$$

Obviously, (10) is a special case of (11), and (4) is a special case of (11) with $w_j = K$. If the widths of Q-predicates are unequal, (11) can be modified as in (5). The choice (11) again guarantees the convergence property (9).

2. *The Atomistic Postulate*

Hintikka's interrogative model of inquiry is based on the following idea. An Inquirer is trying to derive from a given theory T an answer to the question whether C is true or not. Besides deductive steps the Inquirer may ask questions about the related domain or universe U. Answers to such interrogative moves are provided by Nature. The logical complexity of Nature's answers then defines a hierarchy of more and more complex methodological situations (see Hintikka 1987, 1988).

As the interrogative model allows only deductive inferences, and does not mention induction at all, Hume's problem seems to disappear from the "serious theory of the scientific method" (Hintikka 1992, p. 25). This general conclusion may be questioned by pointing out that, in the terms of Lakatos, there is a "whiff of inductivism" in the assumptions that Nature's answers are true (*ibid.*, p. 24) or true with a certain probability (*ibid.*, p. 25). But what Hintikka has in mind here is a more specific conception of inductive inference as non-deductive generalization from singular instances.

Hintikka argues that, in a typical controlled experiment with two quantitative variables x and y, Nature's answers provide the Inquirer with a function that links every x to some y. This quantitative law or curve is then used as such (its "uncertainty" notwithstanding) for further purposes in scientific reasoning.[2] Hence, the logic of experimental science has at least $\forall\exists$-complexity.

Contrast this with the situation which satisfies the *Atomistic Postulate*: Nature's answers are restricted to negated or unnegated atomic propositions. Hintikka (1988) claims that this has been assumed as a "super-dogma" "in virtually all of the recent philosophy of science".[3]

The concept of atomism has earlier been used to characterize "Carnapian" universes where no universal laws are true. In this sense, the universe U is *atomistic* with respect to L if and only if it instantiates all Q-predicates of L, i.e., (6) is true in U. Similarly, Carnap's λ-continuum is atomistic, since it assigns the prior probability one to constituent C^K, i.e., to statement (6).

Hintikka's claim that all inductive logic has been based upon dogmatic "atomism" has to be understood in the new interrogative sense. Standard inductive logic indeed gives gist to Hintikka's mill here. In its usual application, the desired conclusion C is a singular prediction (Carnap) or a universal generalization (Hintikka), theory T vanishes, and Nature's answers are summarized in the singular evidence statement e_n. The Atomistic Postulate is presupposed.

However, the applications of inductive logic need not be restricted to simple situations concerning uncertain inferences from indubitable singular premises (*ibid.*, p. 186).[4] Indeed, there is a whole monograph *Theoretical Concepts and Hypothetico-Inductive Inference* (1973), by Niiniluoto and Tuomela, where inductive inference is taken to proceed from theoretical *cum* observational evidence. More precisely, this work studies probabilities of the form $P(g/e_n\&T)$, where e_n is singular observational evidence as above, g is a universal observational generalization, and T is a theory in a language enriched by new theoretical concepts. Calculation of these probabilities is given in Hintikka's generalized combined system.

One might say that this situation still fits the Atomistic Postulate, since it is not theory T but singular e_n that plays the role of observational or experimental evidence (i.e., Nature's answers). I think three replies are in order here.

First, if probabilities of the form $P(g/e_n\&T)$ can be determined within inductive logic, the same holds of probabilities like $P(g/T)$ (without atomistic evidence) or $P(g/g'\&T)$ (where the singular e_n has been replaced by a generalization g'). Such probabilities depend on the number of cells claimed to be empty by g but not by T (and g').[5] The calculation of these probabilities is possible in Hintikka's system precisely for the reason that it is not "atomistic" in the Carnapian sense.

Secondly, when T is a sufficiently strong homogeneity condition, knowledge of one instance is sufficient to warrant a generalization (Niiniluoto and Tuomela 1973, p. 240). In such cases, relative to T, a singular statement is equivalent to a universal generalization, and the difference between A_0-complexity (atomistic case) and $\forall$-complexity collapses.

Such strong homogeneity assumptions may be valid in controlled experiments – and may thus explain why the traditional Humean problem of inductive generalization is not relevant in such contexts. For example, a physicist studying the thermal expansion of metals assumes that all pieces of iron behave in the same way, and the problem is not one of generalizing from instances but rather that of finding a curve fitting measured values of temperature and length. But inductive generalization remains as an important issue in those contexts where such strong background assumptions cannot be made.

Thirdly, one may argue that, in a typical curve-fitting problem, the function f relating the variables x and y is not given by Nature but has to be discovered by the Inquirer. More precisely, even though the controlled experiment may seem to give a reply to a $\forall\exists$-question, a curve going through the observed points would "overfit" the data (see Forster and Sober 1994). The function f (and its expression in analytic form) has to be supplied by the Inquirer. This agrees with William Whewell's Kantian formulation in 1840: in every induction, there is a conception "superinduced" by the human mind on the observed facts. This insight has been emphasized by theorists of induction from Aristotle to Whewell, Peirce, and Hintikka (see Niiniluoto 1995). It is thus possible to support the Atomistic Postulate by stressing the role of human mind (*vis-à-vis* that of Nature) in concept and theory formation.

In my view, Kant and contemporary nominalists go too far in claiming that all lawfulness is mind-involving. A scientific realist admits that generality (or Peirce's Thirdness) is something that belongs to Nature independently of human thinking. But this point means the world is not atomistic in the Carnapian sense. It is another matter how we gain knowledge about such lawful generalizations in inquiry.

Finally, one might add that the important macrolevel of scientific reasoning – i.e., deductive and inductive inferences between statements with at least ∃- or ∀-complexity – has been extensively studied under such methodological titles as explanation, reduction, and prediction (rather than confirmation).

For these reasons, I think Hintikka is right in emphasizing the importance of both micro- and macrolevel methodological contexts. But there is nothing that prevents inductive logic to serve as a philosopher's tool in both areas.

3. Inductive Logic with Observational Error

Standard inductive logic assumes that observations are error-free. This can be seen clearly in Hintikka's relativized probability assignment (8): the probability of finding an individual in a cell Q_i is non-zero only for those cells that are instantiated by constituent C_j. This means that observations cannot be non-veridical. For example, if C_j tells that there are only yellow, brown, and black ants in the world, the observed evidence will not include orange, red, or grey ants.

For this reason, inductive logic cannot be directly compared with the Bayesian treatments of parameter estimation or curve-fitting. In these applications, there is a probability distribution $f(x/\theta)$ for observation x given the value of the parameter θ. This may be, e.g., a Gaussian normal distribution which concentrates around θ and decreases when the distance between x and θ grows. If $g(\theta)$ is the prior distribution for θ, then Bayes' Theorem gives the posterior distribution $g(\theta/x)$ for θ given evidence x:

$$(12) \quad g(\theta/x) = \frac{g(\theta)f(x/\theta)}{\int g(t)f(x/t)dt}.$$

A similar approach applies to curve-fitting, where again a probability distribution of errors in data is defined relative to a curve.[6]

Thus, (12) shows how we can learn from errors: the prior distribution $g(\theta)$ is transformed to the posterior distribution $g(\theta/x)$ by the observed or measured value x, where x is known to deviate from the truth.

In inductive logic, the only proposal for dealing with uncertain evidence is due to Richard Jeffrey (1983).[7] To illustrate Jeffrey's idea, suppose that language L contains only one family of color predicates $Q_1,\ldots Q_K$, and let

$P_0(Q_i(a))$, $i=1,\ldots,K$, be our initial probabilities about the color of an individual a. Assume then that we see object a dimly by candlelight, and let $P(Q_i(a))$, $i=1,\ldots,K$, be the new probabilities after the observation. Then, according to *Jeffrey conditionalization*, for any statement h in L,

$$(13) \quad P(h) = \sum_{i=1}^{K} P(Q_i(a)) P_0(h / Q_i(a))$$

if the Rigidity Condition

$$(14) \quad P(h / Q_i(a)) = P_0(h / Q_i(a))$$

holds for all $i=1,\ldots,K$. If we learn that a really is Q_m, say, then $P(E)=1$, where $E=Q_m(a)$, and (13) reduces to the principle of simple conditionalization:

$$(15) \quad P(h) = P_0(h / E).$$

Bayes' formula in the form (12) is, so to say, in the middle between standard inductive logic (conditionalization with respect to a veridical evidence statement e_n) and Jeffrey conditionalization (change of degrees of belief without an evidence statement). In (12) there is conditionalization relative to a value x, obtained by measuring a random variable, and the deviation of x from the truth is governed by a probabilistic error distribution.

It is somewhat surprising that the strategy corresponding to (12) has not been used for treating uncertain evidence within inductive logic. This is what I shall sketch below.

First, we need a distinction between *descriptive* statements ('This ant is brown') and *perceptual* statements ('This ant seems brown to me', 'This ant is observed to be brown', 'This ant is seen as brown'). For the former, we use the vocabulary of language L in Section 1. Thus, $Q_i(a)$ says that *a is* Q_i. For the latter, we let $S_i(a)$ be the corresponding observational statement that *a seems* to be Q_i. We take observation to be uncertain, so that $S_i(a)$ does not imply $Q_i(a)$.[8]

Secondly, it is natural to assume that inductive probabilities for the descriptive language (with $Q_1,\ldots,Q_K$) constitute a *SH*-system. For simplicity, we assume that the descriptive Q-predicates have equal widths:

$$(16) \quad P(Q_i(a)) = 1/K.$$

The probability $P(Q_i(a_{n+1})/e_n\&C_j)$ is given by (11).

Thirdly, we need a probabilistic model of observation. Such models are of course highly context-dependent. Perception may be systematically biased in many different ways. Observation by candlelight makes red to appear brown, blue to appear grey, etc. A colour blind person cannot discriminate between red and green.

Here I shall restrict my attention to *error probabilities* that are counterparts to the normal distribution in the quantitative case. Let d be a distance function on the set $\mathbf{Q}$ of Q-predicates (cf. Section 1). Then a conditional probability P on $\mathbf{Q}\times\mathbf{Q}$ is said to *agree with d* if and only if

(17) The probability $p_{ij}=P(S_i(a)/Q_j(a))$ is a decreasing function of the distance $d_{ij}=d(Q_i, Q_j)$.

We shall assume that the probability of correct observation is the same for all cells:

(18) $p_{ii} = P(S_i(a)/Q_i(a)) = \beta$ for all $i=1,\ldots,K$.

The constant β is called *observational reliability*. Hence, when $r_{ij}=1-d_{ij}$ is the degree of similarity between cells Q_i and Q_j,

(19) $$\begin{aligned} p_{ij} &= \beta \quad \text{iff} \quad i=j \\ p_{ij} &< p_{kj} \quad \text{iff} \quad r_{ij} < r_{kj}. \end{aligned}$$

The standard application of inductive logic without observational error corresponds to the assumption that

(20) $$\begin{aligned} p_{ij} &= 1 \quad \text{if} \quad i=j \\ &= 0 \quad \text{if} \quad i\neq j. \end{aligned}$$

Let us denote, for $i=1,\ldots,K$,

(21) $$p_i = \sum_{j=1}^{K} p_{ij}.$$

Then

$$\sum_{i=1}^{K} p_i = \sum_{i=1}^{K}\sum_{j=1}^{K} p_{ij} = K.$$

Assumptions (16) and (17) entail now that

$$(22)\quad P(S_i(a)) = \sum_{j=1}^{K} P(Q_j(a))P(S_i(a)/Q_j(a)) = \frac{1}{K}\sum_{j=1}^{K} p_{ij} = p_{i\cdot}/K.$$

By Bayes' Theorem,

$$(23)\quad P(Q_j(a)/S_i(a)) = \frac{P(Q_j(a))P(S_i(a)/Q_j(a))}{P(S_i(a))} = p_{ij}/p_{i\cdot}.$$

Hence, by (17), the posterior probability of $Q_j(a)$, given that a is observed in cell S_i, decreases with the distance d_{ij}.

Fourthly, we should be able to specify probabilities of the form

$$(24)\quad P(S_i(a_{n+1})/e_n^S \,\&\, C_j)$$

where sample e_n^S is like (2) but Q_j's replaced with S_j's. Unlike in (11), we have to allow here that a_{n+1} will be erroneously observed in a cell S_i outside those in $\mathbf{CT}_j$ allowed by constituent C_j. Similarly, the sample e_n^S may contain observations outside $\mathbf{CT}_j$.

One way of proceeding is to make a "resilience" or invariance assumption about error probabilities. Probabilities of errors are independent of each other, e.g.,

$$P(S_i(a_1)\,\&\,S_j(a_2)/Q_k(a_1)\,\&\,Q_l(a_2)) = p_{ik}\cdot p_{jl}.$$

And they are not affected by additional information about other individuals or the constitution of the universe, e.g.,

$$P(S_i(a)/Q_k(a)\,\&\,C_j) = p_{ik}$$

for any constituent C_j and any $Q_k \in \mathbf{CT}_j$.[9]

For each predicate S_i $(i=1,\ldots,K)$ and constituent C_j, define

$$(25)\quad \gamma_i^j = \frac{1}{K}\sum_{m\in CT_j} p_{im}.$$

In particular, $\gamma_i^K = 1/K$. The sum of these error factors over all $i=1,\ldots,K$ is w_j/K. Hence, by resilience,

$$P(S_i(a)/C_j) = \sum_{k \in CT_j} P(Q_k(a)/C_j)P(S_i(a)/Q_k(a)\&C_j)$$

$$= \frac{1}{w_j} \sum_{k \in CT_j} p_{ik} = \frac{\gamma_i^j}{w_j/K}.$$ [10]

More generally,

$$P(S_i(a_2)/Q_k(a_1)\&C_j) = \frac{p_{ik}+\lambda\gamma_i^j}{1+\lambda w_j/K}$$

and

$$(26) \quad P(S_i(a_{n+1})/e_n\&C_j) = \frac{\alpha_i^j(e_n)+\lambda\gamma_i^j}{n+\lambda w_j/K},$$

where e_n is a descriptive sample and

$$(27) \quad \alpha_i^j(e_n) = \sum_{m \in CT_j} n_m p_{im}.$$

Resilience implies that probability (24) has the general form:

$$\frac{g(e_n^S)+\lambda\gamma_i^j}{n+\lambda w_j/K}$$

where $g(e_n^S)$ depends on the error probabilities and on all the ways in which the observations in e_n^S could have arisen from their true origins in $\mathbf{CT}_j$. In particular,

$$(28) \quad P(S_i(a_2)/S_k(a_1)\&C_j) = \frac{\dfrac{\sum_{m \in CT_j} p_{km}p_{im}}{\sum_{m \in CT_j} p_{km}}+\lambda\gamma_i^j}{1+\lambda w_j/K}.$$

Similarly, probability $P(Q_k(a_{n+1})/e_n^S\&C_j)$ has the form

$$\frac{h(e_n^S)+\lambda/K}{n+\lambda w_j/K}.$$

In particular,

$$(29)\qquad P(Q_k(a_2)/S_i(a_1)\&C_j) = \frac{\dfrac{p_{ik}}{\sum\limits_{m\in CT_j} p_{im}}+\lambda/K}{1+\lambda w_j/K}.$$

The possibility of inductive generalization depends now on the error probabilities. If $p_{ij}>0$ for all i and j, the universe appears atomistic in the Carnapian sense, and all constituents (except C^K) will have the probability zero. The situation may be otherwise for specific error distributions. We shall also see, in the next section, that the chances of inductive generalization may be improved by the repetition of error-laden observations.

4. *Repetition of Observations*

Hintikka (1987) has noted that, in the interrogative model of inquiry, it makes sense to repeat the same question. This standard requirement of methodology, viz. repetition of experiments, has not played any role in inductive logic. I think this is due to the assumption that observation is error-free. In our treatment of induction with observational error, it makes sense to repeat the examination of the same individual. Let $a_i^1, a_i^2, \ldots$ refer to repeated independent observations of the same individual a_i. Then

$$\begin{aligned}(30)\qquad & P(S_{i_1}(a^1)\&\ldots\&S_{i_m}(a^m)/Q_j(a))\\ &= p_{i_1j}\cdot\ldots\cdot p_{i_mj}\\ &= p_{1j}^{m_1}\cdot\ldots\cdot p_{Kj}^{m_K},\end{aligned}$$

where $m_1,\ldots,m_K$ are the frequencies of $Q_1,\ldots,Q_K$ in $\{a^1,\ldots,a^m\}$. Hence,

$$(31)\qquad P(Q_j(a))/S_{i_1}(a^1)\&\ldots\&S_{i_m}(a^m)) = \prod_{i=1}^{K} p_{ij}^{m_i} / \sum_{q=1}^{K}\prod_{i=1}^{K} p_{iq}^{m_i}.$$

In particular, if a is repeatedly found in cell Q_j (i.e., $i_k=j$ for all k),

$$(32)\quad P(Q_j(a)/S_j(a^1)\&\ldots\&S_j(a^m)) = \frac{p_{jj}^m}{\sum_{q=1}^{K} p_{jq}^m} = \frac{\beta^m}{\beta^m + \sum_{q\neq j} p_{jq}^m}$$

As, by (19), $\beta > p_{jq}$ for $q \neq j$, the probability (32) approaches the limit one when $m \to \infty$.

In the quantitative case, a series of repeated measurements is often replaced by their average; for the normal error curve, this average is an unbiased estimate of the unknown parameter. In our case, the value of (31) is maximized by that choice of Q_j which gives the largest value to (30). Let us denote this *maximum likelihood estimate* by $Q_*(a)$. When the observational reliability β is sufficiently large, Q_* will be that cell which is most frequently exemplified in the sample $\{a^1,\ldots,a^m\}$. When m grows without limit, so that the relative frequencies m_i/m remain approximately constant, the posterior probability of the maximum likelihood estimate $Q_*(a)$ approaches the limit one.

If β and m are sufficiently large, the maximum likelihood estimate $Q_*(a)$ relative to a given constituent C_j will belong to one of the cells $\mathbf{CT}_j$ allowed by C_j. If e_n^* is a n-fold sample, where each a_i is associated with its maximum likelihood estimate $Q_*(a_i)$, probabilities of the form

$$P(S_i(a_{n+1})/e_n^* \& C_j)$$

are determined by (26). When n grows, this probability will approach the limit zero for any cell S_i outside $\mathbf{CT}_j$. Asymptotically these probabilities behave like in Hintikka's original system, or in the *SH*-system, which guarantees that the posterior probabilities relative to evidence e_n^* satisfy the fundamental condition (9) concerning inductive generalization.

Department of Philosophy
P.O. Box 24
00014 University of Helsinki
Finland

NOTES

1 A good summary of this programme is given by Theo Kuipers in this volume. See also Niiniluoto (1987), Ch. 2.

2 According to Hintikka (1992), such tasks include the widening of the scope of conditional generalizations by extrapolation to longer intervals and the reconciliation of partial generalizations (for the latter problem, see also Forster and Sober 1994). As Hintikka (*ibid.*) shows, this gives interesting new light on Newton's method.

3 Hintikka (1992, p. 27) states that "the Atomistic Postulate was not adopted in the earlier tradition of philosophy and science from Aristotle to Newton (inclusive)". However, I think this postulate was clearly expressed by Francis Bacon, whose method is advertised as establishing "progressive stages of certainty", "starting directly from the simple sensuous perception" (Bacon 1960, pp. 33–4). According to Aphorism XIX of *Novum Organum*, "the true way" of "searching into and discovering truth" "derives axioms from the senses and particulars, rising by gradual and unbroken ascent, so that it arrives at the most general axioms last of all" (*ibid.*, p. 43).

4 Zamora Bonilla (1992), working in the theory of truthlikeness, has assumed that the empirical evidence is given directly by a set of accepted empirical laws.

5 See the formula (3.5) in Niiniluoto and Tuomela (1973), p. 37.

6 See Forster and Sober (1994).

7 See also Howson and Urbach (1989) and Earman (1992).

8 Perceptual statements ($S_a p \equiv a$ sees that p) without a success condition ($S_a p \supset p$) are analyzed in Hintikka's logic of perception. See Niiniluoto (1983).

9 Hintikka's logic of perception (see note 8) allows "conscious illusions", where sensory impressions are not affected by background knowledge.

10 Rosenkrantz (1980) defines this probability by

$$(1-\alpha)(1/w_j)+\alpha(1/K), \text{ if } S_i \in \mathbf{CT}_j$$
$$\alpha(1/K) \qquad , \text{ if } S_i \notin \mathbf{CT}_j,$$

where α is the probability of erroneous classification. He does not employ the concept of distance between Q-predicates.

REFERENCES

Bacon, F. (1960). *The New Organon*. Bobbs-Merrill: Indianapolis.

Carnap, R. (1952). *The Continuum of Inductive Methods*. Chicago: The University of Chicago Press.

Earman, J. (1992). *Bayes or Bust? A Critical Examination of Bayesian Confirmation Theory*. Cambridge, MA: The MIT Press.

Forster, M. and Sober, E. (1994). How to Tell when Simpler, More Unified, or Less *ad hoc* Theories will Provide More Accurate Predictions. *The British Journal for the Philosophy of Science* **45**, 1–35.

Hintikka, J. (1966). A Two-Dimensional Continuum of Inductive Methods. In: J. Hintikka and P. Suppes (Eds.), *Aspects of Inductive Logic*. Amsterdam: North-Holland, pp. 113–32.

Hintikka, J. (1987). The Interrogative Approach to Inquiry and Probabilistic Inference. *Erkenntnis* **26**, 429–42.

Hintikka, J. (1988). What is the Logic of Experimental Inquiry?. *Synthese* **74**, 173–90.

Hintikka, J. (1992). The Concept of Induction in the Light of the Interrogative Approach to Inquiry. In: J. Earman (Ed.), *Inference, Explanation, and Other Frustrations*. Berkeley: University of California Press, pp. 23–43.

Howson, C. and D. Urbach (1989). *Scientific Reasoning: The Bayesian Approach*. La Salle: Open Court.

Jeffrey, R. (1983). *The Logic of Decision*, 2nd ed. New York: McGraw-Hill.

Kuipers, T. (1978). *Studies in Inductive Probability and Rational Expectation*. Dordrecht: D. Reidel.

Niiniluoto, I. (1977). On a K-Dimensional System of Inductive Logic. In: F. Suppe and P.D. Asquith (Eds.), *PSA 1976*, Vol. 2. East Lansing: Philosophy of Science Association, pp. 425–47.

Niiniluoto, I. (1983). Remarks on the Logic of Perception'. In: I. Niiniluoto and E. Saarinen (Eds.), *Intensional Logic: Theory and Applications*. Helsinki: *Acta Philosophica Fennica* **35**, 116–29.

Niiniluoto, I. (1988). Analogy and Similarity in Scientific Reasoning. In: D. Helman (Ed.), *Analogical Reasoning*. Dordrecht: Kluwer, pp. 271–98.

Niiniluoto, I. (1995). Hintikka and Whewell on Aristotelian Induction. *Grazer philosophische Studien* **49**, 49–61.

Niiniluoto, I. and Tuomela, R. (1973). *Theoretical Concepts and Hypothetico-Inductive Inference*. Dordrecht: D. Reidel.

Rosenkrantz, R.D. (1980). Measuring Truthlikeness. *Synthese* **45**, 463–87.

Zamora Bonilla, J.P. (1992). Truthlikeness Without Truth: A Methodological Approach. *Synthese* **93**, 343–72.

Poznań Studies in the Philosophy of the Sciences and the Humanities
1997, Vol. 51, pp. 133–146

Arto Mutanen

THEORY OF IDENTIFIABILITY

1. Introduction

Questions of definability and identifiability of concepts have occupied a central place in the methodology and philosophy of science. Traditional accounts of the notions are often vague and obscure. In the literature there is extensive discussion about the notion of definability. In mathematical logic there is a rich theory of definability which is well known among logicians and philosophers of science. It has been instructive to use exact logical notions of definability in philosophy of science. Unfortunately the notion of definability (in its exact logical sense) does not capture all the relevant features that are subsumed under the notion in the methodology and philosophy of science.

The notion of identifiability was introduced to the literature by econometricians. The notion was used to resolve some problems in the actual methodology of econometrics. In fact the notion, and its generalizations (see Simon 1977), captured the features that are incorporated in the notion of definability quite well. So the notion of identifiability has an important role in the methodology of empirical sciences. Unfortunately there is no (well known) general theory of identifiability.

In the literature there exists very many different meanings of the term 'identifiability'. There is nothing wrong in this situation, but it may cause confusions if we do not recognize these different meanings of the term, or if there is no clear and distinct meaning of the term at all. In fact the use of the terms 'identifiability' and 'definability' is messy. Sometimes writers prefer the term 'definability' to the term 'identifiability', sometimes vice versa. Take for example the controversy between Suppes and Simon: Suppes uses the notion of definability (identifiability) as a synonym of definability in its exact logical sense, and Simon uses the notion of definability (identifiability) as a generalized notion of identifiability which is studied by econometricians. So it would be instructive to study general theory of identifiability. In what follows I try to sketch some aspects of such a theory by following Hintikka's ideas. In fact, our restricted treatment can be seen as a special case of Hintikka's theory of identifiability.

We will construct theory of identifiability parallel to the theory of definability. This means that we will define analogous of identifiability in analogy with the different kinds of definability, and try to prove theorems analogous to those in the theory of definitions. The main differences between notions of definability and identifiability are: (i) The notion of definability is defined independently of any model or relative to all the models of the theory, whereas the notion of identifiability is defined relative to a single model or relative to single model of the theory. And (ii) in the theory of definability intended definition is determined by the theory alone, and in the theory of identifiability intended (pseudo)definition is determined by the theory together with some auxiliary information. A natural framework for the theory of identifiability is offered by Hintikka's notion of an interrogative model of inquiry.

2. *Interrogative Model of Inquiry*

The interrogative model can be formulated in the form of a game. This game is two-person game, where the players are called 'the Inquirer' and 'Nature'. The Inquirer is an active player, and he or she has two kinds of moves: logical moves and interrogative moves. Nature is a passive player, and comes to the picture mainly as a source of answers to the Inquirer's questions. In logical moves the Inquirer can make deductive inferences from the premises, and in interrogative moves he or she may put a question to Nature. Nature's answers can be used as additional premises in the Inquirer's logical moves.

There can be different kinds of restrictions on the Inquirer's questions. A very natural restriction says that the Inquirer has to establish the *presupposition* (see Hintikka 1976) of a question before he or she may ask it from Nature. When the Inquirer asks an allowed question Nature must comply. In this paper we assume that all Nature's answers are true (in the underlying model). Answers may contain names of individuals from the domain. So answers are formulated in the extended language which contains names of individuals from the domain.

In specifying an interrogative game we need to know what restrictions are put on Nature's answers. The most important restriction that can be imposed on Nature's answers concerns their quantificational complexity.

These restrictions impose an AE-hierarchy of different kind of answers. If all the answers that Nature can give are atomic or negated atomic sentences the case is called $A(0)$-case. This restriction will be called the *atomistic assumption*. The next step in the hierarchy is the A- or $A(1)$-case. In this case Nature's possible answers can be universally quantified, i.e., answers are of the form $\forall x_1 \ldots x_n S(x_1,\ldots,x_n)$, where $S(x_1,\ldots,x_n)$ is quantifier-free. After this case we have AE- or $A(2)$-case. In this case possible answers can have a string of universal quantifiers followed by a string of existential quantifiers, i.e., they are of the form $\forall x_1 \ldots x_n \exists y_1 \ldots y_m\, S(x_1,\ldots,x_n,y_1,\ldots,y_m)$, where $S(x_1,\ldots,x_n,y_1,\ldots,y_m)$ is quantifier-free. After this we have $A(3)$-case, then $A(4)$-case and so on. Finally, we have the unlimited case, where no restrictions are put on the degree of quantification. Notice that for all $n,m \in \omega$, if $n \leq m$ then, if a sentence is at level $A(n)$ then it is also at level $A(m)$.

In pure observational science it seems to be a natural assumption that Nature cannot tell the Inquirer what the case is or is not always and everywhere, but merely how things are or are not in one particular case, in the particular situation in which the Inquirer is making his or her observation. That means that Nature's answers can be formulated by atomic and negated atomic sentences. So the atomistic assumption is a characteristic assumption in pure observational science. In experimental science the Inquirer is trying to find out functional dependencies. In an experiment the Inquirer can vary one variable (controlled variable) and observe the changes of the other variable (observed variable). So if an experiment is seen as a question of the Inquirer Nature gives the dependence between two variables as an answer. In mathematical language these kind of dependencies can be expressed by functions. So the logical form of Nature's answers in experimental science is at least on level $A(2)$.

We need some logical and model theoretical notation to introduce notions of definability and identifiability in an explicit way. Our notation is a standard one. Let us mention the following: Let L be a first-order language and let $L(P)$ be the extension of L with a one-place predicate symbol P. Let $T(P)$ be a theory in $L(P)$. Let $\mathbf{A}$ be a model of $T(P)$ and let A be the domain of $\mathbf{A}$. We denote by $L(P,A)$ the extension of $L(P)$ with names of individuals from the domain A.

Let F be a formula of $L(P,A)$. If the Inquirer wins a game associated with F in $\mathbf{A}$ against Nature then we say that the Inquirer has *derived F interrogatively from $T(P)$ in $\mathbf{A}$*. If the Inquirer wins no matter what Nature does, so that the Inquirer has a winning strategy in a game associated with F in $\mathbf{A}$, we say that F is *interrogatively derivable from $T(P)$ in $\mathbf{A}$*. We denote this by the following formula

(1) $\mathbf{A}:T(P) \vdash F$.

Let F be a formula of $L(P)$. In this case the notion of interrogative derivability is between notions of deductive derivability and truth in a model. If we do not use any information from the model, then (1) reduces to

(2) $T(P) \vdash F$.

If there is no restriction on information that we can get from the model $\mathbf{A}$, then (1) reduces to

(3) $\mathbf{A} \vDash F$.

Restrictions concerning the available information from the model must be spelled out. It is quite interesting that our results hold largerly independently of what restrictions are imposed on this information from the model, as long as it remains constant throughout the argument. In this paper we suppose that the atomistic assumption holds which means that the set of available information is included in the diagram D of the model. So we may 'stabilize' the interrogative process by supposing that the set of available information is the diagram D of the model. In this case (1) reduces to

(4) $T(P) \cup D \vdash F$.

Because of compactness we have the formula (4) if and only if there is a finite subset D' of D such that

(5) $T(P) \cup D' \vdash F$.

Notice that every sentence in the set D' is atomic or negated atomic sentence of $L(P,A)$ which is true in $\mathbf{A}$.

If F would be a formula of $L(P,A)$ then the only difference in (2)−(5) would be that in (2) we should quantify universally over individuals from A. So F is of the form $F(a_1,\ldots,a_n)$ where $a_1,\ldots,a_n$ are all the individuals from A that occurs in F. Let $x_1,\ldots,x_n$ be new variables. Then in (2) we should replace F by the formula $\forall x_1\ldots x_n F(x_1/a_1,\ldots,x_n/a_n)$.

3. Basics of Definability

In the following we state some basic notions and results of definability. We say that P is *explicitly definable* on the basis of $T(P)$ if and only if an explicit definition of P of the form

$$(6) \quad \forall x[Px \leftrightarrow D(x)]$$

is deducible from $T(P)$, where $D(x)$ is a formula of L and x is the only free variable in $D(x)$.

We say that P is *piecewise definable* on the basis of $T(P)$ if and only if a finite disjunction of explicit definitions of the form (6) is deducible from $T(P)$.

Explicit definability can be seen as a uniform piecewise definability which results when there is only one disjunct in the piecewise definition of P. In a complete theory explicit and piecewise definability coincide.

Let Q be a new one-place predicate symbol. Let $T(Q)$ be the theory in $L(Q)$ obtained from $T(P)$ by substituting Q for P in $T(P)$. We say that P is *implicitly definable* on the basis of $T(P)$ if and only if

$$(7) \quad T(P) \cup T(Q) \vdash \forall x[Px \leftrightarrow Qx].$$

We say that P is *definable in a model* $\mathbf{A}$ if and only if an explicit definition of P of the form (6) is true in $\mathbf{A}$.

In following we give some basic results of definability. (Theorems 1–4 are proved in Rantala 1977.)

Theorem 1: P is piecewise definable on the basis of $T(P)$ if and only if it is definable in every model of $T(P)$.

Theorem 1 states that there cannot be more than one interpretation of P once the interpretation of the other concepts of the theory are fixed. If P is explicitly definable on the basis of $T(P)$ then P is definable by the same formula in every model of $T(P)$, and if P is piecewise definable on the basis of $T(P)$ then definition can be different in different model of $T(P)$.

Theorem 2: P is explicitly definable on the basis of $T(P)$ if and only if it is implicitly definable on the basis of $T(P)$.

This very deep result of definability, known as Beth's definability theorem, is a consequence of Craig's interpolation theorem. The result does not hold for all logics and it is one of the characteristic properties of first-order logic.

We say that P is *restrictedly definable* on the basis of $T(P)$ if and only if there are formulas $D_1(x,y_1,\ldots,y_k),\ldots,D_n(x,y_1,\ldots,y_k)$ $(n < \omega)$ of L such that

(8) $T(P) \vdash \bigvee_{i=1}^{n} \exists y_1 \ldots y_k \forall x[Px \leftrightarrow D_i(x,y_1,\ldots,y_k)]$

where $x,y_1,\ldots,y_k$ are the only free variables in each $D_i(x,y_1,\ldots,y_k)$ $(i = 1,\ldots,n)$.

The following theorem characterizes the notion of restricted definability.

Theorem 3: P is restrictedly definable on the basis of $T(P)$ if and only if for every infinite model $\mathbf{A}$ of L, $\mathbf{A}$ has at most $crd(\mathbf{A})$ expansions $(\mathbf{A},X)$ to the language $L(P)$ such that $(\mathbf{A},X)$ is a model of $T(P)$, where $crd(\mathbf{A})$ is the cardinality of $\mathbf{A}$.

We say that P is *finitely definable* on the basis of $T(P)$ if and only if there are formulas $F(y_1,\ldots,y_k)$ and $D_1(x,y_1,\ldots,y_k),\ldots,D_n(x,y_1,\ldots,y_k)$ $(n < \omega)$ of L such that

(9) $T(P) \vdash \exists y_1 \ldots y_k F$ and
(10) $T(P) \vdash \forall y_1 \ldots y_k (F \rightarrow \bigvee_{i=1}^{n} \forall x[Px \leftrightarrow D_i])$

where $y_1,\ldots,y_k$ are the only free variables in F and $x,y_1,\ldots,y_k$ are the only free variables in each D_i $(i = 1,\ldots,n)$.

The following theorem characterizes the notion of finite definability.

Theorem 4: Predicate P is finitely definable on the basis of $T(P)$ if and only if for every model $\mathbf{A}$ of L, $\mathbf{A}$ has at most n expansions $(\mathbf{A},X)$ to the language $L(P)$ such that $(\mathbf{A},X)$ is a model of $T(P)$.

As a corollary of theorem 4 we have the following: If n in (9) and (10) is 1, then P is explicitly definable on the basis of $T(P)$. The result follows immediately from theorem 1 and the note after it. In this case the explicit definition of P would be the formula $\forall x[Px \leftrightarrow \exists y_1 \ldots y_k(F \,\&\, D_1)]$.

4. Definition of Identifiability

Identifiability can be seen as a generalization of definability similarily as interrogative derivability can be seen as a generalization of deductive derivability. So it is possible to define analogous notions of identifiability as there are notions of definability. In following let **A** be a model of *T(P)* such that *crd*(**A**) $\geq \omega$.

We say that *P* is *explicitly identifiable* on the basis of *T(P)* in a model **A** if and only if a pseudodefinition of the form

(11) $\forall x[Px \leftrightarrow D(x,a_1,\ldots,a_n)]$

is interrogatively derivable on the basis of *T(P)* in **A**, where $D(x,y_1,\ldots,y_n)$ is a formula of L, $x,y_1,\ldots,y_n$ are the only free variables in $D(x,y_1,\ldots,y_n)$ and individuals $a_1,\ldots,a_n$ are from A ($n < \omega$). Auxiliary information may not contain *P*. This restriction means, because of the atomistic assumption, that we have as an auxiliary information the diagram *D* of **A** which is restricted to the language *L(A)*. So *P* is explicitly identifiable on the basis of *T(P)* in a model **A** if and only if

(12) $T(P) \cup D \vdash \forall x[Px \leftrightarrow D(x,a_1,\ldots,a_n)]$

where $D(x,y_1,\ldots,y_n)$, $a_1,\ldots,a_n$ and *D* are as above.

We say that *P* is *piecewise identifiable* on the basis of *T(P)* in a model **A** if and only if a finite disjunction of explicit pseudodefinitions of the form (11) is interrogatively derivable on the basis of *T(P)* in **A**, i.e, if and only if

(13) $\mathbf{A}{:}T(P) \vdash \vee_{i=1}{}^{n} \forall x[Px \leftrightarrow D_i(x,a_1,\ldots,a_n)]$.

Auxiliary information may not contain *P*.

We say that consistent theory *T(P)* is *model complete* if every embedding between its models is elementary. Model completeness has analogous properties in an interrogative model with the atomistic assumption as completeness in the deductive case: If *T(P)* is model complete then explicit and piecewise identifiability coincide.

Let *Q* and *T(Q)* be as above. We say that *P* is *implicitly identifiable* on the basis of *T(P)* in a model **A** if and only if

(14) $\mathbf{A}{:}(T(P) \cup T(Q)) \vdash \forall x[Px \leftrightarrow Qx]$.

Auxiliary information may not contain *P* or *Q*.

We say that *P* is *weakly implicitly identifiable* on the basis of *T(P)* in a model **A** if and only if for every individual *b* in *A* either

(15) $\mathbf{A}{:}T(P) \vdash Pb$ or
(16) $\mathbf{A}{:}T(P) \vdash \sim Pb$.

Auxiliary information may not contain P.

5. Basic Results of Identifiability

The following theorem is the very basic theorem in the theory of identifiability. Hintikka calls this theorem the Extended Interpolation Theorem. Let C be a sentence of $L(P,A)$.

Theorem 5: Assume that $T(P)$ is a consistent theory such that
- (i) $\mathbf{A}{:}T(P) \vdash C$ and
- (ii) not: $T(P) \vdash C$.

Then there is an interpolant $I(a_1,\ldots,a_n)$ such that
- (a) each nonlogical constant of $I(a_1,\ldots,a_n)$ occurs in both T and C except for finite number of individual constants $a_1,\ldots,a_n$,
- (b) $\mathbf{A}{:}T(P) \vdash I(a_1,\ldots,a_n)$,
- (c) $I(a_1,\ldots,a_n) \vdash C$.

As a consequence of theorem 5 we have the following theorem, which Hintikka calls Extended Beth's Theorem.

Theorem 6: P is explicitly identifiable on the basis of $T(P)$ in $\mathbf{A}$ if and only if P is implicitly identifiable on the basis of $T(P)$ in $\mathbf{A}$.

Theorems 5 and 6 are proved in Hintikka and Harris (1988). Theorem 6 helps us to use the techniques of interrogation in the theory of identifiability just as Beth's theorem helps us in the theory of definability. So the theorem has great practical value.

It is easy to see that if P is explicitly identifiable on the basis of $T(P)$ in $\mathbf{A}$ then P is definable in $\mathbf{A}$ with parameters. The converse does not hold generally. For model complete theories explicit identifiability and definability in a model with parameters coincide.

To see the interconnection between definability and identifiability we have to the generalize definition of the notion of identifiability. We say that P is *generally explicitly identifiable* on the basis of $T(P)$ if and only if P is explicitly identifiable on the basis of $T(P)$ in every model of $T(P)$. If the pseudodefiniens is the same formula in every model of $T(P)$, then we say that P is *strongly generally identifiable* on the basis of $T(P)$, otherwise we say that P is *weakly generally identifiable* on the basis of $T(P)$.

We have the following theorems.

Theorem 7: If P is strongly generally identifiable on the basis of $T(P)$ then P is explicitly definable on the basis of $T(P)$ in the extended language $L(P, c_1, \ldots, c_n)$.

Proof: Let P be strongly generally identifiable on the basis of $T(P)$. Then there is a formula $D(x, y_1, \ldots, y_n)$ of L such that for every model $\mathbf{A}$ of $T(P)$ there are individuals $a_1, \ldots, a_n$ such that

(17) $\mathbf{A}:T(P) \vdash \forall x[Px \leftrightarrow D(x, a_1, \ldots, a_n)]$.

Now we have

(18) $\mathbf{A} \models \forall x[Px \leftrightarrow D(x, a_1, \ldots, a_n)]$.

Let $c_1, \ldots, c_n$ be new constants which will be interpreted as $a_1, \ldots, a_n$ respectively. Now we have

(19) $(\mathbf{A}, a_1, \ldots, a_n) \models \forall x[Px \leftrightarrow D(x, c_1, \ldots, c_n)]$.

This happens in every model of $T(P)$, and so we have the intented result. □

Theorem 8: If P is weakly generally identifiable on the basis of $T(P)$ then P is piecewise definable on the basis of $T(P)$ in the extended language $L(P, c_1, \ldots, c_n)$.

Proof: By compactness there are only finitely many different formulas $D_i(x, y_1, \ldots, y_n)$ of L which are used as pseudodefiniens. The result follows as in the proof of theorem 7. □

If P is weakly implicitly identifiable on the basis of $T(P)$ in a model $\mathbf{A}$, then for every a in A there is a sentence F of $L(A)$ (without quantifiers) such that

(20) $T(P) \vdash F \rightarrow Pa$.

Formula (20) follows easily from formula (5).

We have the following theorem which is proved in its general form in Hintikka and Harris (1991).

Theorem 9: If P is weakly implicitly identifiable on the basis of $T(P)$ in every model of $T(P)$ and there is a fixed finite upper bound to individuals in different sentences F in the formula (20), then P is finitely definable on the basis of $T(P)$.

Proof: Let **A** be a model of $T(P)$. Now P is weakly implicitly identifiable on the basis of $T(P)$ in **A**. From restrictions on n (number of individuals in F) and the available information (the atomistic assumption) it follows by compactness that there are only a finite number m of candidates for the role of F for different $a \in A$. Let them be

(21) $F_i(x,y_1,\ldots,y_n),\ i = 1,\ldots,m.$

Each $a \in A$ satisfies condition for some F_i and for some $a_{i1},\ldots,a_{in}$. These individuals can be chosen independently of a. So we have

(22) $\mathbf{A}{:}T(P) \vdash \exists y_{11}\ldots y_{1n}\, \exists y_{21}\ldots y_{2n} \ldots \exists y_{m1}\ldots y_{mn}$
$\forall x \bigvee_{i=1}{}^{m} F_i(x,y_{i1},\ldots,y_{in}).$

For every F_i we have $\mathbf{A}{:}T(P) \vdash F_i \rightarrow Pa$ or $\mathbf{A}{:}T(P) \vdash F_i \rightarrow \sim Pa$, so we can formulate the following sets

(23) $\Sigma^+ = \{F_i : \mathbf{A}{:}T(P) \vdash F_i \rightarrow Pa\}$ and
(24) $\Sigma^- = \{F_i : \mathbf{A}{:}T(P) \vdash F_i \rightarrow \sim Pa\}.$

Let $crd(\Sigma^+) = p$ and $crd(\Sigma^-) = r$, where $crd(A)$ is the cardinality of A. In this case $p + r = m$. Formula (20) gives us the following

(25) $T(P) \vdash \forall y_1\ldots y_n\, \forall x\, [F^+(x,y_1,\ldots,y_n) \rightarrow Px],$

where $F^+ \in \Sigma^+$. So we have

(26) $T(P) \vdash \forall y_1\ldots y_n\, \forall x\, [\bigvee_{i=1}{}^{p} F_i^+(x,y_1,\ldots,y_n) \rightarrow Px],$

where each $F_i^+ \in \Sigma^+$ $(i = 1,\ldots,p)$. Similarly we have

(27) $T(P) \vdash \forall y_1\ldots y_n\, \forall x\, [\bigvee_{i=1}{}^{r} F_i^-(x,y_1,\ldots,y_n) \rightarrow \sim Px],$

where each $F_i^- \in \Sigma^-$ $(i = 1,\ldots,r)$. By definition we have that

(28) $T(P) \vdash \forall y_1\ldots y_n\, \forall x\, [\sim F_i^+(x,y_1,\ldots,y_n) \rightarrow F_j^-(x,y_1,\ldots,y_n)]$

for every $F_i^+ \in \Sigma^+$ and for every $F_j^- \in \Sigma^-$. So we have that

(29) $T(P) \vdash \forall y_1 \ldots y_n (\forall x \vee_{i=1}{}^m F_i(x,y_1,\ldots,y_n) \rightarrow \forall x[\sim \vee_{i=1}{}^p F_i^+(x,y_1,\ldots,y_n) \rightarrow \vee_{j=1}{}^r F_j^-(x,y_1,\ldots,y_n)])$.

Now from (26), (27) and (29) we get

(30) $\mathbf{A} \vDash \forall y_1 \ldots y_n (\forall x \vee_{i=1}{}^m F_i(x,y_1,\ldots,y_n) \rightarrow \forall x[Px \leftrightarrow \vee_{i=1}{}^p F_i^+(x,y_1,\ldots,y_n)])$.

We put $F(y_1,\ldots,y_n) = \forall x \vee_{i=1}{}^m F_i(x,y_1,\ldots,y_n)$ and $D(x,y_1,\ldots,y_n) = \vee_{i=1}{}^p F_i^+(x,y_1,\ldots,y_n)$.

This happens in every model of *T(P)*. Let us suppose that *T(P)* is given in the distributive normal form (see Rantala 1977). The situation described above happens in every constituent of *T(P)* at some, not necessarily the same, finite depth. In each completion of *T(P)* we get formulas *F* and *D* described above. By compactness there are only finitely many of them. Put disjunction of different *F*'s as an existential sentence in formula (9) and different *D*'s can be differerent definiens in formula (10). So we have the intented result. (For details, see Rantala 1977; Hintikka and Harris 1991.)□

6. *Conditions of Definability and Identifiability*

In the theory of definability there are some conditions that types of definitions are assumed to acquire. In logical accounts of definability usually conditions that are studied are *eliminability*, *noncreativity* and *Ramsey-eliminability*. We say that *P* is *eliminable* in *T(P)* if and only if for each formula *F* of *L(P)* there is a formula F_L of *L* such that

(31) $T(P) \vdash F \leftrightarrow F_L$.

We have the following theorem, which is proved in Rantala (1991).

Theorem 10: *P* is explicitly definable on the basis of *T(P)* if and only if *P* is eliminable in *T(P)*.

Let *T(P)* be the set of all sentences of *L(P)* deducible from a set $T \cup \{M\}$, where *T* is a set of sentences of *L* and *M* is a definition (of some kind) of *P* in terms of *L*. So *T(P)* is of the form

(32) $Cn(T \cup \{M\})$

where $Cn(X) = \{F : X \vdash F\}$.

Let $T(P)$ be of the form (32). We say that definition M is *noncreative* if and only if for any sentence F of L such that $T(P) \vdash F$, it also holds that $T \vdash F$. So if M is noncreative then all new consequences that can be obtained by means of M are in the language $L(P)$, and they contain P.

We say that P is *Ramsey-eliminable* in $T(P)$ if and only if every model of T can be expanded to a model of $T(P)$. As a consequence of the completeness of first-order logic we have that, if P is Ramsey-eliminable then P is noncreative. The converse does not hold generally. For example explicit, piecewise, finite and restricted definitions are Ramsey-eliminable.

We say that P is *eliminable in a model* $\mathbf{A}$ if for every formula $F(x_1,\ldots,\ldots,x_n)$ of $L(P)$ there is a formula $F_L(x_1,\ldots,x_n)$ of L such that

(33) $\mathbf{A} \vDash \forall x_1\ldots x_n[F(x_1,\ldots,x_n) \leftrightarrow F_L(x_1,\ldots,x_n)]$.

We have the following theorem which is proved in Rantala (1991).

Theorem 11: P is piecewise definable on the basis of $T(P)$ if and only if P is eliminable in every model of $T(P)$.

Of cource, if P is explicitly definable on the basis of $T(P)$ then P is eliminable in every model of $T(P)$. On this, see notes after the definition of piecewise definability and theorem 1 above. Now clearly we have the following results.

Theorem 12: If P is identifiable on the basis of $T(P)$ in $\mathbf{A}$ then P is eliminable in the extended model $(\mathbf{A},a_1,\ldots,a_n)$.

Proof: Theorem follows immediately from theorems 7 and 10.□

Theorem 13: If P is generally identifiable on the basis of $T(P)$ then P is noncreative.

Proof: It is enough to prove that P is Ramsey-eliminable. The theory $T(P)$ is of the form (32), where M is a finite disjunction of pseudodefinitions of the form (11). Let $\mathbf{A}$ be a model of L such that $\mathbf{A} \vDash T$. Let D_i be one of the pseudodefiniens of P. Let X be a following set

(34) $\{b \in A : \mathbf{A} \vDash D_i(b,a_1,\ldots,a_n)\}$.

So we have that

(35) $(\mathbf{A},X) \models M$.

We have that $\mathbf{A} \models T$, and so we have that

(36) $(\mathbf{A},X) \models T(P)$,

which is intended result. □

Department of Philosophy
P.O. Box 24
00014 University of Helsinki
Finland

REFERENCES

Chang, C.C. and H.J. Keisler (1973). *Model Theory*. Netherlands: North-Holland.

Earman, J. (Ed.) (1992). *Inference, Explanation, and Other Frustrations, Essays in the Philosophy of Science*. Oxford: University of California Press.

Hintikka, J. (1976). *The Semantics of Questions and The Questions of Semantics*. Amsterdam: North-Holland Publishing Company.

Hintikka, J. (1988). What Is The Logic of Experimental Inquiry. *Synthese* vol. **74**, 173–90.

Hintikka, J. (1987). The Interrogative Apporoach to Inquiry and Probabilistic Inference. *Erkenntnis* **26**, 429–42.

Hintikka, J. (1991). Towards a General Theory of Identifiability. In: J.H. Fetzer, D. Shatz and G. Schlesinger (Eds.), *Definitions and Definability: Philospohical Perspectives*. Netherlands: Kluwer Academic Publishers, pp. 161–83.

Hintikka, J. (1992). Theory-Ladenness of Observations as a Test Case of Kuhn's Approach to Scientific Inquiry. *PSA 1992*, vol. 1, pp. 277–86.

Hintikka, J. and I. Halonen (1994). Quantum Logic as a Logic of Identification: In: P. Humphreys (Ed.), *Patrick Suppes: Scientific Philosopher*. Netherlands: Kluwer Academic Publishers, pp. 125–43.

Hintikka, J. and S. Harris (1988). On The Logic of Interrogative Inquiry. *PSA 1988* vol. 1, pp. 233–40.

Hintikka, J. and S. Harris (1991). Identifiability in First-Order Theories: Some Basic Results, forthcoming.

Hintikka, J. and M. Hintikka (1982). Sherlock Holmes Confronts Modern Logic. In: Barth E. M. and Martens J. L. (Eds.), *Argumentation: Approaches to Theory Formation*. Amsterdam: Benjamins, pp. 55–76.

Hodges, W. (1993). *Model Theory*. Cambridge: Cambridge University Press.

McKinsey, J.C.C., A.C. Sugar and P. Suppes (1953). Axiomatic Foundations of Classical Particle Mechanics. *Journal of Rational Mechanics and Analysis*, vol 2, pp. 253–72.

Rantala, V. (1977). *Aspects of Definability*. Amsterdam: North-Holland.
Rantala, V. (1991). Definitions and Definability. In: J.H. Fetzer, D. Shatz and G. Schlesinger (Eds.), *Definitions and Definability: Philospohical Perspectives*. Netherlands: Kluwer Academic Publishers, pp. 161–83.
Simon, H. (1970). The Axiomatization of Physical Theories. *Philosophy of Science*, vol. **37**(1), 16–26.
Simon, H. (1977). Identifiability and the Status of Theoretical Terms. In: Butts and Hintikka (Eds.), *Basic Problems in Methodology and Linguistics*. Netherlands: D. Reidel Publishing Company, pp. 43–61.
Suppes, P. (1969). *Studies the Methodology and Foundation of Science*. Dordrecht-Holland: D. Reidel.
Tuomela, R. (1973). *Theoretical Concepts*. Wien, NY: Springer-Verlag.

III

QUESTIONS IN INQUIRY: THE INTERROGATIVE MODEL

Poznań Studies in the Philosophy of the Sciences and the Humanities
1997, Vol. 51, pp. 149–163

Sylvain Bromberger

NATURAL KINDS AND QUESTIONS

1. Introduction

This paper is based on the tenet that reasoning about natural kinds is intimately linked to reasoning with questions. If right, it should be viewed as amplifying Jaakko Hintikka's rich contributions to our understanding of the central role that questions play in scientific thought.

The term *natural kind* is a philosopher's term. It was presumably used first by Venn in his work on induction and probability. But terms with similar connotations have been used by others. Mill in his treatises on logic (Mill 1875) writes of *Kinds* and of *Real Kinds*. Duhem (1914) writes about *classifications naturelles*. And one finds references to the same idea in many earlier philosophers concerned with issues of metaphysics, epistemology, or philosophy of natural history. Nor is the notion primarily a philosophic or scientific one. Recent work in cognitive science suggests that it is part of our innate intellectual endowment. It is certainly part of the common sense of many people. Thus a column in my daily newspaper *The Boston Globe,* once contained the following question and answer:

> Q. How many kinds of animals are there on earth, and at what rate are they disappearing?
> A. No one really knows how many types of animals there are or how many of those species are disappearing. *The World Book Encyclopedia* says scientists have classified almost a million animals ranging in size from the microscopic to the gargantuan. Each year hundreds of new animals are discovered. The more than 800,000 varieties of insects make up the largest group. There are about 21,000 kinds of fishes, 8,600 species of birds, 6,000 types of reptiles, 3,000 kinds of amphibians and 4,000 species of mammals.

In spite of its familiarity and the ease with which we rely on it, the notion that there are natural kinds is replete with problems that call for philosophic scrutiny. In what follows, I will first review some of these problems. Any adequate account of Natural Kinds should help us solve or dissolve them. I will then sketch a conjecture about what the notion that there are natural kinds comes to. The conjecture will put us in a position to revisit the problems from a different perspective, though I will postpone such a revisit for an occasion where much more space is available. I will also concentrate on what is implicit in the belief that there are natural kinds, and not on so-called natural kind terms, though I will finish with a brief comment about such terms, a family of lexical items that has received much attention in philosophic and linguistic literature.

2. *Some Classical Issues about the Notion of Natural Kinds*

A. *Methodological Issues and Rational Taxonomy*

Statements like the one quoted from *The Boston Globe* presume that animals – that is, individual animals – can be systematically and unequivocally sorted out into species, types, classes, varieties, and so on. But of course, like any set of individuals, they can be sorted in indefinitely many ways. There is a whole discipline, so-called systematics, devoted to the study of systems of classification and evaluation of such systems. It is based on the plausible view that some classifications are better, more effective, more convenient, more natural, perhaps closer to the facts for the purposes of natural history than others. But what are the features by which good classifications differ from bad ones? That turns out to be a deeply controversial and often divisive issue, as is attested by the literature on taxonomy in e.g., entomology, or the debates that surrounded Lavoisier's proposals on the classification of chemical substances, or the controversies that accompanied the transition from Linnean taxonomies to cladistic ones[1].

B. *Ontological Issues*

What is the ontological status of Natural Kinds? Are there so-called Real kinds, or are there only Nominal kinds? Note that the *Globe* reporter wrote of kinds and species and types as if they were part of the furniture of the world, as if the world contained not only individuals but also kinds, as if we come across not only new individuals but new kinds. But this cannot be literally true. We never actually come across a kind, but only across individuals: since kinds are not the sort of things that can have any causal efficacy, they cannot affect our senses, be seen, heard, felt, tasted, smelled, or be bumped into. But even so, talk about kinds does not come down to talk about individuals. For instance, the enumeration of kinds is crucially different from the enumeration of individuals since it is one thing to know (or to be ignorant about) the number of kinds of animals there are and an altogether different thing to know how many animals there are. Nor is the appearance and disappearance of kinds simply the appearance or disappearance of individuals. But if it is not literally true that Natural Kinds exist as entities over and above their members, how is it true, how is one to understand the claim that kinds in general exist, or even that some specific kind exists?

Some people are actually willing to say that one must understand these claims literally. After all, there are not only individuals, but also sets of individuals, and natural kinds are but sets of individuals. One has to be a very finicky nominalist indeed, to deny the existence of sets. However, this attitude raises a problem of its own. So-called natural kinds may be sets, but they can't be just arbitrary sets. They presumably are special sets, sets that are distinct from random sets. Their members have been grouped by nature, so to say. But exactly what does that mean? What facts about the world, about nature, about reality, about God for that matter, does the distinction of sets into natural kinds and non-natural kinds (nominal kinds) register? Or – put differently – what sort of new fact about the world does the discovery of a new natural kind represent? How is it related to, how is it like, and how is it unlike the discovery of new individuals?

Of course some people think that such worries are misguided, are the price of unclear thinking which fails to acknowledge that so-called natural kinds are human intellectual artifacts. We, according to them, not nature, are the agents who parcel things into kinds. If there is a problem here at all, it is the problem of making clear to ourselves the ends for which we classify. Good classifications are not discovered but contrived. On such a view there are no right or wrong classifications, only more useful or less useful ones, and the ontological problem reduces to the methodological one.

Others rightly reply that no classification of natural objects can be convenient, or useful, unless it is grounded on facts about these objects, facts by virtue of which they fall into kinds that we may choose to acknowledge or not acknowledge, but which are not up to us to invent. A classification can be effective only if it, as the saying goes, "cuts nature at the joints". So this way of looking at things, this pragmatic descent, does not really dissolve the ontological issue.

C. Epistemological Issues

There are, to begin with, problems that are the obverse of the metaphysical problems and the issues in systematics. Thus the history of science is replete with claims that new species have been discovered. (Think of Darwin's trip on the *Beagle*.) But how is the claim that a new kind has been discovered ever warranted? When does evidence justify it, and how? What is the nature of the reasoning, of the inferences from observation to conclusions that justifies such claims or such beliefs? It obviously is not pure deductive reasoning from observational premises. Nor is it simple induction by enumeration. Nor does it seem, offhand, to be induction to the best explanation (or if it is, explanation of what?).

The problem about evidence arises in other contexts brought to our attention initially by Nelson Goodman (1965) but about which there now exists an enormous literature. When we infer from observed cases to unobserved ones we rely on classifications of individuals into something like natural kinds. So suppose that I observe that some elephant has a heart with four cavities, and suppose that I classify objects into sets, one of which contains my elephant, the elephants in the Bronx Zoo, the snakes in South America and the flies in New Jersey, and suppose that I call the members of that set *shmiers*. It would clearly be a mistake to infer from the fact that one or more shmiers has a heart with four cavities that all shmiers do. Shmiers don't make up a kind across which inductive projections are licensed. On the other hand, if I classify things so that there is a class that contains all and only elephants, then the inference, though not conclusive, will be at least rational, plausible. Or, to take an example with a more respectable Goodmanesque philosophic lineage, suppose that I define *emphire* as the class of all things that are either emeralds and have been observed by some human being before March 3, 2000, or are sapphires, but have not been observed before March 3, 2000. All the emphires observed so far have been green. So have all the emeralds observed so far. In fact every emphire observed so far has been an emerald and vice versa. Yet the induction to the conclusion that all emphires are green is unwarranted, whereas the induction to the conclusion that all emeralds are green is warranted. Even the inference that all emphires are grue (green if observed before March 3, 2000, blue otherwise) seems abnormal. What is the difference between the species called emphire and the species called emeralds that underlies and generates this difference? What is there about natural kinds that licenses projection from individual members to further individual members?

D. Semantic Issues

The notion of natural kinds has come up in contemporary philosophic literature in relation to a set of problems raised initially by Saul Kripke (1980) and Hilary Putnam (1975), but about which there now exists a vast literature and innumerable doctoral dissertations. There is a view in semantics, traced usually to the German mathematician Frege (1893), according to which singular terms, i.e., names (and descriptions) do their linguistic job by virtue of being attached, on the one hand, to something called a *sense* (*Sinn*) and on the other hand to something called an extension or *denotation* (*Bedeutung*). The extension is the set of things to which the name applies. Thus the extension of the word 'dog' is the set of all dogs that have, do, and will exist. The sense of the name is something more subtle and not easily characterized, but roughly it is that which speakers know about the word by virtue of which they can recognize members of its extensions, that learnable fact tied to the phonological (or orthographic) and syntactic features of a sign by virtue of which the sign is used to refer to members of its specific extension. Some hold that English and other languages (possibly all natural languages) contain so-called natural kind terms, that is, terms whose extensions by semantic necessity make up natural kinds. Those who hold this view must thus believe that the senses of natural kind terms share features by virtue of which their bearers qualify as natural kind terms. It becomes a task of semantics to characterize that feature. Kripke and Putnam have advanced powerful arguments that cast deep doubts on this view. They deny that names of natural kinds have Fregean senses. They hold that such names are *rigid designators* and get their extension, not through a semantic content, but by being tacked directly and deliberately to some member of the extension in a sort of baptism. The word *water*, for instance, on this view, has its extension because something like the following happened: someone at some time pointed to a sample of water and said "This stuff is going to be called *water*", or words to that effect, and thereby initiated the convention that all samples standing in the relation of being "the same substance" to this sample may be referred to by utterances of the sound *water*.

Kripke and Putnam's way of picturing things may free us from the problems raised by the theory that names of kinds have a sense that share a distinctive feature, but only at the price of a new problem, namely, that of determining what constitutes *being the same substance*, or, in the case of animals, *being of the same species*. And that problem is as recalcitrant as the problem raised by the Fregean assumption. In short, Kripke and Putnam do not deny that natural kind terms themselves constitute some sort of natural kind, that is, they still hold that the extensions of some terms constitute natural kinds and that this is something that must be known (or at least believed) by anyone who can be said to understand the term, but they leave us in the dark about the content of that knowledge (or belief). I shall come back to this issue in the last section.

3. A Different Approach to the Notion of Natural Kind

Whatever the problems that burden the notion of natural kind, the presumption that the world contains such kinds is unquestionably a very deeply ingrained presumption, so deeply ingrained, in fact, as to suggest that it not only springs from our native cognitive endowment, but that it does so in a way that shapes how we learn about the world, think about it, cope with it, and talk about it. We should therefore not be surprised that it is at once transparent, utterly compelling, and yet elusive.

In what follows I want to sketch a theory about what the presumption that the world contains natural kinds comes to. Note that I speak here of a presumption. I don't propose to show that there are natural kinds. I don't know whether there are. But I think that much of our intellectual commerce with the world is inescapably based on the presumption that there probably are. We approach the world and inquire about it with expectations built on that presumption. Science would be impossible without it. But what is its content?

I have spoken of a presumption, as if there were a single one. But the analysis I am about to propose assumes that belief in natural kinds is actually constituted of a family of presumptions in which more complex ones rely on simpler ones, but in which simpler ones do not require commitment to the more complex ones. I will therefore divide the discussion as if the presumption implicated different levels of commitment, and speak of *levels* of that presumption.

I. THE FIRST LEVEL OF THE PRESUMPTION: The world contains sets of objects that stand as exact models to each other.

The word *model* has many meanings in contemporary English. I use it here in a sense on which I have relied before (Bromberger 1992, p. 178). Examples of the sort of things I have in mind are familiar: small models of real airplanes, plastic models of DNA molecules, models of the Eiffel tower, maps of territories, and so on. What makes some of these things models of others? The fact that we can find out things about the object modeled by investigating the object that serves as model. Thus counting the number of engines of a model airplane will tell us the number of engines of the actual airplane, counting the twists on the DNA model will tell us the number on an actual molecule, computing the ratio of height to that of the base of a mock-up of the Eiffel tower will tell us what that relation is in the case of the real Eiffel tower.

The relevant notion of a model on which I rely is thus the following: *M is a model of O relative to a quadruple* $<Q_m, Q_o, P, A>$ *if and only if (a) M and O are numerically distinct, (b) in that quadruple* Q_m *is a non-empty set of questions about M,* Q_o *is a non-empty set of questions about O, P pairs members of* Q_m *with members of* Q_o*, and A is an algorithm that translates answers to any member of* Q_m *into answers to the member of* Q_o *paired with it by P, and translates correct answers to the former into correct answers to the latter.*

This notion of a model is obviously a relational notion. Nothing is intrinsically a model. More importantly, it is a relativised notion. No two objects stand or fail to stand in the relation of model to modelee *tout court*. They may stand in that relation relative to some quadruple and not relative to other quadruples. In short, it is built on a three-term relation (actually a six-term relation), not merely a two-term one.

To fix ideas, think of a map, which is a form of model under this definition. A question in Q_m for the relevant quadruple might be "What is the distance between dots *a* and *b* on the map?" and in the corresponding Q_o "What is the distance between city *A* and city *B* in the territory?" where the pairing is provided by the mapping conventions and the algorithm by the scale. Other questions about the map will not enter into any such quadruples, for instance, "What is the chemical composition of the paper?" or "What is the name of the publisher printed at the bottom?"

An *exact* model is defined as follows: *M is an exact model of O if and only if, in the quadruples relative to which they are models, (a) members of* Q_m *are paired by P only with members of* Q_o *with which they are identical but for the replacement of references to M in* Q_m *by references to O in* Q_o, *and (b) A matches answers to each that are identical but for a similar replacement of references.*

So, for instance, any sphere S_1 is an exact model of any identical sphere S_2 relative to a quadruple that includes in Q_m "What is the radius of S_1?", "What is the circumference of S_1?", "What is the area of S_1?", and that includes in the paired Q_o the same questions but with S_2 mentioned instead of S_1, and that has as *A* the equality function.

So the first level of the presumption comes down to this: *Objects in the world fall into sets of indefinite size whose members are exact models of each other relative to some known or discoverable sets of questions.*

That is a bit abstract, so let us look at two examples that support the presumption.

Samples of mercury are exact models of each other relative to the questions "What is the boiling point of *m*?" (356.6°C), "What is the freezing point of *m*?" (–38.87°C), "What is the molecular weight of *m*?" (200.6).

Utterances of the expression *Massachusetts Institute of Technology* are exact models of each other relative to the questions "How many words does it contain?", "What does it refer to?", "What are the morphemic constituents of each word?", "What is its constituent structure?".

The first level of the presumption may not seem very weighty, but it is not a tautology. Is it a necessary truth? That is a complex issue which cannot be considered without delving into the nature of the relevant modalities. It does not seem to be a necessary truth under most construals. But even if it is not a necessary truth, its import for science is not that of a hypothesis to be confirmed or disconfirmed. As a hypothesis it is rather thin and dull. Its import is that of a ground for inquiries or what a Kantian might pompously call a Postulate of Pure Erotetic Empirical Reason. If there are sets of objects that stand as exact models of each other in nature (artifacts are less interesting), we should want to know what they are, or we should at least want to know as many of them as possible, and perhaps how they came to be. Satisfying such a want is not a simple matter. One does not discover such sets by simply looking at things: one has to equip oneself with questions. Take, as a case in point, the first of our examples. The discovery that there are samples of matter that are exact models of each other in these specific respects could not have been made before the erotetic notions of temperature, of boiling point, and of molecular weight had been established, and that happened rather late in our conceptual history and required enormous preparation.

In what follows, I will use the expression *Projectible Questions for the kind K* to refer to the questions that make some set *K* a confirming instance of this first-level presumption. I shall call such confirming instances "Minimal Natural Kinds". The first presumption can therefore be restated as follows: *The world contains minimal natural kinds.*

II. THE SECOND LEVEL OF THE PRESUMPTION: The world contains sets of objects that constitute Quasi-Natural Kinds.

A quasi-natural kind is a minimal natural kind *N* of which the following is also true: *There is a non-empty set of questions Q_w: (a) Whose presuppositions are satisfied by every member of N; (b) To which different (though not necessarily all) members of N bear different answers, i.e., Q_w is not a* *set of projectible questions for N; (c) Whose right answer – for each member – is the object of a sound why-question; (d) At least one answer to that why-question in each case follows from common nomological principles, instantiated by contingencies peculiar to the case, but of a sort applicable to all.*

Let us refer to such questions as *w-projectible questions over the kind.*

To make this more concrete, let us again consider samples of mercury. They constituted a minimal natural kind by virtue of the question and common answer pairs: "What is the boiling point of *m*? 356.6 °C"; "What is the freezing point of *m*? –38.87 °C"; "What is the molecular weight of *m*? 200.6." And they furthermore constitute a quasi-natural kind by virtue of the *w*-projectible questions such as the following, askable of each sample (but not always supporting the same answer for each sample): "What was its temperature on July 4, 1987 at noon?" That question is *w*-projectible because its right answers for each sample are individually themselves the topic of a sound why-question "Why was the temperature *T* on July 4, 1987?", one of whose answers satisfies (d) above if we assume, as seems plausible, that the temperature can be accounted for in each case by appeal to the same appropriate thermodynamical theory and "boundary conditions" of the sort selected by these principles, e.g., details about the degrees of freedom of the particles in the sample and distributions of energy among these degrees of freedom.

In short: *A class of objects that constitutes a minimal natural kind also constitutes a Quasi-Natural Kind relative to a set* $<Q_p/A_p, Q_w>$ *-- in which the* Q_i/A_i*'s are non-empty sets of questions paired with their true answers -- if and only if it is the largest set of objects for which those in* Q_p *are projectible, and those in* Q_w *are w-projectible.*

The presumption that there are quasi-natural kinds, like the presumption that there are minimal natural kinds, does not represent a necessary truth. And its import for science too is that of a regulative principle of inquiry. If there are quasi-natural kinds, we want to know what they are, or we may want to discover as many of them as possible, and we may want to know how they came to be.

Here too, it is worth noting that the discovery of quasi-natural kinds requires much more than mere observation of things: it too may require the conception of new questions. But it may require even deeper reconceptualizations. The presumption entails that the world sustains a distinction between brute facts and explainable facts, or, more exactly, between facts that are legitimate objects of sound why-questions and those that are not. The discovery of quasi-natural kinds may require a relocation of the division between these two kinds of facts.

To see what is involved here, think of Lavoisier after he had performed the experiments that convinced him that water is a compound of Hydrogen and Oxygen. Suppose that at that point he has raised the question "Why is water a compound of Hydrogen and Oxygen?" There would have been something blatantly nonsensical about that question, something calling for a rhetorical "Why should water not be composed of Hydrogen and Oxygen?" Every fact is not necessarily the topic of a sound why-question. At the time, the question "Why do Hydrogen and Oxygen combine at all?" would have seemed equally hopeless. So the status of a why-question can change from pointless to significant as new knowledge is acquired.

But how can one tell whether a fact is – or could become – the topic of a sound why-question? Often one simply can't. I have argued elsewhere (Bromberger 1992, pp. 145 ff.) that why-questions can put one in the very peculiar predicament of not only not knowing the answer and not knowing how to find the answer and not being able to come up with a non-objectionable answer but of not even knowing what one needs to know in order to know whether the question has an answer at all. Why-questions are very different from other questions in this respect[2]. In the case of other wh-questions one need only establish whether their presuppositions are true to establish whether they have an answer, and their presuppositions are inferable from non-elliptic formulations. Thus neither "What is the age of the King of the U.S.?" nor "What is the square root of the King of the Belgians?" has an answer. The first suffers from failure of referential presupposition (there is no King of the U.S), the second from failure of attributive presupposition (the King of the Belgians is not the sort of thing that has a square root). Why-questions can have referential presuppositions inferable from non-elliptic formulations. They can also have propositional presuppositions inferable from non-elliptic formulations ("Why does hydrogen combine with oxygen?" has the propositional presupposition that hydrogen combines with oxygen.). But they have no attributive presuppositions inferable from non-elliptic formulations. Nor is the truth of their referential and attributive presuppositions sufficient to guarantee that they have a true answer. This is due to the fact that why-questions are governed by assumptions of a special sort, to the effect, (roughly) that one condition (at least) is unsatisfied from a set of conditions that are jointly sufficient and individually necessary for some contrary of the propositional presupposition to obtain. But there is no way to infer from mere knowledge of the propositional presupposition, whether the assumption is true, and what its content might be in a case at hand. There may not be such a set of jointly necessary and individually sufficient conditions in the case at hand, or there may be more than one such set, and there is no way of knowing in advance (that is, from mere understanding of the question) which ones, if any, can fail to be satisfied (Bromberger 1992, pp. 75 ff.).

III. THE THIRD LEVEL OF THE PRESUMPTION: The world contains biological kinds.

A biological kind, for present purposes, is *a non-empty set which: (a) includes a subset whose members constitute a quasi-natural kind as defined above. Let us call that subset the "stereotypical core" of the biological kind; (b) also includes the preponderance of any of its members' descendants; (c) if Q is a projectible question for the stereotypical core, but Q is not projectible across all the descendants, then Q is at least w-projectible across both the stereotypical core and all the descendants.*

A biological kind is thus the union of a stereotypical core and the descendants of that core[3]. Note that the definition is designed to accommodate not only stereotypical tigers, but also the exceptional albino and three-legged ones that are descendants of stereotypical ones.

The notion of descendant required for this definition is simply the notion captured by Frege's classical definition. Thus if *P* stands for the relation of parenthood then *b* is an ancestor of *a* if and only if

$$(\forall A)(a \in A \;\&\; (\forall y)(\forall x)(x \in A \;\&\; yPx :\rightarrow y \in A)) \rightarrow b \in A))$$

This notion of descendant raises an intriguing possibility. It is built on the relation of parenthood. But we know that ancestor-like relations (ancestrals) can be built on other relations. Frege, for instance, defined the set of natural numbers by means of an ancestral-like relation built on the notion of successor. Could the presumption that there are biological kinds be extended along similar lines?

More explicitly, parenthood is a causal relation. Roughly, *a* and *b* are parents of *c* if *a* and *b* are causally and biologically responsible for *c*'s earthly existence. Could there be other causal relations whose Fregean ancestral plays a similar role in shaping our curiosity about the world?

IV. THE FOURTH AND FINAL LEVEL OF THE PRESUMPTION: The world contains CATEGORIES that are the domain of descriptive and explanatory laws.

A category is simply a set of distinct quasi-natural kinds that are all relativized to the same projectible, *w*-projectible, and individuating questions. So, for instance, samples of mercury form a category with samples of gold, water, iron, hydrochloric acid, and other substances. More specifically, as we noted before, samples of mercury form a quasi-natural kind by virtue of the questions "What is its boiling point?", "What is its molecular weight?", "What is its freezing point?" and so on, for which they all get the same answers. But samples of (pure) water are also subject to these questions and they too all share the same answers. Of course, the answers are different for water and for mercury. The same similarities and differences apply to samples of gold, iron, hydrochloric acid, and other chemical substances. That is how chemical substances therefore constitute a category. More specifically then, *A set of quasi-natural kinds forms a category relative to a set* $< Q_p, Q_w >$ *if and only if it is the largest set of quasi-natural kinds for each of whose members the questions in* Q_p *are projectible and those in* Q_w *are w-projectible.*

When is a category the domain of a descriptive law and of an explanatory law?

To see what is involved here, we must first return to projectible questions. A set of projectible questions characteristic of a category (the Q_p in the definition above) constitute the dimensions of a space. In the case of our example of chemical substances the space is defined by the questions "What is its boiling point?", "What is its freezing point?", "What is its molecular weight?", and so on. The spaces that concern us are spaces defined by such projectible questions.

The points of such a space are tuples whose members are possible answers to the defining questions. Some of these points correspond to actual substances, some don't. So, for instance, in the case of our example, the point where the answer to "What is the boiling point?" is 356 °C, that to "What is the freezing point?" is –38.87 °C and that to "What is the molecular weight?" is 200.6 corresponds to mercury, the point where the answer to "What is the boiling point?" is 2807 °C, that to "What is the freezing point?" is 1064 °C and that to "What is the molecular weight?" is 197 corresponds to gold. Each of the other substances is located in this fashion at a corresponding point. On the other hand, most of the points defined by arbitrary answers to these questions do not correspond to any substance at all. Returning then to our question, *A category is the domain of a descriptive law if there is a law that is the locus of all the members of that category, i.e., a law that partitions the points of the space into two subsets, those whose points are answers for actual attested or attestable cases and those whose elements don't add up to a description of any actual cases. A category is the domain of explanatory laws when every quasi-natural kind in it has w-projectible questions whose answers depend on the same explanatory principles.*

We can now summarize all this succinctly: *Natural kinds are quasi-natural kinds that are members of categories which are themselves the domain of descriptive and explanatory laws*. And the presumption that there are natural kinds thus also implicates the presumptions that there is more than one and that there are descriptive and explanatory laws.

The discovery of natural kinds is a many faceted enterprise. Elsewhere I listed some of the tasks required for their discovery:

> Observation of phenomena is only the most obvious among them. Questions, answers that fit them, concepts with which to embed them into laws, expressions with which to process them, must be thought of; those that are projectible across significant numbers and varieties of things must be recognized; .... regularities must be spotted, canonical usages that guard against the misleading associations of the current vernacular must be introduced, etc. Since, in the performance of these tasks, conceptual and terminological creativity plays an enormous part, one might be tempted to conclude that scientists, rather than facts about the world, are responsible for the existence of categorial spaces in which laws operate. But that is hardly plausible. Surveyors label and map the territory; they sometimes divide it; but they do not create it. Gloves are human creations, but it does not follow from this that the hands they are designed for are also human creations, or that the shape of those hands springs from the glovemaker's mind. (Bromberger 1992, pp. 199–200.)

4. A Word about so-called Natural Kind Terms

Much of the recent discussion about natural kinds has actually been discussion about so-called natural kind *terms* carried out in what Carnap called "the material mode." But are there such terms, are there terms whose meaning – however we think of meaning – implies that the corresponding extension makes up a natural kind? Are there terms whose understanding requires that competent users construe them as having a natural kind as extension?

The answer, it would seem, must be both "yes" and "no"; it depends on what part of the vocabulary one is talking about.

Any person's vocabulary, lexicon, can be divided into two distinct sets. Part of the vocabulary – let us call this *the native vocabulary* – is made up of words and morphemes acquired through exposure, under conditions of paucity of stimulus, with relatively little negative feedback, i.e., in the course of natural linguistic development. There is no reason to believe that the native vocabulary does or can include natural terms. It may, of course, turn out to contain terms whose extensions happen to constitute natural kinds relative to some set of questions, but that would be a coincidence, a rather amazing coincidence. Part of the vocabulary – at least in some cultures – consists of terms acquired not through mere exposure but through tutoring, instruction that, at least initially, requires prior mastery of a native vocabulary. Let us call that part of the vocabulary *nomenclature*. There is such a thing as scientific nomenclature. It is subject to explicit stipulations and deliberate conventions, it is codified in canonical texts and it is under the governance of experts and professional groups to whom others must defer. That part of the vocabulary does contain terms whose extension is presumed, by convention, to constitute natural kinds. Its terms are introduced at specific historical moments and with stipulations designed to give them that character. Terms in the nomenclature have a history of their own which is quite distinct from the history of the rest of the language. They evolve along lines and under conditions that have little to do with the way other words evolve. (Compare the history of such English words as 'shoulder' or 'meat' which go back to Indo-European origins and are acquired without definitions, and the history of the words 'atom' or 'mass' or 'heat' as used in physics and chemistry. The introduction of the latter can be dated fairly precisely, and was subject to explicit datable revisions acknowledged by professional organizations. Many such terms do not even belong to any particular language, are recorded as ideographs, have no phonology proper, occur in formulae that need not be translated to become accessible to speakers of different languages.)

Conflation of these two classes is due to at least two facts: a) Some phonological shapes of items in the native vocabulary have come to also encode items in the nomenclature. How and why such double assignment ever occurs are interesting questions to be left for another occasion. But by fixing on such an ambiguous phonetic shape while thinking that one is fixing on the term itself, one can easily come to overlook the shape's double life. b) Many items in the nomenclature can be taught and are taught. But the process by which they are taught is not always recognized as utterly different from the processes through which lexical items in the native vocabulary are acquired through natural development. By slighting the difference between these two processes one is easily led to disregard crucial differences among their end results.

Whether samples of water – where I am using the word as an item in the native vocabulary – constitute a natural kind was a matter to be discovered. It turns out that they don't, though a subset of them does. Whether samples of H_2O constitute a natural kind was never a matter open to discovery. That they are follows from the meaning of H_2O.

Massachusetts Institute of Technology

NOTES

1 For an important discussion of some of these problems and a useful bibliography see Dupré (1993).

2 Though not from all. "What is ..." questions, for instance, can share the same peculiarity.

3 This definition does not make allowance for the possibility of speciation. It represents therefore only a first approximation. The further conditions required to accommodate speciation can probably not be formulated at the present level of abstraction since they need to take into account the various – often conflicting – interests that regulate division into species and other taxa.

REFERENCES

Bromberger, S. (1992). *On What We Know We Don't Know: Explanation, Theory, Linguistics, and How Questions Shape Them.* Chicago: The University of Chicago Press and Stanford: CSLI.

Duhem, P. (1914). *La Théorie Physique: Son Objet, Sa Structure.* Paris: Marcel Rivière & Cie.

Dupré, J. (1993). *The Disorder of Things. Metaphysical Foundations of the Disunity of Science.* Cambridge: Harvard University Press.

Frege, G. (1893). Über Sinn und Bedeutung. *Zeitschrift für Philosophie und Philosophische Kritik* **100**, 25–50.

Goodman, N. (1965). *Facts, Fiction, and Forecast.* 2d ed. Indianapolis: Bobbs-Merrill.

Kripke, S. (1980). *Naming and Necessity.* Cambridge: Harvard University Press.

Mill, J.S. (1875). *System of Logic.* 8th ed. London: Longmans.

Putnam, H. (1975). The Meaning of 'Meaning'. *Philosophical Papers, ii: Mind. Language and Reality.* Cambridge: Cambridge University Press, 215–71.

Poznań Studies in the Philosophy of the Sciences and the Humanities
1997, Vol. 51, pp. 165–180

Scott A. Kleiner

THE STRUCTURE OF INQUIRY IN DEVELOPMENTAL BIOLOGY

1. Introduction

A recent issue of *Science* (28 October, 1994) is devoted largely to a review of the current state of developmental biology. The introductory article includes a survey of 100 prominent developmental biologists who were asked what they thought were the most important unanswered questions in their discipline. They were also asked where they expected the most rapid progress over the next five years. There were 66 respondents, and the responses gave the following order of importance:

Unanswered Questions

Q1. What are the molecular mechanisms of morphogenesis?
Q2. What is the connection between development and evolution?
Q3. How do cells become committed to a particular fate?
Q4. What is the role of cell-cell signaling in development?
Q5. How are patterns established in the early embryo?
Q6. How do neurons establish their specific connections?
Q7. How do cells know when to divide and when to die in the sculpting of organs and tissues?
Q8. How do transcription factors control tissue differentiation?

This ranking was generated by ordering the number of respondents voting for each. These numbers ranged from 36 for the first to 15 for the last question. By a similar procedure the questions for which progress is expected over the next five years were ordered, and interestingly the ordering was Q4,Q1,Q3,Q8,Q6,Q5,Q7 and Q2. Also three areas of promise beyond the above questions were included with the following ranking:

T3. Vertebrate development, aided by mouse and zebrafish genetics.
T6. The identification of developmentally important genes.
T9. The molecular mechanisms of tissue induction.
T12. Plants as models for understanding development.

These topics can also be construed as areas of inquiry and briefly described as focusing on the following questions:

Q9. How do vertebrates develop? (The choice of mice and zebrafish as models is part of a heuristic for the search for answers to this question.)
Q10. What genes are developmentally important?
Q11. How does tissue induction take place?
Q12. What important insights into development and related areas might plants provide?

I have argued elsewhere (Kleiner 1993) that an important but neglected component of the logic of scientific discovery consists of normative principles according to which questions or research problems should be given preferential order. Scientists are always endowed with limited intellectual and technological resources for generating knowledge, and principles underlying the rational allocation of these resources should be of great interest to philosophers of science. The aim of this paper is to set forth principles underlying the choice of these and related questions concerning development as foci for the research efforts of embryologists, geneticists and evolutionary theorists. These principles should provide an ordering of questions into two hierarchies, one of subordinate and superordinate questions and a second of questions of more or less epistemic importance. Such a 'rational' ordering should replace the subjective and intuitive orderings of questions reported in *Science*.

Question Q is subordinate to (a sub-question) of Q' if Q is believed to serve as a means to answering Q'. Question Q is more important than Q' epistemically if answers to Q are expected to contribute more or more significantly to the enterprise of knowing a subject than are the answers to Q'. Questions arranged in these hierarchies would constitute a more detailed representation of a scientific research program than is usually offered (Lakatos 1978).

Practical as well as epistemic interests are involved in the question preferences of developmental biologists. This paper will focus upon epistemic interests, the valuing of these questions as instruments for the advancement of knowledge for its own sake.

2. *Models of Question Choice*

Professor Hintikka has offered a model of question preference which is based on the 'method of analysis' in deduction (Hintikka and Hintikka 1982, Kleiner 1993). This method has been interpreted in a variety of ways in the history of philosophy and science, but one important interpretation involves the search for premises in an effort to deduce a given conclusion. In this procedure Professor Hintikka distinguishes deductive from interrogative moves. The interrogative moves are appropriate in the search for information that will complete what is needed for the deduction of a conclusion, provided that the question's presupposition has been affirmed in the inquiry at hand.

To apply this scheme to natural science, a plausible first step would be to consider scientific explanation. Explanation is one of the central epistemic objectives of scientific research, and the metascientific analysis of explanation might be expected to provide a framework for the organization of many important scientific research programs. If scientific explanations could all be construed as deductions, as in the case of the Hempel–Oppenheim model of explanation in the empirical sciences, Professor Hintikka's model of inquiry would be singularly appropriate for modeling the scientist's search for explanations. Deductive explanations consist of covering laws with empirical content, boundary and initial conditions, all of which serve as premises in the deduction of a law or phenomenon to be explained. As Professor Hintikka points out, various conceptual and cognitive elements make such deductions possible, e.g. concepts in terms of which credible lawlike statements can be formulated, covering laws, specifications of systems, their components, initial states and influences across their boundaries. The conceptualization schematized in Professor Hintikka's deontic-epistemic logic of questions tightens the deductive-nomological conception of explanation insofar as it excludes some of the counterexamples to this conception based on the properties of classical quantificational logic (Kitcher 1981, Hintikka 1981, Kleiner 1993). Also reasoning from the conclusion or explanandum back to what it would take to deduce that conclusion is an effective strategy for identifying missing elements whose presence or determination would complete the deduction. Thus one might focus upon the search for a concept that would permit the formulation of a covering law, or for the conception and description of a system which will complete the initial and boundary conditions required for a deductive explanation. Such questions concerning covering laws, system composition, initial and boundary conditions are subquestions to the overall question demanding the explanation. As such they inherit the importance attributed to the superordinate question (Kleiner 1993).

However, this determination of questions by deductive requirements works only in the context of particular background beliefs that the scientist brings to a given problem situation: If the covering laws are known, then the search might be directed to appropriate conceptions of systems and initial conditions. Alternatively a search might be directed to concepts or laws that will provide explanatory connections within a known system or between the system and its environment.

A number of writers (e.g. Brody 1972, Schaffner 1993) have pointed out that many scientific explanations do not take this deductive form. They are not even reasonably regarded as 'explanation sketches' which when fully elaborated fit the Hempelian form. Rather, many explanations in the biological sciences simply describe with varying generality biological structures at several levels of organization and processes by which states of their components and their organization change over time. The descriptions may also specify causal 'mechanisms' by which the components interact in sequence over time. 'Mechanisms' need not all be 'mechanical' in the sense of Archimedes, Galileo or Newton; in contemporary biology they include paradigmatic biochemical processes such as DNA replication, transcription and the translation of RNA messages into proteins, as well as various interactions between proteins and DNA, catalytic digestion of RNA, protein clipping, antibody reactions, and so forth. The scope of explanatory 'mechanisms' also includes processes at other levels of organization, such as the mechanical action of valves in the veins or the heart, or they may be interlevel, as in the process of selectively transporting proteins across cell membranes.

In general what count as 'mechanisms' or fundamental causal processes is stipulated in the scientific research program (cf. Kuhn 1970). Sometimes these processes are described in detail in some model organism which is chosen because it is well known or easy to investigate, but is also supposed to embody processes similar to those embodied in other related organisms. At other times these mechanisms are described in a generalized way that abstracts from the particularities of individual models, e.g. the description of mitosis and meiosis for cells in general. This account of explanation and its implications for scientific inquiry will be further spelled out in the discussion of developmental biology to follow.

3. A Model of Development

The principles of 'laws' behind current models of development in multicellular sexual organisms include the following: All cells in an individual organism, both somatic and precursors to germ cells, descend by mitotic division from the fertilized (or sometimes unfertilized) ovum. In mitotic division each paired chromosome in the parent cell is duplicated and one of each pair is given to each daughter cell, so each descendent cell has the same genes as has the fertilized ovum. Assuming that the genes program all activities and characteristics of each cell, the principal genetic problem in development is to explain the differentiation and specialization of the various cell types that emerge in the developing embryo and the differential timing of these developmental episodes. One would expect on these genetic assumptions that all cells of immediate common ancestry should be alike. This problem is epistemically important because it calls into question the two principal assumptions of Mendelian genetics, (i) that all somatic and diploid germ line cells have the same genes, and (ii) that all heritable characteristics of cells are genetically determined. It is also interesting because it is a recent specification of the old problem of life and matter: What explains the information of matter manifest in life? 'Information' refers to the development of functional structures, their maintenance through life by the transformation of nutritive substances and their transmission to offspring. For ancient thinkers such as Aristotle this problem is one of specifying for various living organisms the formal, material, teleological and occasional factors that govern their generation, nutrition and senescence. For modern thinkers of mechanistic proclivities, the problem is highlighted because the simple primary qualities and the several interactions envisioned for the atoms do not seem sufficient to generate organic form and function.

The problem of development is further specified in the standard descriptive account of cell division, migration and specialization in cytological models of embryonic development. In one 28 year old biology text (Keeton 1967) the early development of the chordate *amphioxus* is described at the cellular level. Initial cleavages of the fertilized egg involve little growth and result in the *morula*, a cluster of cells the same size as the egg. As cleavage continues, cells become arranged into a hollow sphere called the *blastula*. Next there occurs a series of cell movements that eventuate in a definite shape and pattern of precursors to more specialized cells, early stages of *morphogenesis*. In animal embryos a grove eventually appears in the surface of the blastula at the *vegetal pole*, and the outer cell layers fold inward and join the opposite inner surface of the blastula, producing a two layered structure called the *gastrula*. The new cavity thus produced is the *archenteron*, which eventually becomes the cavity of the digestive tract. Also in gastrulation two primitive cell layers are produced, the outer *ectoderm* and the inner *endoderm*. A third intermediate layer, the *mesoderm* is produced from pouches formed in the endoderm. The ectoderm eventually gives rise to the outermost layers of the body, the epidermal part of the skin and to further derived structures, hair, nails, eye lens, certain glands, neural tissue etc. Similarly the endoderm gives rise to the epithelial lining of the digestive tract, the lungs, liver, pancreas, thyroid and the bladder. The mesoderm is the origin of tissues between endodermal and ectodermal, e.g. muscle, blood, connective tissue and bone.

This model holds for some but not all vertebrates. For example there are significant differences in patterns of gastrulation in vertebrates whose eggs have yolks, though models for development of these latter embryos are homologous in stage and in tissue structure with those just described. Also different embryos differ in the timing of the differentiation of component cells: In the ascidian *styela* embryo the two daughter cells are immediately *determined*. Killing one of these cells results in an incomplete embryo in which subsequent development produces only neural plate and muscle cells. On the other hand, the nuclei from cells in the gut of the amphibian *Xenopus* tadpoles remain *totipotent* at this late stage of development in the sense that they could be induced into all of the different specializations manifest in the adult. Ennucleated egg cells with these nuclei implanted can develop into adult frogs. In *Drosophila melanogaster* the egg's nucleus divides eight times without the formation of daughter cell membranes. The cells then migrate to the periphery and only when there are about 6000 nuclei do cell membranes form.

The account given so far fits Kenneth Schaffner's description of biomedical 'theories of the middle range':

> Theories of the middle range fall between biochemistry at one extreme and evolutionary theory at the other extreme on the continuum of levels of aggregation, from molecules to populations.... Components of such theories are usually presented as temporal models, namely as collections of entities that undergo a process. The entities that appear in the model are idealized representations in the sense that they are partially abstracted from the full richness of biological details (at each entity's own level). The models as a whole are usually interlevel. In the characterization of such a middle-range theory, not all possible (or even existing) model variants are usually described explicitly. Rather , a subset of variants of the models are selected to constitute *prototypes* that represent the most *typical* of the wild-type and mutant models. It is these prototypes that "overlap" with respect to their assumptions. (1993, p. 98)

The model just sketched refers to organs, tissues, cells, and chromosomes and hence is interlevel. Its temporality is evident in the temporal sequencing of the several developmental stages described. Several variants are mentioned for different classes or phyla, and the overlap is apparent in homologous structures, such as the *blastocoel* or the hollow at the center of the blastula. Laws governing cell division are components of the model, but equally important components include the prototypical developing structure and the component processes, such as determination and morphogenesis. The laws constrain these processes, e.g. determination involves no change in the genes present in the cells and mitotic cell division in morphogenesis produces daughter cells genetically identical to the parent cells. However, there is no deduction of developmental stages or processes from these laws and initial and boundary conditions describing some initial state of the organism.

Let us now consider what kind of structuring our theory of early embryonic development gives to the questions Q1–Q12 listed at the beginning of this paper. As to the question

Q5. How are patterns established in the early embryo?

the theory appears to be a direct answer insofar as it describes the more notable stages of these patterns, e.g. blastulation, gastrulation, and could go on to segmentation and so forth. That is, under one interpretation Q5 asks for the kind of descriptive model just elaborated. We shall see that this is not the intended interpretation of the question drawn from *Science*. Nonetheless the descriptive theory provides a conceptual framework that is necessary for initially interpreting and ordering the questions Q1–Q12. It gives theoretical meaning to terms in these questions such as 'development' in Q2, 'patterns' in Q5, 'cell determination' in Q3, 'tissue differentiation' in Q8, 'morphogenesis' in Q1, 'vertebrate development' in Q9, and 'tissue induction' in Q11. This cytological developmental model extends beyond the account I have given and assigns theoretical meaning to other terms in the questions, including 'cell-cell signaling' in Q3, and 'neuronal connections' in Q6.

Also the subject matter of Q5, pattern development, embraces the subject matter of a number of the other questions: tissue induction (Q11), tissue differentiation (Q8), cell determination (Q3), cell growth and death (Q7) are all components of pattern formation or morphogenesis in early and later stages of development. These considerations would place Q11, Q8, Q3, and Q7 as subquestions to Q5, that is questions whose answers, according to the model of development, are necessary for answering Q5. (See Figure 1.)

Figure 1

From the Cytoplasmic Model

HIERARCHY OF SUB-QUESTIONS (SUPERORDINATE→SUBORDINATE):

Q1 (morphogenesis)

/------------------------

Q5 (pattern development)

/-----------------------------------
Q11 (tissue induction)
Q8 (tissue differentiation)
Q3 (cell determination)
Q7 (cell growth and death)
Q4 (cell-cell signaling)

I should hasten to add, however, that a portion of the terminology of these questions, e.g. 'transcription factors' (Q8) and 'molecular mechanisms' (Q1) do not appear in the cytological model of development. Hence the ranking of the questions just given should be regarded as only provisional awaiting incorporation of these terms into a different model, a genetic model of development. We shall see in the next section that the genetic model adds a further level of organization to developmental theory, the molecular level.

5. Models of Gene Regulation

F. Jacob's and J. Monod's *lac* operon model for the prokaryote (cells without nuclei) *Escherichia coli* was offered in 1961 and exemplifies part of what is sought in asking for 'mechanisms' of morphogenesis (Q1). Extensions of this model to eukaryote cells (cells with nuclei) provide the prototype needed for understanding or equivalently specifying the goal states proposed by the several 'how' questions Q3,Q5,Q6,Q7,Q8,Q9 and Q11 and specify the possible mechanisms sought under Q1.

The *lac* operon consists of a string of three genes for enzymes needed for lactose metabolism. These coding regions are preceded (in the direction in which messenger RNA is transcribed) by regulatory regions that contain binding sites for proteins. These sites include a *promoter* site for RNA polymerase which catalyzes transcription, and sites for proteins which positively or negatively regulate the transcription. Negative regulation occurs when the regulating protein prevents transcription when bound to an *operator* site located just downstream (in the direction of transcription) from the promoter. The regulating protein is the product of another gene, in this case the *lacI* gene, a regulator as opposed to a structural gene. The *lac* genes can be turned on by allolactose, a product of lactose metabolism that binds to the repressor causing it to be released and thereby permitting transcription of the genes in the operon. A model for positive control was offered in 1965 (Englesberg et al. 1965) for the *ara* operon, which encodes enzymes that degrade arabinose. In this case no transcription occurs unless the activator protein is bound to the operator site. The *lac* operon also is positively regulated by the CAP/cAMP promoter which responds to the absence of glucose and binds just upstream from the *lac* promoter. Hence the same genes or operons can be both positively and negatively regulated. A further variant is the *trp* operon, which encodes for an enzyme that builds up the substance tryptophan. The production of tryptophan is regulated by its presence; tryptophan combines with a repressor that shuts off the *trp* operator. Thus the model provides for regulation by feedback loops of varying length and composition as well as stimulation and repression by external and genetically coded substances. Further regulation is provided by attenuators that halt transcription and specificity factors, *sigma* factors, that attach to the polymerase and determine various kinds of promoter at which transcription is to be initiated.

Eukaryotic cells possess three different kinds of DNA polymerase, one each for the production of messenger RNA (*m*RNA), transfer RNA (*t*RNA) and ribosomal RNA (*r*RNA). Eukaryotic genes have three regulatory sites. The *core promoter* is closest to coding regions and is the binding site for RNA polymerase II, the enzyme responsible for producing *m*RNA. A second set of regulatory elements, called *enhancers* are generally located some thousands of base pairs up or downstream from the core promoter. A third set, *silencers*, to which proteins that inhibit transcription bind, are at comparable distances from the core promoter. Although there is considerable consensus among genes regarding the location and the sequence of the core promoter, individual genes generally differ in their combination of enhancers and silencers. These latter regulatory regions are often tissue specific in the sense that genes with given enhancers function only in certain kinds of cell. Gene expression is controlled by a complex of several proteins, including several *basal factors* that are essential to transcription, one of which binds directly to an element within the core promoter called the TATA box. *Activators* bind at enhancer regions and control the rate of transcription, and *repressors* bind to silencer regions and slow transcription. Finally there are at least six *coactivators* that mediate between activators and repressors, on the one hand, and basal factors on the other. Furthermore eukaryotic cells possess chromosomes in which the DNA is variously bound up by proteins thereby providing a means of regulation not found in prokaryotes. The transcription complex acts by bending the DNA so that the enhancer and repressor regions are brought close to the promoter site. Various ways in which the DNA is bound and wound with proteins in the chromosomes may facilitate or block transcription (Tjian 1995).

The mechanisms for controlling gene expression just described are at the level of transcription, which is believed to be the predominant means of regulation. Paradigms of control at this level include the production of globins in red blood cells but not in oviduct cells and conversely the production of ovalbumin in oviduct but not in blood cells. The specialized production of these products is correlated with the presence of globin *m*RNA's only in the blood cells and ovalbumin *m*RNA's only in the oviduct cells. However control can occur at other levels: Postranscriptional control can occur when cells selectively degrade or vary in the rate of degradation of *m*RNA, where the quantity of *m*RNA present determines the amount of product that can be produced. The concentration of *m*RNA's for the milk protein *casein* increases because the hormone *prolactin* increases the stability of the casein *m*RNA. Control at the level of translation can occur when maternal *m*RNA's are stored in the egg cytoplasm. The translation of these messages may controlled by their being complexed with various proteins, and their products support episodes in egg development without expression of the embryo's own genes. Translation could also be regulated by the amount of *t*RNA and hence by regulators affecting the activity of RNA Polymerase III which catalyzes the synthesis of *t*RNA. At the postranslational level proteins may be clipped by various proteases or may be coupled to other compounds. Other points of possible control include the product's crossing the endoplasmic reticulum and its being packaged for export by the Golgi apparatus in the cytoplasm (Weaver and Hedrick 1992, pp. 283–87).

6. Conclusion

The *lac* operator model and its eukaryotic extension have provided a theoretical definition for 'transcription factors' in Q8: These are the proteins that promote and inhibit transcription and they may be coded in the embryo's genome or they may be provided externally. Genes coding for transcription factors are among those that are developmentally important (Q10). Q10 can also be construed as asking for the identity of these genes, viz. their location, function and sequence. The models also exemplify the kind of mechanisms sought in Q1,Q3,Q5,Q6,Q7, and Q8. For example in the light these models

Q3. How do cells become committed to a particular fate?

is more informatively paraphrased as:

Q3′. What mechanism of gene regulation determines the fate of differentiating cells.

The gene regulation models exemplify the possible answers envisioned for this question and thereby give the question content. Since tissue differentiation is composed of cell differentiation for which determination is necessary,

Q8. How do transcription factors control tissue differentiation?

in the light of cell theory entails

Q8′. By what mechanism do transcription factors control cell determination?

According to the model, there should be several components of this question:

Q8″. By what mechanism are transcription factors (determiners and hormones) distributed to the appropriate regions of control?
Q8‴. By what mechanism within the cell do transcription factors determine the fate of cells?

Q8‴ is equivalent to a specification of Q3′, that is, it differs from Q3′ only in that it contains the added information that transcription factors are involved in internal mechanisms of cell determination. This added information is provided by the genetic model. Thus, because informationally more specific questions for a given topic should be preferred, Q8‴ should replace Q3′ in the research program for developmental genetics.

Also, according to cell theory and the genetic model, the topic of Q11, tissue induction, should be explained or further illuminated by that of Q4, cell-cell signaling. Thus the question in the cytological model

Q11. How does tissue induction take place?

should be replaced by these in the genetic program:

Q11′. What transcription factors (hormones) cause cell-cell signaling.
Q11″. By what mechanism do transcription factors induce different characteristics in different cells.

Q11″ is just a restatement of Q8‴ and hence should be replaced by it. Q11′ seems rather similar to Q8″; However, they differ in topic because Q8″ concerns intracellular gradients and polarizations in the egg and in the early stages of blastulation and Q11 concerns intercellular interactions. Inter- and intracellular interactions should be kept distinct because of the activities of cell membranes, particularly in selectively transmitting transcription factors. The following summarizes these shifts in question:

QUESTION REPLACEMENT, RESTATEMENT, DIVISION:

Q3 (determination)
is replaced by:
Q3′ (regulative mechanism for cell determination)

Q8 (control of tissue differentiation)
is replaced by:
Q8′ (regulative mechanism of cell determination)

Q8′ (regulative mechanism of cell determination)
splits into:
Q8″ (spatial distribution of transcription factors) and
Q8‴ (intracellular regulative mechanism for cell determination)

Q8‴ (intracellular regulative mechanism for cell determination)
replaces:
Q3′ (regulative mechanism for cell determination)

Q11 (tissue induction)
splits into:
Q11′ (transcription factors in cell-cell signaling)
Q11″ (induction of differentiation by transcription factors)

Q11″ (induction of differentiation by transcription factors)
reduces by equivalence to:
Q8‴ (intracellular mechanism for cell determination)

The genetic model gives the following ordering to the questions constituting the genetic program: As in the cytological model, understanding pattern development entails understanding cell determination, cell growth and death, and cell-cell signaling, as well as other characteristics such as cell migration and adhesion. Thus

Q5. How are patterns established in the early embryo?

divides into subquestions Q8″,Q8‴ concerning intracellular mechanisms of cell determination, and Q11′ concerning intercellular signaling. Q7 when elaborated in terms of the role of transcription factors in causing cell growth and death also becomes a subquestion of Q5, for reasons found in the cytoplasmic model. Q10, which asks for the identity of genes coding for transcription factors, is a means to specific answers for Q8″, Q8‴ and Q11″, and hence is a subquestion to all of these. Finally, Q1, which is a question that can only be asked in the light of the genetic model, is the molecular genetic variant of Q5. Also embryonic pattern formation consists of early stages of morphogenesis, so Q1 is broader in scope than Q5. Q5 remains a subquestion of Q1, but should be replaced by

Q5′. What is the molecular mechanism of pattern formation in the early embryo?

Thus the transformation of traditional cytological embryology into its contemporary molecular genetic form consists in part in grafting molecular models of gene expression onto the traditional conceptual framework. This graft gives content to new epistemic objectives for developmental biology, as can be seen in the introduction of new terms into the basic questions of development and in the possible answers to the questions exemplified by the model. It also implies some reduction, merging and rearranging of superordinate and subordinate questions in developmental biology. Still some of the ordering of traditional objectives, particularly some of the components of morphogenesis, is carried into the newer discipline. (For a summary see Figure 2.)

Figure 2

HIERARCHY OF SUB-QUESTIONS IN THE MOLECULAR MODEL:

Q1 (mechanisms of morphogenesis)
/——————————————————\

Q5 (mechanisms of pattern development in early embryo)

/——————————————————\

Q8″ (spatial distribution of transcription factors)
Q8‴ (intracellular mechanism for cell determination)
Q11′ (transcription factors in cell-cell signaling)
Q7 (trascription factors in cell growth and death)
/——————————————————\
Q10 (genes coding for transcription factors)

We have considered how some of the questions of contemporary developmental biology relate in conceptual order, but we have not reflected upon the overall value of these questions in the economy of allocating epistemic resources in science. To this we now turn.

1. Classical Mendelian genetics and embryology were practiced in the first part of this century in virtual isolation from one another. Experimental embryologists such as Ross Harrison, Joseph Needham and Paul Weiss hardly mention genetics and conversely geneticists and evolutionary theorists such as T.H. Morgan, R.H. Fisher, T. Dobzhansky and E. Mayr offer models that include nothing about development. This separation is largely a consequence of there being very little known about what genes are and how they are affect the morphology of the organism. These questions could not be answered on the morphological, tissue, cytological or sub-cytological levels of organization. Only when conceived and formulated on the molecular level did there appear a promise of explaining the gene and its activities (Maull 1977). Particularly, the models of gene expression imply that it is only on the molecular level that we can expect an answer to the basic question of developmental genetics,

How do genetically identical cells become differentiated?

Thus selection of the molecular genetic program for development is warranted as the only apparent means of solving an important problem in genetics.

2. The inability to solve this problem separated genetics and embryology, but the promise of its solution has brought them into relations of mutual relevance. Thus the epistemic desideratum of unified science through discovering relations between fields and levels of organization also warrants embracing the molecular program for development (Laudan 1977, Maull 1977, Darden and Maull 1977).

3. Certain implications of molecular theories of development for evolutionary theories also should support epistemic interest in the molecular program (cf. Q2). First of all, most models of evolution in the neo-Darwinian synthesis assign fitness values to individual genes. However it is often the case that selection acts at the organismal level, whether at the embryonic or at later stages of growth. Hence there is a gap between neo-Darwinian models of evolution, e.g. mathematical models of genetic change within a population, and the equally Darwinian belief that resistance to disease, predators, environmental change, the ability to find food and mates, etc. are the characters of levels of organization other than that of the gene, and these characters are the ones that contribute to an organism's fitness. The gap is the missing link between the genes and the phenotypic traits that are exposed to selective forces. Another indication of this gap is the disparity in rate between molecular and phenotypic evolution. Also the phenotypic stasis that appears in most lineages at one time or another is problematic given the improbability that the environment causing selective pressures remains correspondingly static. The suggestion here is that developmental mechanisms may be highly robust, but they must be understood in detail to see how this robustness is brought about. Furthermore there are powerful objections against the neo-Darwinian belief that the variation that feeds selection is 'random'. These objections concern the enormous improbability of useful mutations occurring randomly in an organismic context in which many independent mechanisms must coordinate to produce a functioning system. A promising line of inquiry into this problem would be the study of co-evolutionary processes in the developing organism. (Kauffman 1993, pp. 18ff, Ho and Sanders 1984, Ho and Fox 1988.)

College of Arts & Sciences, Philosophy Department
The University of Georgia

REFERENCES

Brody, B. (1972). Towards an Aristotelian Theory of Scientific Explanation. *Philosophy of Science* **39**, 20–31.

Bowler, P. (1989). *The Mendelian Revolution.* Baltimore MD: Johns Hopkins University Press.

Darden, L. and Maull, N. (1977). Interfield Theories. *Philosophy of Science* **43**, 44–64.

Englesberg, E., Irr, J., Power, J. and Lee, N. (1965). Positive Control of Enzyme Synthesis by Gene C in the L-arabinose System. *Journal of Bacteriology* **90**, 946–55.

Hines, Pamela J., Marx, Jean, and Parks, Suki (Eds.) (1994). Frontiers in Biology: Development, *Science* **286**, 28 October, 561–614.

Hintikka, J. (1981). On the Logic of an Interrogative Model of Scientific Inquiry. *Synthese* **47**, 69–83.

Hintikka, J. and Hintikka, M.B. (1982). Sherlock Holmes Confronts Modern Logic: Toward a Theory of Information-Seeking Through Questioning. In: E.M. Barth and J.L. Martens (Eds.), *Studies in Language: Companion Series.* Vol. **8**. Amsterdam: John Benjamins B.V.

Ho, M.W. and Sanders, P.T. (Eds.) (1984). *Beyond Neo-Darwinism: Introduction to the New Evolutionary Paradigm.* London: Academic Press.

Ho, M.W. and Fox, S.W. (Eds.) (1988). *Evolutionary Processes and Metaphors.* New York NY: John Wiley and Sons.

Kauffman, S. (1993) *The Origins of Order.* Oxford: Oxford University Press.

Keeton, W.T. (1967). *Biological Science.* New York NY: W.W. Norton.

Kitcher, P. (1981). Explanatory Unification. *Philosophy of Science* **48**, 507–31.

Kleiner, S.A. (1993). *The Logic of Discovery: A Theory of the Rationality of Scientific Research.* Boston MA: Kluwer.

Kuhn, T.S. (1970). *The Structure of Scientific Revolutions.* Second Edition. Chicago IL: University of Chicago Press.

Lakatos, I. (1978). Falsification and the Methodology of Scientific Research Programmes. In: J. Worrall and G. Currie (Eds.), *The Methodology of Scientific Research Programmes: Philosophical Papers* Vol. **1**, 8–101.

Laudan, L. (1977). *Progress and its Problems.* Berkeley and Los Angeles: University of California Press.

Maull, N. (1977). Unifying Science without Reduction. *Studies in the History and Philosophy of Science* **8**, 143–62.

Schaffner, K.F. (1993). *Discovery and Explanation in Biology and Medicine.* Chicago IL: University of Chicago Press.

Tijan, R. (1995). Molecular Machines that Control Genes. *Scientific American* **272** 2, 55–61.

Weaver, R.F. and Hedrick, P.W. (1992). *Genetics.* Second Edition. Dubuque IA: William C. Brown.

Wilkins, A.S. *Genetic Analysis of Animal Development.* Second Edition. New York NY: Wiley-Liss.

Poznań Studies in the Philosophy of the Sciences and the Humanities
1997, Vol. 51, pp. 181–211

Andrzej Wiśniewski

SOME FOUNDATIONAL CONCEPTS OF EROTETIC SEMANTICS*

The aim of this paper is to define some semantical concepts which may be useful in the analysis of questions and questioning. Some of them have been already introduced elsewhere (see the papers Wiśniewski 1989, 1991, 1994a, 1994b, and Buszkowski 1989); we shall give here more extended motivation for the proposed definitions. Some concepts we introduce are in principle borrowed from Belnap's erotetic semantics (cf. Belnap and Steel 1976, Chapter 3); yet, contrary to Belnap, in our analysis we make use of the concept of multiple-conclusion entailment. In general, this concept is one of the main tools applied throughout this paper.

We shall be considering here a class of formalized languages whose meaningful expressions are either declaratives or questions; a language of the analyzed kind consists of the assertoric part being a standard first order language with identity and the erotetic part with questions as the key expressions. Yet, first order languages can be supplemented with questions in many ways and it is not the case that the concepts we are going to introduce are applicable regardless of the way of constructing a question-and-answer system. So let us start with some general information about the existing logical theories of questions and answers.

1. Questions and answers

1.1. Questions: reductionism vs. non-reductionism

As we read in one of the very few monographs on the logic of questions, "Different authors developing logical theories of questions accept different answers to the question 'What is a question?'" (Kubiński 1971, p. 97). This statement, written down more than twenty five years ago, still gives us a realistic description of the situation within erotetic logic.

To speak generally, the approaches to questions proposed by different logicians and formal linguists can be divided into *reductionist* and *non-reductionist* ones. Inside the reductionist approach, in turn, the *radical* and *moderate* standpoints can be distinguished.

1.1.1. *Radical reductionism.* According to the radical view, questions are not linguistic entities. The reduction of questions to sets of sentences or propositions is most often adopted here. Sometimes any set of sentences is allowed to be a question, but usually questions are identified with sets of answers of some distinguished category. Stahl (cf., e.g., Stahl 1962) identifies questions with sets of their sufficient answers; these answers are declarative formulas of a strictly defined kind. Hamblin (cf. Hamblin 1973) identifies a question with the set of its possible answers, whereas Karttunen (cf. Karttunen 1978) identifies questions with sets of their true answers; in both cases the relevant answers are propositions in the sense of some intensional logic. Questions are also identified with functions defined on possible worlds; the set of values of a function of this kind consists of truth-values, or of sets of individuals, or of sets of sets of individuals (cf., e.g., Tichy 1978, and Materna 1981). Also in this case some intensional logic serves as the basis of analysis. An analysis of questions and interrogatives in terms of (some versions of) Montague intensional logic is to be found e.g. in Groenendijk and Stokhof (1984). Some linguists developed the so-called categorial approach to questions: according to this view, questions are to be considered as functions from categorial answers to propositions (cf., e.g., Hausser 1983); a categorial answer may be a full sentence, but may also be a part of it, e.g. a noun phrase, an adverb, etc. There are philosophers of language who tend to identify questions with speech acts rather than with expressions. Let us stress, however, that in all of the above cases a distinction is made between an interrogative (or an interrogative sentence) and a question: whereas interrogatives are linguistic entities, questions are claimed not to be.

Let us finally add that some linguists (cf. Keenan and Hull 1973, and Hiż 1978) proposed theories in which the semantically meaningful units are not questions, but question-answer pairs. Sometimes questions are also analyzed as ordered pairs consisting of interrogative terms and statements expressing the relevant presuppositions.

1.1.2. *Moderate reductionism.* The moderate reductionist view considers questions as linguistic entities which, however, can be reduced to expressions of some other categories. To be more precise, it is claimed here that every question can be adequately characterized as an expression which is synonymous (or synonymous to some reasonable degree) to a certain expression of a different syntactical category. Or, to put it differently, each question can be adequately paraphrased as an expression belonging to some other syntactic category and then formalized within some logic which, although not primarily designed as the logic of questions, can thus be regarded as providing us with the foundations of erotetic logic.

Some theorists propose the reduction of questions to declarative formulas of strictly defined kind(s). Sometimes questions are identified with declarative formulas having free variables, that is, with sentential functions (cf., e.g., Cohen 1929). But questions are also identified with sentences, that is, declarative formulas with no free variables. According to the early proposal of David Harrah (cf. Harrah 1961, 1963) whether-questions are to be understood and then formalized as declarative sentences having the form of exclusive disjunctions, whereas which-questions should be identified with existential generalizations.

Questions are also identified with imperatives of a special kind. The imperative-epistemic approach is the most popular here with Lennart Åqvist and Jaakko Hintikka as its most eminent representatives.

According to Åqvist (cf., e.g., Åqvist 1971, 1972, 1975), a question can be paraphrased as an imperative-epistemic expression of the form "Let it (turn out to) be the case that φ", where φ is a formula which describes the epistemic state of affairs which should be achieved. Pragmatically, a question is thus understood as an imperative which demands of the respondent to widen the questioner's knowledge. Questions are formalized within the framework of some imperative-epistemic logic; on the level of formal analysis we deal with interrogatives. Each interrogative consists of an interrogative operator and its arguments. Interrogatives are defined as abbreviations of certain formulas of the language of the considered imperative-epistemic logic.

The imperative-epistemic approach to questions is also adopted by Jaakko Hintikka in his theory of questions and answers (cf., e.g., Hintikka 1974, 1976, 1978, 1983). Hintikka interprets questions as requests for information or knowledge: according to his view, each question can be paraphrased as an expression which consists of the operator "Bring it about that" followed by the so-called *desideratum* of the question. The desideratum describes the epistemic state of affairs the questioner wants the respondent to bring about. Although the main ideas of Åqvist and Hintikka are similar, they are elaborated on in different ways.

As far as the moderate reductionist view is concerned, the imperative-epistemic approach is the most widely developed one. Yet, there are also other proposals. In particular, there is an old idea (which goes back at least to Bolzano) that the paraphrase of a question should contain an optative operator. (It is worth noticing that Hintikka sometimes calls the operator "Bring it about that" an "imperative or optative operator.") Apostel (cf. Apostel 1969) claims that questions can be reduced to expressions which contain epistemic, deontic and alethic operators as well as the assertion operator; yet, Apostel also claims that the deontic operator "ought to" used by him should be replaced by some optative or imperative operator when a more adequate analysis of such operators will be available. In a short note Åqvist (cf. Åqvist 1983) sketched an outline of the imperative-assertoric analysis of questions; according to this proposal, questions should be paraphrased (and then formalized) as imperatives which contain the assertoric operator "You tell me truly that" instead of the epistemic operator "I know that." Let us stress, however, that none of the above proposals has been elaborated on.

1.1.3. *Non-reductionism.* According to the non-reductionist approach, questions are specific expressions of a strictly defined form; they are not reducible to expressions of other syntactic categories.

The most widespread proposal here is to regard a question as an expression which consists of an interrogative operator and a sentential function. This view is accepted, among others, by Ajdukiewicz (cf., e.g., Ajdukiewicz 1974), Hiż (cf. Hiż 1962), Kubiński (cf., e.g., Kubiński 1960, 1971, 1980), Koj (cf. Koj 1972), Leszko (cf., e.g., Leszko 1983), and Harrah in his later papers (cf., e.g., Harrah 1975, 1984). Some prominent authors (e.g. Carnap, Reichenbach, Cresswell) who only incidentally paid attention to questions also shared this view. Yet, the above idea is most widely elaborated on in the books and papers of Tadeusz Kubiński.

Kubiński's analysis is mainly a syntactical one: questions of a formalized language are defined as expressions which consist of interrogative operators and sentential functions. Interrogative operators, in turn, consist of both constants and variables. The only free variables in the sentential functions which occur in questions are the variables of the corresponding interrogative operators; these variables are "bound" by the interrogative operators. The variables which occur in questions may belong to various syntactical categories. Roughly, the categories of variables indicate the (ontological) categories of objects which are asked about. For example, a question whose interrogative operator contains only individual variables asks about individuals. If the relevant variables run over sentential connectives, then the corresponding questions are about either the existence of some state(s) of affairs or some connection(s) between states of affairs. Questions with predicate variables, in turn, ask about properties or relations. When a question contains only sentential variables, it is a question about logical values (truth and falsehood). Kubiński considers also "mixed" questions, that is, questions whose interrogative operators contain variables belonging to two or more different categories.

Although the "interrogative operator-sentential function" view is shared by most of the adherents of the non-reductionist approach to questions, there are also other proposals. Among them special attention should be paid to Nuel D. Belnap's theory of questions and answers (cf., e.g., Belnap 1963, and Belnap and Steel 1976).

Belnap distinguishes between: (a) natural language questions, (b) interrogatives and (c) questions understood as abstract (set-theoretical) entities. Interrogatives are expressions of some formalized languages. They are not only formal counterparts of natural language questions, but they also express questions understood as abstract entities.

A simple interrogative consists of the question mark ?, the lexical subject and the lexical request. The question mark is interpreted as a sign of the function which assigns to the lexical subject and the lexical request of a simple interrogative the corresponding (abstract) question; such a question consists, in turn, of the abstract subject and the abstract request. A compound interrogative can be obtained from simple interrogatives by performing some (logical of Boolean) operations on them or their lexical subjects.

The basic idea of Belnap's approach is that an interrogative "presents" a set of alternatives together with some suggestions or indications as to what kind of choice or selection among them should be made; the situation is analogous in the case of the corresponding questions. The function of the lexical subject of an interrogative is to offer the relevant (nominal) alternatives, whereas the role of the lexical request is to characterize the required kind of selection. The lexical request consists of three parts: the lexical selection-size specification, the lexical completeness-claim specification, and the lexical distinctness-claim specification. Roughly, the lexical selection-size specification informs how many (that is, how many exactly, how many at least and/or how many at most) of the alternatives offered by the lexical subject of the interrogative are called for. The lexical completeness-claim specification, in turn, informs about the amount of true nominal alternatives called for; there are interrogatives which call for all the true alternatives presented by them, but there are also interrogatives which do not demand so much. Finally, the lexical distinctness-claim specification tells whether the alternatives called for should be semantically different.

1.2. Answers

Most logical theories of questions pay at least as much attention to answers to questions as to questions themselves. It is usually assumed that a question can have many answers; the phrase "answer to a question" *is not* used synonymously with "the true answer to a question." In other words, the analyzed answers are usually *possible* answers: their logical values are not prejudged. Yet, it is not the case that all possible answers are equally interesting to erotetic logicians. The standard way of proceeding is to define some *basic* category of possible answers. They are called *direct answers* (Åqvist, Belnap, Harrah, Kubiński), *proper answers* (Ajdukiewicz), *sufficient answers* (Stahl), *conclusive answers* (Hintikka and his associates), *indicated replies* (Harrah in his later papers), etc. Those "principal" possible answers (let us use this general terms here) are supposed to satisfy some general conditions, usually expressed in pragmatic (in the traditional sense of the word) terms. For example, direct answers in Kubiński's sense are "these sentences which everybody who understands the question ought to be able to recognize as the simplest, most natural, admissible answers to this question" (Kubiński 1980, p. 12). Direct answers in Belnap's sense are the answers which "are directly and precisely responsive to the question, giving neither more nor less information than what is called for" (Belnap 1969a, p. 124). Direct answers in Harrah's sense are replies which are complete and just-sufficient answers (cf. Harrah 1963, p. 26 et al.). In the light of Hintikka's theory a reply is called conclusive just in case it completely satisfies the epistemic request of the questioner, that is, brings about the epistemic state of affairs the questioner wanted to be brought about.

Let us stress that although the above conditions are formulated in pragmatic terms, some of the logical theories of questions define the principal possible answers to questions or interrogatives of *formalized languages* in terms of syntax and/or semantics. The questions (interrogatives) of formalized languages, however, are usually formalizations of natural language questions. Consequently, the principal possible answers are usually defined in such a way that the natural language sentences which correspond to them are those answers to the analyzed natural language questions which have the above-mentioned pragmatic properties. Or, to be more realistic, to define them in such a way is the aim of the enterprise. Yet, there are natural language questions which admit many readings and thus many formalizations. There are also natural language questions which seem to have no well-defined sets of answers; why-questions are often recalled in this context. Moreover, most theories admit questions or interrogatives which have no counterparts in natural languages, but nevertheless have well-defined sets of the "principal" possible answers.

It is not the case that only the principal possible answers are of interest to erotetic logicians. Most theories provide us also with definitions of other kinds of possible answers: partial answers, complete answers, just-complete answers, incomplete answers, corrective answers, etc. These answers are usually defined in terms of the "principal" answers; yet, it also happens that they (or some of them) are defined independently. Their definitions differ from theory to theory; no well-established terminology has been elaborated yet.

Let us finally add that sometimes replies which are not statements (e.g. noun phrases, nods, grunts) are regarded as answers; in most cases, however, answers are assumed to be statements and replies of other kinds are regarded as abbreviations of the corresponding statements.

2. *General assumptions*

As the above sketchy presentation shows, first-order languages (as well as other formalized languages) can be supplemented with questions in different ways. Yet, regardless of the way how the extension goes on, it requires, first, the introduction of new symbols to the vocabulary of some basic extensional formalized language. Questions (or interrogatives) are made up of the new symbols and the old ones. Then the assignment of the principal possible answers to questions takes place; this is sometimes done in purely syntactical terms, but also on the level of semantics or/and pragmatics. The principal possible answers are usually expressions of the basic extensional language; in most cases they are declarative well-formed formulas of it. The result is a question-and-answer system.

In what follows we will be adopting the following general assumptions. First, we assume that some first-order language with identity $\mathcal{L}_0$ is given. Second, we assume that the vocabulary of $\mathcal{L}_0$ is supplemented with some new symbols in order to constitute the vocabulary of some new (formalized) language $\mathcal{L}$. By *terms* and *declarative well-formed formulae* (d-wffs for short) of $\mathcal{L}$ we shall mean those of $\mathcal{L}_0$. We assume that the language $\mathcal{L}$ contains some new category of meaningful expressions (built up, int.al, by means of the new symbols) which are called *questions*; yet, we do not forejudge the way of constructing questions of $\mathcal{L}$. We then assume that to each question of $\mathcal{L}$ there is assigned an at least two-element set of principal possible answers, which are sentences (d-wffs with no free variables) of $\mathcal{L}$. For brevity we shall call them *direct answers* to the question.

Although the above conditions are rather general, it cannot be said that they are uncontroversial; it even cannot be said that they are fulfilled by each existing question-and-answer system. The motivation for some of them is philosophical. We require each question to have at least two principal possible (i.e. direct) answers because we think that a necessary condition of being a question is to present at least two "alternatives" or conceptual possibilities among which some selection can be made. The "Hobson choice" questions are thus excluded, but rhetorical questions are allowed – the selection need not be rational. Some logical theories of questions (for example, Belnap's theory or Kubiński's theory) allow questions which have only one direct answer, but it seems that this step is motivated rather by the pursuit of generality than other reasons. Concerning the condition according to which each direct answer is a sentence: in the light of semantics we are going to propose the free variables are interpreted in the "generalizing" manner (not as "dummy" names): a sentential function of the form Ax_i is true if and only if, to speak generally, it is satisfied by all the possible values of x_i. Under this interpretation a sentential function expresses a condition which may be satisfied by some object and not satisfied by others; on the other hand, we want the principal possible answers to be answers in the very serious sense of the word. But the most controversial is of course the assumption according to which each question *has* a set of principal possible answers. It cannot be said that each question analyzed in any logical theory of questions fulfills this condition; it also cannot be said that any natural-language question fulfills it. Yet, some of them do, and most question-and-answer systems distinguish questions having this property. So the above assumption may be viewed as restricting the range of applicability of the concepts we are going to propose. In other words, our further considerations pertain to first-order languages with identity enriched with questions in such a way that each question has at least two-element set of principal possible answers; these answers are sentences of the basic extensional language. Yet, let us stress, the details of the extension procedure are irrelevant from the point of view of applicability of the concepts we are going to propose.

We will be using the letters A, B, C, ..., possibly with subscripts, as metalinguistic variables for d-wffs, and the symbols X, X_1, ..., Y, Y_1, ..., Z, Z_1, ... as metalinguistic variables for sets of d-wffs. The symbols Q, Q_1, ... will be used as metalinguistic variables for questions. The set of (all) direct answers to a question Q will be referred to as dQ. On the metalanguage level we assume the von Neumann–Bernays–Gödel version of the set theory (we choose this version because we want to have the possibility of speaking about both sets and classes). We shall use the standard set-theoretical terminology and notation. The expression "iff" is an abbreviation of "if and only if."

3. *Basic concepts*

3.1. *Interpretations, satisfaction and truth*

According to what has been said above the language $\mathcal{L}$ results from the first-order language with identity $\mathcal{L}_0$. To speak generally, our first step is to supplement the language $\mathcal{L}_0$ with a standard model-theoretical semantics. To be more precise, we shall first define some concepts which do not pertain to questions of $\mathcal{L}$.

Let us temporarily assume that the vocabulary of $\mathcal{L}$ contains apart from some predicate symbol(s) also some function symbol(s) and individual constant(s). By non-logical constants of $\mathcal{L}$ we mean below individual constants, function symbols and predicate symbols of this language.

DEFINITION 1. An *interpretation* of the language $\mathcal{L}$ is an ordered pair $<\mathbf{M}, \mathbf{f}>$, where $\mathbf{M}$ is a non-empty set and $\mathbf{f}$ is a function defined on the set of non-logical constants of $\mathcal{L}$ which fulfills the following conditions:

(i) for each individual constant a_i, $\mathbf{f}(a_i) \in \mathbf{M}$,

(ii) for each n-argument function symbol F_i^n, $\mathbf{f}(F_i^n)$ is a n-argument function defined on the set $\mathbf{M}$ and whose values belong to the set $\mathbf{M}$,

(iii) for each n-place predicate symbol P_i^n, $\mathbf{f}(P_i^n)$ is a n-ary relation in $\mathbf{M}$.

If $<\mathbf{M}, \mathbf{f}>$ is an interpretation, the set $\mathbf{M}$ is called the *domain* of this interpretation, whereas the function $\mathbf{f}$ is called the *interpretation function*. Note that the new symbols which enable to form questions of $\mathcal{L}$ are not arguments of an interpretation function!

When speaking about interpretations we will be using the symbols $\mathfrak{I}$, $\mathfrak{I}'$,... .

The definition of the concept of interpretation should be adjusted in an obvious way if the vocabulary of $\mathcal{L}$ contains no function symbols or no individual constants.

Let $\mathfrak{I} = \langle \mathbf{M}, \mathbf{f} \rangle$ be an arbitrary but fixed interpretation of $\mathcal{L}$. A $\mathfrak{I}$-*valuation* is a denumerable sequence of elements of the domain of the interpretation $\mathfrak{I}$ (by "denumerable" we mean here and below "countably infinite"). The concept of *value* of a term of $\mathcal{L}$ in the interpretation $\mathfrak{I}$ with respect to a given $\mathfrak{I}$-valuation **s** is defined in the standard way. Similarly, the concept of *satisfaction* of a d-wff in the interpretation $\mathfrak{I}$ by a $\mathfrak{I}$-valuation **s** is defined in the usual manner.

A d-wff A of $\mathcal{L}$ is *true in an interpretation* $\mathfrak{I}$ of $\mathcal{L}$ if and only if A is satisfied in $\mathfrak{I}$ by each $\mathfrak{I}$-valuation. If a d-wff A is true (is not true) in $\mathfrak{I}$, we write $\mathfrak{I} \models A$ (we write $\mathfrak{I}$ *non* $\models A$). By a *model* of a set of d-wffs X we mean any interpretation of the language in which all the d-wffs in X are true. If an interpretation $\mathfrak{I}$ is a model (is not a model) of a set of d-wffs X, we write $\mathfrak{I} \models X$ (we write $\mathfrak{I}$ *non* $\models X$).

Let us stress that the concepts of satisfaction and truth do not pertain to questions of $\mathcal{L}$.

3.2. Normal interpretations and consistency

Let us now assume that the class of interpretations of $\mathcal{L}$ includes a non-empty subclass (not necessarily a proper subclass) of *normal interpretations*. The reasons for distinguishing normal interpretations from the remaining ones can be various. For instance, one may intend to construe some non-logical constant(s) in a way that complies with some intuitions; in this case normal interpretations can be defined as those which make true some definition(s) or meaning postulate(s) worded in $\mathcal{L}$. Normal interpretations can also be defined as models of some first-order theory worded in (the declarative part of) $\mathcal{L}$. They can also be defined as those models of such a theory which fulfill some additional conditions; the so-called standard models of Peano's arithmetics give us a simple example here. Another possibility lies in defining normal interpretations as those which make true some sentences which are regarded as "laws of science", that is, first-order counterparts of some scientific laws (expressed in $\mathcal{L}$ which serves as the language of logical analysis). Normal interpretations can also be identified with the "intended models" in the sense of philosophy of science. But normal interpretations can also be distinguished for purely "erotetic" reasons. For example, in Belnap's theory of questions special attention is paid to those interpretations of the basic "assertoric" languages in which, to speak generally, the objects assigned to terms that belong to the nominal category determined by a given category condition are among the objects that satisfy this category condition[1]. When questions about objects that satisfy some sentential function or functions are considered, it seems natural to call normal interpretations only those interpretations in which each element of the domain has a name: by doing so we can avoid the situation that there are objects which satisfy the appropriate sentential function(s), but nevertheless the analyzed questions have no true direct answers.

It seems impossible to define the general concept of "normalness" of interpretation: this concept varies from language to language. As far as the language $\mathcal{L}$ is concerned we only assume that the class of normal interpretations of it exists and is non-empty. Moreover, the remaining semantic concepts pertaining to the language $\mathcal{L}$ will be defined here by means of the concept of normal interpretation of $\mathcal{L}$. Thus their definitions remain schematic until the concept of normal interpretation of $\mathcal{L}$ will be defined in detail. But $\mathcal{L}$ is assumed to be an arbitrary but fixed formalized language which fulfills some general conditions and the situation is analogous in the case of its semantics. To speak generally, we thus leave room for different possibilities. Let us stress that we even do not assume (but also do not deny) that the class of normal interpretations of $\mathcal{L}$ is a *proper* subclass of the class of all interpretations of $\mathcal{L}$: it may happen that all interpretations of some language are regarded as normal.

A d-wff A of $\mathcal{L}$ is called a *tautology* of $\mathcal{L}$ if and only if A is true in each normal interpretation of $\mathcal{L}$. A d-wff A of $\mathcal{L}$ is said to be a *contradictory d-wff* (or *contradiction*) of $\mathcal{L}$ if and only if there is no normal interpretation $\mathfrak{I}$ of $\mathcal{L}$ such that some $\mathfrak{I}$-valuation satisfies A in $\mathfrak{I}$. A d-wff A of $\mathcal{L}$ is a *synthetic d-wff* of $\mathcal{L}$ just in case A is neither a tautology of $\mathcal{L}$ nor a contradictory d-wff of $\mathcal{L}$.

By a *normal model* of a set of d-wffs X of $\mathcal{L}$ we mean any normal interpretation of $\mathcal{L}$ being a model of X. A set of d-wffs X of $\mathcal{L}$ is said to be *consistent* just in case there is a normal model of X; otherwise X is said to be *inconsistent*. We shall use the symbol ***Inc*** for the family of all inconsistent sets of d-wffs of $\mathcal{L}$.

4. Entailment and multiple-conclusion entailment

4.1. Entailment

Let us assume again that the class of normal interpretations of $\mathcal{L}$ is defined in some way or another. The semantic concept of *entailment in a language* can now be defined as follows:

DEFINITION 2. A set of d-wffs X of $\mathscr{L}$ *entails in* $\mathscr{L}$ a d-wff A of $\mathscr{L}$ iff A is true in each normal interpretation of $\mathscr{L}$ in which all the d-wffs in X are true.

Let us stress that the above concept of entailment is relativized to the class of normal interpretations of the considered language. Thus entailment in a given language need not be tantamount to logical entailment; of course these concepts coincide if each interpretation is regarded as normal. The concept of logical entailment we have in mind here is defined as follows: X *logically entails* A iff A is true in each interpretation of the language in which all the d-wffs in X are true.[2] In general, the stronger the conditions we impose on normal interpretations the wider becomes the range of entailment in the language. But sometimes this is the effect we intended to achieve: if we intend to reflect in a formal language some non-logical implicatures which are already present in the natural language which is the subject of formalization, the simplest way is to define entailment according to the pattern presented by Definition 2 and to define normal interpretations in the appropriate manner: as those which make true some meaning postulates, or some theory, or some "laws", etc.[3]

We shall use the symbol $\models$ for entailment in a language.

We say that two d-wffs A, B of $\mathscr{L}$ are *equivalent* if and only if B is entailed in $\mathscr{L}$ by A and A is entailed in $\mathscr{L}$ by B.

Let us now introduce the concept of compactness of entailment. A relation $\models$ of entailment in a language is said to be *compact* if whenever $X \models A$ there exists a *finite* subset Y of X such that $Y \models A$. We neither assume nor deny here, however, that entailment in $\mathscr{L}$ is compact; there are languages of the considered kind in which entailment is compact, but there are also languages in which it is not. The compactness of entailment depends on the conditions imposed on the class of normal interpretations.

4.2. Multiple-conclusion entailment

When we are dealing with questions whose sets of direct answers are well-defined, we may think about the direct answers as offering some possibilities or "alternatives" among which some selection should be made. Thus some notion of, to speak generally, "entailing a set of possibilities" is needed. There is a logic, however, within which such a notion has been elaborated on: it is multiple-conclusion logic (cf. Shoesmith and Smiley 1978; see also Scott 1974, and Zygmunt 1984).

Multiple-conclusion logic generalizes the concept of entailment: now a *set* of conclusions is allowed. Any such set is regarded as, intuitively speaking, setting out the field within which the truth must lie if the premises are all true. Let us then introduce the concept of *multiple-conclusion entailment*.

DEFINITION 3. A set of d-wffs X of $\mathcal{L}$ *multiple-conclusion entails in* $\mathcal{L}$ a set of d-wffs Y of $\mathcal{L}$ if and only if the following condition holds:

(*) whenever all the d-wffs in X are true in some normal interpretation of $\mathcal{L}$, then there exists at least one d-wff in Y which is true in this interpretation of $\mathcal{L}$.

To speak generally: X multiple-conclusion entails Y just in case Y must contain at least one true d-wff if all the d-wffs in X are true; "must" in the sense that it is impossible that all the d-wffs in Y are not true in any normal interpretation of the language which makes true all the d-wffs in X.

We will be using the term "mc-entailment" instead of the long expression "multiple-conclusion" entailment.

We shall use the symbol $\Vdash$ for mc-entailment in a language. If Y is not mc-entailed by X, we write X *non* $\Vdash Y$. If a set of d-wffs Y is mc-entailed by a singleton set $\{A\}$, we simply say that Y is mc-entailed by the d-wff A.

The concept of multiple-conclusion entailment is a generalization of the standard (i.e. "single-conclusion") concept of entailment. As an immediate consequence of Definition 2 and Definition 3 we get:

COROLLARY 1. *$X \models A$ iff $X \Vdash \{A\}$.*

Corollary 1 says that a set of d-wffs X entails a d-wff A just in case the set X mc-entails the singleton set whose element is the d-wff A. Thus entailment is definable in terms of mc-entailment. On the other hand, in the general case mc-entailment is not definable in terms of entailment. It happens that a set of d-wffs X mc-entails some set of d-wffs Y, but does not entail (in the standard sense of the word) any d-wff in Y. The sets of the form $\{A \vee \neg A\}$ and $\{A, \neg A\}$, where A is an atomic sentence, give us a simple example here.

The basic properties of mc-entailment are characterized by the following corollaries ($\Vdash$ stands below for mc-entailment in $\mathcal{L}$; let us recall that $\mathcal{L}$ is an arbitrary but fixed language that fulfills the conditions listed above):

COROLLARY 2. *If $X \cap Y \neq \emptyset$, then $X \Vdash Y$.*
COROLLARY 3. *If $X \subseteq X_1$, $Y \subseteq Y_1$ and $X \Vdash Y$, then $X_1 \Vdash Y_1$.*

COROLLARY 4. *If* $X \cup Z_1 \Vdash Y \cup Z_2$ *for any* Z_1, Z_2 *such that* $Z_1 \cap Z_2 = \emptyset$ *and* $Z_1 \cup Z_2 = Z$, *then* $X \Vdash Y$.

Corollaries 2 and 3 are immediate consequences of Definition 3. For the proof of Corollary 4 assume that X *non* $\Vdash Y$. So there exists a normal interpretation $\mathfrak{J}$ of $\mathcal{L}$ such that $\mathfrak{J} \models X$ and for each $B \in Y$, $\mathfrak{J}$ *non* $\models B$. Let Z_1 be the set of all the d-wffs of $\mathcal{L}$ which are true in $\mathfrak{J}$ and let Z_2 be the set of all the d-wffs of $\mathcal{L}$ which are not true in $\mathfrak{J}$. We now have $X \cup Z_1$ *non* $\Vdash Y \cup Z_2$ as required.

Thus mc-entailment in any of the considered languages is a multiple-conclusion consequence in the sense of the monograph Shoesmith and Smiley (1978) (see also below).

A relation $\Vdash$ of mc-entailment is said to be *compact* if whenever $X \Vdash Y$ there exist *finite* subsets X_1 of X and Y_1 of Y such that $X_1 \Vdash Y_1$. We have:

COROLLARY 5. *Mc-entailment in* $\mathcal{L}$ *is compact iff entailment in* $\mathcal{L}$ *is compact.*

Proof:
(⇒) By Corollary 1.
(⇐) Let us first observe that if entailment in $\mathcal{L}$ is compact, then the class of normal interpretations of $\mathcal{L}$ fulfills the following condition:

(:) for each set of d-wffs X of $\mathcal{L}$: if each finite subset of X has a normal model, then the set X has a normal model.

For, if there is a set of d-wffs, say, X_1, such that each finite subset of X_1 has a normal model, but X_1 has no normal model, then X_1 is an infinite set; moreover, X_1 entails some contradictory sentence, say, A, which is not entailed by any finite subset of X_1. It follows that entailment in $\mathcal{L}$ is not compact; so if entailment in $\mathcal{L}$ is compact, the condition (:) holds.

Since the compactness of entailment yields the condition (:), it suffices to prove that if $X \Vdash Y$ and the condition (:) holds, then there are finite sets X_1, Y_1 such that $X_1 \subseteq X$, $Y_1 \subseteq Y$ and $X_1 \Vdash Y_1$. Moreover, it suffices to consider the cases in which X or Y are infinite sets.

Assume that Y is an infinite set. Let $uc(Y)$ be the set of universal closures[4] of the d-wffs of Y. Let us designate by $\neg uc(Y)$ the set of negations of the sentences of the set $uc(Y)$. Suppose that $X \Vdash Y$. Hence $X \cup \neg uc(Y) \in$ ***Inc***. By condition (:) we get that there exists a finite and non-empty subset Z of the set $X \cup \neg uc(Y)$ such that $Z \in$ ***Inc***. There are three possibilities: (a) $Z \subseteq X$, (b) $Z \subseteq \neg uc(Y)$, (c) $Z \subseteq X \cup \neg uc(Y)$, where $Z \not\subseteq X$ and $Z \not\subseteq \neg uc(Y)$. If the possibility (a) holds, then for each finite subset Y_1 of the set Y we have $Z \models\!\!\!\models Y_1$; at the same time Z is a finite subset of the set X. If the possibility (b) takes place, then some finite subset of the set Y is mc-entailed by the empty set and thus also by each finite subset of the set X. It is obvious that if the possibility (c) holds, then some finite subset of the set Y is mc-entailed by some finite subset of the set X.

Assume that X is an infinite set. If $Y = \emptyset$, then $X \in \boldsymbol{Inc}$. By condition (:) we get that there exists a finite subset X_1 of the set X such that $X_1 \in \boldsymbol{Inc}$. Thus $X_1 \models\!\!\!\models \emptyset$; on the other hand, $\emptyset$ is a finite subset of Y. If $Y \neq \emptyset$, we proceed analogously as in the case in which Y is an infinite set. □

Thus mc-entailment in a language is compact just in case entailment in this language is compact. Let us stress that Corollary 5 speaks of any language of the considered kind. Since we neither assume nor deny here that entailment in $\mathcal{L}$ is compact, the same holds true in the case of mc-entailment in $\mathcal{L}$: we leave room for different possibilities.

By a sentence we mean here a d-wff with no free variables; otherwise a d-wff is said to be a sentential function. One can easily prove:

COROLLARY 6. *If* $A_1, \ldots, A_n$ *are sentences, then* $\{A_1 \vee \ldots \vee A_n\} \models\!\!\!\models \{A_1, \ldots, A_n\}$.

COROLLARY 7. *If* $A_1, \ldots, A_n$ *are sentences, then:* $X \models\!\!\!\models \{A_1, \ldots, A_n\}$ *iff* $X \models A_1 \vee \ldots \vee A_n$.

According to Corollary 6, a finite and non-empty set of sentences of a given language is mc-entailed in this language by a disjunction of all its elements (to be more precise, by a singleton set which contains this disjunction). Corollary 7 says that a finite and non-empty set of sentences is mc-entailed by a set of d-wffs X just in case the set X entails some disjunction of all the elements of this set. Let us stress that the corollaries 6 and 7 describe some general properties of mc-entailment: they are true with respect to any language of the considered kind. Let us also stress that the assumption that $A_1, \ldots, A_n$ are sentences is essential: in the case of sentential functions the situation is different. For, let us consider a sentential function $P(x_i)$ (where P is a one-place predicate symbol) of some language and let us assume that there is a normal interpretation $\mathfrak{J}$ of the language such $P(x_i)$ is satisfied in $\mathfrak{J}$ only by some $\mathfrak{J}$-valuations, but not by all of them. The set $\{P(x_i) \vee \neg P(x_i)\}$ does not mc-entail in the analyzed language the set $\{P(x_i), \neg P(x_i)\}$, but does entail the sentence $P(x_i) \vee \neg P(x_i)$. To give a more concrete example: let us imagine that P stands for "is a prime" and the domain of the interpretation consists of all the natural numbers.

The universal closure[5] of a d-wff A is referred to as $\bar{A}$. The following is a consequence of the corollaries 3, 5 and 7:

COROLLARY 8. *If entailment in $\mathcal{L}$ is compact, X is a set of d-wffs of $\mathcal{L}$ and Y is a non-empty set of d-wffs of $\mathcal{L}$, then $X \Vdash Y$ iff either there is $A \in Y$ such that $X \models \bar{A}$ or there are $A_1, \ldots, A_n \in Y$ such that $X \models \bar{A}_1 \vee \ldots \vee \bar{A}_n$.*

Thus if entailment is compact, mc-entailment of a non-empty set of d-wffs Y reduces either to entailment of the universal closure of a single d-wff of Y or to entailment of some disjunction of universal closures of d-wffs of Y. It does not mean, however, that the concept of mc-entailment is superfluous: there are languages of the considered kind in which entailment (and thus mc-entailment as well) is not compact.

Historical note. The idea of multiple-conclusion consequence goes back to Gentzen (1934); one of the possible ways of looking at a valid sequent of the form $A_1, \ldots, A_m \vdash B_1, \ldots, B_n$ is to construe it, to speak generally, as stating multiple-conclusion entailment of the set made up of the formulas referred to by $B_1, \ldots, B_n$ from the set made up of the formulas referred to by $A_1, \ldots, A_m$. Under this interpretation the turnstile $\vdash$ is a relation symbol and a calculus of sequents is a (single-conclusion) metacalculus for a multiple-conclusion object-calculus. Yet, there is also another possibility: a sequent $A_1, \ldots, A_m \vdash B_1, \ldots, B_n$ is a notation for a formula $A_1 \wedge \ldots \wedge A_m \rightarrow B_1 \vee \ldots \vee B_n$ and Gentzen's calculi of sequents are variants of the corresponding conventional calculi. Shoesmith and Smiley claim that Gentzen interpreted his calculi of sequents in this latter way. If this is so, it is Carnap who for the first time introduced the concept of multiple-conclusion entailment (cf. Carnap 1943; Carnap uses the term "involution"). The concept of multiple-conclusion consequence was incorporated into the general theory of logical calculi by Dana Scott (1974). Multiple-conclusion consequence and related concepts (multiple-conclusion calculus, multiple-conclusion rules, etc.) are analyzed in detail in the monograph Shoesmith and Smiley (1978); for developments see also the book Zygmunt (1984). Let us add that the approach presented by Shoesmith and Smiley is much more general than ours: it is not restricted to first-order languages supplemented with model-theoretical semantics. Assume that L is a formalized language and let us designate by *Form*(L) the set of formulas of L; the only condition imposed on *Form*(L) is the non-emptiness clause. A relation $\vdash$ between sets of formulas of L is a called a *multiple-conclusion conse-*

quence if and only if $\vdash$ fulfills the following conditions for any $X, Y, Z \subseteq Form(L)$:

(C$_1$)(*Overlap*): *If* $X \cap Y \neq \emptyset$, *then* $X \vdash Y$.
(C$_2$)(*Dilution*): *If* $X \subseteq X_1$, $Y \subseteq Y_1$ *and* $X \vdash Y$, *then* $X_1 \vdash Y_1$.
(C$_3$)(*Cut for sets*): *If* $X \cup Z_1 \vdash Y \cup Z_2$ *for any* Z_1, Z_2 *such that* $Z_1 \cap Z_2 = \emptyset$ *and* $Z_1 \cup Z_2 = Z$, *then* $X \vdash Y$

This is the syntactical concept of multiple conclusion-consequence. But there is also a semantical one. Assume that L is supplemented with some semantics rich enough to define some (relativized) concept of truth for formulas. A *partition of* L is an ordered pair $<T, U>$ such that $T \cap U = \emptyset$ and $T \cup U = Form(L)$; we may think of the elements of the set T of a partition $<T, U>$ as of consisting of truths in the sense of the underlying semantics and of the set U as of consisting of untrue formulas. Let Δ be a class of partitions of L. A relation $\vdash$ between sets of d-wffs of L is the *multiple-conclusion consequence relation characterized by* Δ iff for each partition $<T, U> \in \Delta$ and for each $<X, Y> \in \vdash$, $X \cap U \neq \emptyset$ or $Y \cap T \neq \emptyset$: this is the semantical concept of multiple-conclusion consequence. It may be proved that each relation being a multiple-conclusion consequence in the syntactical sense of the word is a multiple-conclusion consequence in the semantical sense, that is, a multiple-conclusion consequence characterized by some class of partitions of the language. It can also be proved that for each class of partitions of the language, the multiple-conclusion consequence characterized by this class is a multiple-conclusion consequence in the syntactical sense of the word, that is, fulfills the conditions (C$_1$), (C$_2$) and (C$_3$). For proofs, see Shoesmith and Smiley (1978), p. 30.

What we have called above multiple-conclusion entailment in $\mathcal{L}$ would presumably be called by Shoesmith and Smiley multiple-conclusion consequence characterized by the class of normal interpretations of $\mathcal{L}$, or, to be more precise, by the class of partitions of the set of d-wffs of $\mathcal{L}$ determined by the class of normal interpretations of $\mathcal{L}$.

By means of the concept of mc-entailment we can define certain useful erotetic concepts in a simple, general and natural way; as a matter of fact, this concept will be the most useful tool of our further analysis. To our best knowledge, the idea of applying the concept of mc-entailment in erotetic logic appeared for the first time in the papers Buszkowski (1987) and Wiśniewski (1987)[6] and in a marginal form in the dissertation Wiśniewski (1986); see also the papers Buszkowski (1989) and Wiśniewski (1989) for more extended expositions. For the application of this concept in erotetic logic see also the books Wiśniewski (1990a) and (1995) and the papers Wiśniewski (1990b), (1991), (1994a) and (1994b).

5. Erotetic concepts

5.1. Soundness, safety and riskiness

The semantic concepts introduced so far do not pertain to questions. Let us now define the semantical erotetic concepts the introduction of which is the main goal of this paper.

The first important step is a negative one: we do not assign here truth and falsehood to questions. The reason is that it is doubtful whether all questions can express thoughts and describe states of affairs. Yet, we shall introduce here a more neutral semantic concept of *soundness* of a question in a given interpretation of a language.

DEFINITION 4. A question Q of $\mathcal{L}$ is *sound in an interpretation* $\mathfrak{I}$ of the language $\mathcal{L}$ iff at least one direct answer to Q is true in $\mathfrak{I}$.

The basic idea of this definition was suggested by Sylvain Bromberger (cf. Bromberger 1992, p. 146). Soundness understood in the above sense is called by Belnap (nominal) truth (cf. Belnap and Steel 1976, p. 119); yet, we prefer to use here the more neutral term.

There are questions which are sound in each normal interpretation of the language, questions which are sound only in some such interpretation(s) and questions which are not sound in any normal interpretation of the language. Following Belnap (cf. Belnap and Steel 1976, p. 130; we omit the relativization to a set of quasiformulae) we shall introduce here the concepts of *safety* and *riskiness* of a question.

DEFINITION 5. A question Q of $\mathcal{L}$ is said to be *safe* iff Q is sound in each normal interpretation of $\mathcal{L}$; otherwise Q is said to be *risky*.

Thus a safe question is a question which has at least one true direct answer in each normal interpretation of the language. Let us observe that safe questions might also have been defined as questions whose sets of direct answers are mc-entailed by the empty set; we can easily prove:

COROLLARY 9. *A question Q of $\mathcal{L}$ is safe iff the set of direct answers to Q is mc-entailed in $\mathcal{L}$ by the empty set.*

Simple yes–no questions, that is, questions whose sets of direct answers consist of a sentence and its negation, are paradigmatic examples of safe questions. Let us stress, however, that they are not the only safe questions. For instance, any questions having at least two direct answers which contradict each other and any questions which have tautologies among their direct answers are safe questions. Yet, it can be proved that each safe question is reducible to a set of questions made up of simple yes–no questions; it can also be proved that if Q is a safe question which has a finite number of direct answers or entailment in the language is compact, then Q is reducible to a finite set of simple yes–no question. For that and related results see Wiśniewski (1994b).

5.2. Just-complete answers and partial answers

If direct answers are defined in syntactic terms, it may happen that a sentence which is equivalent to a direct answer need not be a direct answer; on the other hand, such a sentence may perform the same pragmatic functions as a direct answer. Let us then introduce the semantic concept of just-complete answer; the definition given below is a slight modification of that proposed by Belnap (cf. Belnap and Steel 1976, p. 126).

DEFINITION 6. A sentence A of $\mathcal{L}$ is a *just-complete answer* to a question Q of $\mathcal{L}$ iff there is a direct answer B to Q such that B entails in $\mathcal{L}$ the sentence A and A entails in $\mathcal{L}$ the answer B.

In other words, a just-complete answer is a sentence which is equivalent to some direct answer.

It seems natural to call a partial answer to a question any sentence which is not equivalent to any direct answer to the question, but which is true if and only if a true direct answer belongs to some specified *proper* subset of the set of all the direct answers to the question. In other words, a partial answer is a sentence which is neither direct nor just-complete answer, but whose truth guarantees that a true direct answer can be found in some "restricted area" and whose truth is guaranteed by this fact. By means of the concept of mc-entailment we can express this intuition as follows:

DEFINITION 7. A sentence A of $\mathcal{L}$ is a *partial answer* to a question Q of $\mathcal{L}$ iff A is not a just-complete answer to Q and there exists a non-empty proper subset Y of the set of direct answers to Q such that: (i) Y is mc-entailed in $\mathcal{L}$ by A and (ii) A is entailed in $\mathcal{L}$ by each element of Y.

Let us stress that our concept of partial answerhood differs from those analyzed by Harrah, Belnap, and Kubiński.

If a question has more than two direct answers, then each disjunction of at least two but not all direct answers which is not equivalent to any single direct answer is a partial answer to the question. If entailment is compact, then each partial answer is either a disjunction of at least two but not all direct answers or a sentence which is equivalent to such a disjunction. The above definition yields, however, that questions with exactly two direct answers have no partial answers (let us recall that each question was assumed to have at least two direct answers). But, looking from the pragmatic point of view, a question which has exactly two direct answers requires a selection of one of them and does not leave room for a partial selection. Moreover, such a question can have answers of other kinds: incomplete, corrective, etc. We will not define here, however, these concepts of answers. Let us finally add that if questions with exactly one direct answer were allowed, no such question would have a partial answer in our sense.

5.3. Presuppositions of questions

The concept of a *presupposition* of a question is defined in various logical theories of questions in different ways. We will define here this concept following the general idea proposed by Belnap (cf. Belnap 1969b; see also Belnap and Steel 1976, p. 119).

DEFINITION 8. A d-wff A of $\mathcal{L}$ is a *presupposition* of a question Q of $\mathcal{L}$ iff A is entailed in $\mathcal{L}$ by each direct answer to Q.

The main advantage of the above definition is that is expresses a clear logical intuition: a presupposition of a question is a d-wff whose truth is necessary for the soundness (having a true direct answer) of the question.

The set of presuppositions of a question Q will be referred to as Pres$Q$.

It can happen that the truth of some presupposition is not only necessary, but also sufficient condition of soundness of the question. Let us then introduce the concept of *prospective presupposition* of a question:

DEFINITION 9. A presupposition A of a question Q of $\mathcal{L}$ is a *prospective presupposition* of Q iff A mc-entails in $\mathcal{L}$ the set of direct answers to Q.

To speak generally, a prospective presupposition is thus a presupposition which, if true, guarantees the existence of a true direct answer to a question.[7] The set of prospective presuppositions of a question Q will be denoted by PPres$Q$.

The set of presuppositions of a question is always nonempty (at least the tautologies of the language belong to it), but it need not be the case with the set of prospective presuppositions. There are languages of the considered kind which contain questions that have no prospective presuppositions; an example will be given below. But each question which has only finitely many direct answers has a non-empty set of prospective presuppositions: by Corollary 6 a disjunction of all the direct answers is a prospective presupposition of it. Let us also add that if a question has prospective presuppositions, all of them are equivalent.

Prospective presuppositions should be distinguished from *maximal presuppositions*.

DEFINITION 10. A presupposition A of a question Q of $\mathcal{L}$ is a *maximal presupposition* of Q iff A entails in $\mathcal{L}$ each presupposition of Q.

A maximal presupposition is thus a presupposition which entails any presupposition. The set of maximal presuppositions of a question Q will be referred to as mPres$Q$.

There is no general reason why each question of any language of the considered kind should have a maximal presupposition. But each question whose set of direct answers is finite does have maximal presuppositions: any disjunction of all the direct answers to it perform this function. Let us also observe that if a question has prospective presuppositions, all of them are maximal. This is due to:

COROLLARY 10. PPres$Q \subseteq$ mPres$Q$.

P r o o f: Assume that $A \in$ PPres$Q$ and that $A \notin$ mPres$Q$. So there is a presupposition B of Q such that A *non* $\models B$; it follows that there is a normal interpretation $\Im$ of the considered language such that $\Im \models A$ and $\Im$ *non* $\models B$. But B is a presupposition of Q; so for each $C \in$ dQ we have $\Im$ *non* $\models C$. Thus A *non* $\Vdash$ dQ and hence $A \notin$ PPres$Q$. We arrive at a contradiction. □

The converse of Corollary 10 need not be true: there are languages of the considered kind which contain questions that have maximal presuppositions which are not prospective presuppositions. Some example may be helpful here. Let us assume that $\mathcal{L}^{\#}$ is a language of the considered kind whose terms and d-wffs are those of some first-order language with identity and infinitely many individual constants; assume also that among questions of $\mathscr{L}^{\#}$ there is a question whose set of direct answers consists of all the sentences of the form $P(t)$, where P is a (concrete) one-place predicate symbol of $\mathscr{L}^{\#}$ and t is an arbitrary closed term of $\mathscr{L}^{\#}$. Let us designate this question by Q^{*}. Assume that the class of normal interpretations of $\mathscr{L}^{\#}$ consists of all the interpretations of the language. Clearly the sentence:

(1) $\exists x\, P(x)$

is a presupposition of Q^{*}, but is not a prospective presupposition of it: there are (normal) interpretations of the language in which the sentence (1) is true, but no sentence of the form $P(t)$ (i.e. no direct answer to Q^{*}) is true. At the same time (1) is a maximal presupposition of Q^{*}. For, let us assume that there is a presupposition, say, A, of Q^{*} which is not entailed in $\mathscr{L}^{\#}$ by the sentence (1). So the set:

(2) $\{\exists x\, P(x),\ \neg \bar{A}\}$

where $\bar{A}$ is the universal closure of A, has a model. But it can be proved that if X is a set of d-wffs of $\mathscr{L}^{\#}$ such that there are infinitely many individual constants of $\mathscr{L}^{\#}$ which do not occur in the d-wffs of X, then X has a model if and only if there is an interpretation $\Im = \langle \mathbf{M}, \mathbf{f} \rangle$ of $\mathscr{L}^{\#}$ which is a model of X and fulfills the following condition:

(i) for each $y \in \mathbf{M}$ there exists a closed term t of $\mathscr{L}^{\#}$ such that for each $\Im$-valuation $\mathbf{s}$, y is the value of t in $\Im$ with respect to $\mathbf{s}$.

(To speak generally, the condition (i) amounts to saying that each element of the domain of $\Im$ has a name in $\mathscr{L}^{\#}$). The proof goes along the lines of the Henkin-style proof of Gödel's theorem of the existence of a model: the only difference is that we use the individual constants which do not occur in the d-wffs of X as the "witnesses". Clearly there are infinitely many individual constants of $\mathscr{L}^{\#}$ that do not occur in the d-wffs of the set (2); so there is an interpretation, say, $\Im'$, of $\mathscr{L}^{\#}$ which fulfills the condition (i) and which is a model of the set (2). It follows that for some fixed sentence of the form $P(t)$, say, $P(t^{*})$, $\Im'$ is also a model of the set:

(3) $\{\exists x\, P(x),\ \neg \bar{A},\ P(t^{*})\}$

But this is impossible, since $P(t^*)$ is a direct answer to Q^* and thus entails the sentence $\bar{A}$ (let us recall that A was assumed to be a presupposition of the question Q^*). We arrive at a contradiction: so the sentence (1) is a maximal presupposition of Q^*. On the other hand, (1) is not a prospective presupposition of the analyzed question.

The example analyzed above is instructive for some other reason as well: it presents a language which contains questions that have no prospective presuppositions. If the question Q^* of $\mathcal{L}^\#$ had a prospective presupposition, this presupposition would be entailed by the maximal presupposition (1). So (1) would be a prospective presupposition of Q^*; since it is not, the question Q^* has no prospective presuppositions. Let us stress, however, that we do not claim here that Q^* and similar questions cannot have prospective presuppositions in any language. The concepts of presuppositions introduced above are defined by means of the concepts of entailment and mc-entailment *in a language* and there are languages of the considered kind in which these questions do have prospective presuppositions.

Let us finally introduce the concepts of factual presupposition and maximal factual presupposition, which may be especially useful in the philosophy of science.

DEFINITION 11. A presupposition A of a question Q of $\mathcal{L}$ is a *factual presupposition* of Q iff A is a synthetic d-wff of $\mathcal{L}$.

DEFINITION 12. A factual presupposition A of a question Q of $\mathcal{L}$ is a *maximal factual presupposition* of Q iff A entails in $\mathcal{L}$ each factual presupposition of Q.

It is not the case that each question has factual presuppositions. Moreover, there is no general reason why each question which does have factual presuppositions should have maximal factual presuppositions.

5.4. Relative soundness. Normal questions and regular questions

Let us now introduce the concept of *relative soundness* of a question, which seems to be of basic importance to erotetic logic. The underlying intuition is that a question Q is sound relative to a set of d-wffs X just in case the question Q *must* have a true direct answer *if* all the d-wffs in X are true. This intuition can be expressed in terms of mc-entailment via

DEFINITION 13. A question Q of $\mathcal{L}$ is *sound relative to* a set of d-wffs X of $\mathcal{L}$ iff the set X mc-entails in $\mathcal{L}$ the set of direct answers to Q.

In other words, Q is sound relative to X just in case there is no normal interpretation of the language in which all the d-wffs in X are true, but no direct answer to Q is true.

If Q is sound relative to a singleton set $\{A\}$, we say that Q is sound relative to the d-wff A.

Let us stress that the concept of relative soundness introduced above must be carefully distinguished from the concept of soundness of a question in an interpretation of the language introduced in Section 5.1: these are different concepts.

A safe question is sound relative to any set of d-wffs. This is the trivial case; in order to distinguish the non-trivial cases let us introduce the following concept:

DEFINITION 14. A question Q of $\mathscr{L}$ is *made sound* by a set of d-wffs X of $\mathscr{L}$ iff X mc-entails in $\mathscr{L}$ the set of direct answers to Q although the set of direct answers to Q is not mc-entailed in $\mathscr{L}$ by the empty set.

Thus Q is made sound by X just in case Q is not sound relative to the empty set, but Q is sound relative to X. It is clear that only risky questions can be made sound by sets of d-wffs.

Each question has a non-empty set of presuppositions; the truth of these presuppositions is a necessary condition for having a true direct answer. Yet, when we look in the opposite direction, the truth of all the presuppositions need not be a sufficient condition of having a true direct answer: there are languages of the considered kind in which some questions do not have this property. The language $\mathscr{L}^{\#}$ mentioned in Section 3.3 is a case in point here: since the sentence (1) is a maximal presupposition of the question Q^* but not a prospective presupposition of it, it happens that all the presuppositions of Q^* are true in some normal interpretation of the language, but no direct answer to Q^* is true in it. In order to distinguish questions whose presuppositions guarantee the existence of a true direct answer from the remaining ones let us introduce the concept of *normal question*:[8]

DEFINITION 15. A question Q of $\mathscr{L}$ is said to be *normal* iff the set of direct answers to Q is mc-entailed in $\mathscr{L}$ by the set of presuppositions of Q.

A normal question is thus a question which is sound relative to the set of its presuppositions. Note that since being a normal question depends on mc-entailment in a language (and thus basically on the conditions imposed on the class of normal interpretations), the same question occurring in one language may be normal in it without being normal in some other language. Again, the question Q^* gives us a simple example here: as far as the language $\mathcal{L}^{\#}$ is concerned, Q^* is not normal in it. It can be shown, however, that the question Q^* would be normal if it occured in a language in which the class of normal interpretations would be defined in such a way that the condition (i) of Section 5.3 would be satisfied by each normal interpretation of it. But any question which has only finitely many direct answers is normal in any language of the considered kind; moreover, each safe question is normal. The remaining questions, however, may or may not be normal: it depends on the semantics of the language.

Let us now introduce a more specific concept of *regular question*. To speak generally, a question Q will be called regular if there is a single presupposition of Q whose truth implies the existence of a true direct answer to Q. To be more precise, we adopt

DEFINITION 16. A question Q of $\mathcal{L}$ is said to be *regular* iff there is a presupposition A of Q such that A mc-entails in $\mathcal{L}$ the set of direct answers to Q.

In other words, a question is regular if it is sound relative to some of its presuppositions. Note that each question whose set of direct answers is finite is regular. We also have:

COROLLARY 11. *A question Q is regular iff Q has prospective presuppositions.*

COROLLARY 12. *Each regular question is normal.*

COROLLARY 13. *If entailment in $\mathcal{L}$ is compact, then each normal question of $\mathcal{L}$ is regular.*

Corollary 11 shows that the concept of regular question can be defined in terms of prospective presuppositions. Corollaries 12 and 13 imply that "being a normal question" and "being a regular question" coincide if entailment in a language and thus also mc-entailment in it are compact. Yet, there are languages of the considered kind in which entailment is not compact.

5.5. Self-rhetoricity and informativeness. Proper questions

Let us now define the concept of a self-rhetorical question.

DEFINITION 17. A question Q of $\mathcal{L}$ is said to be *self-rhetorical* iff there is a direct answer A to Q such that A is entailed in $\mathcal{L}$ by the set of presuppositions of Q.

By proposing the above definition we are not going to explicate the general notion of rhetoricity of a question: this notion is clearly a pragmatic one (in the traditional sense of the word). Our aims are limited: we attempt to explicate the concept of "rhetoricity for logical reasons." The following are examples of self-rhetorical questions: questions having tautologies as direct answers, questions all of whose direct answers are contradictory sentences, questions which have only one direct answer being a synthetic sentence. Note that if questions with exactly one direct answer were allowed, these questions would be self-rhetorical questions.

Each self-rhetorical question is normal in the sense of Definition 15. In order to distinguish normal questions which are not self-rhetorical from the remaining ones we introduce:

DEFINITION 18. A question Q is *proper* iff Q is normal but not self-rhetorical.

Let us finally introduce a certain concept of *informativeness* of a question.

Looking from the intuitive point of view, a question is informative relative to a given set of d-wffs just in case each direct answer to the question conveys some information which cannot be legitimately drawn from the analyzed set of d-wffs. In order to explicate this notion let us first introduce the concept of *content* of a set of d-wffs. Let X be a set of d-wffs of $\mathcal{L}$. The *content* of the set X (in symbols: $\mathbf{Ct}(X)$) is defined as follows:

DEFINITION 19.

$\mathbf{Ct}(X) = \{A$: A is entailed in $\mathcal{L}$ by X and A is not a tautology of $\mathcal{L}\}$

In other words, the content of X consists of all the d-wffs which are non-trivially entailed by X (i.e. are not tautologies).[9]

The relativized concept of informativeness of a question can now be defined as follows:

DEFINITION 20. A question Q is *informative relative to* a set of d-wffs X iff for each $A \in \mathrm{d}Q$, $\mathbf{Ct}(X)$ is a proper subset of $\mathbf{Ct}(X \cup \{A\})$.

Thus a question Q is informative relative to a set of d-wffs X just in case the content of X is a *proper* subset of the content of any set which results from X by adding a direct answer to Q: any such set entails more non-trivial consequences than the initial set X. It is easily seen, however, that Q is informative relative to X if and only if no direct answer to Q is entailed by X.

By proposing the above definition we identify the possibility of extracting information from a set of d-wffs with the entailment of the sentence conveying this information. No doubt, there is plenty of idealization in such an approach; consequently, Definition 20 presents an idealized concept of informativeness of a question. The same holds true in the case of our definition of self-rhetoricity. We can easily prove:

COROLLARY 14. *A question Q is not self-rhetorical iff Q is informative relative to the set of presuppositions of Q.*

Let us finally note:

COROLLARY 15. *A question Q is proper iff Q is normal and Q is informative relative to the set of presuppositions of Q.*

Thus a proper question is a question which is both normal and informative relative to the set of its presuppositions.

Institute of Philosophy
Tadeusz Kotarbiński Pedagogical University
al. Wojska Polskiego
65-762 Zielona Góra
Poland

NOTES

* This paper was completed in February, 1995. Some material from it was then used in the book *The Posing of Questions: Logical Foundations of Erotetic Inferences*, which was completed later.

1 Cf. Belnap and Steel (1976), p. 10 et al.

2 Let us observe that according to this definition a d-wff of the form $P(x_i)$ logically entails its universal generalization, i.e. $\forall x_i\, P(x_i)$. This is intuitively acceptable only if the free variable x_i in $P(x_i)$ is not construed as a name of some unspecified object, but is interpreted in the "generalizing" manner, as referring to any element of the domain. This interpretation of free variables underlies the above definitions of logical entailment

and entailment in a language. Let us add that the system of classical predicate calculus with identity which is complete with respect to logical entailment defined in the above manner can be built, int.al., as follows: (1) *Axioms*: substitution-instances of valid formulas of classical propositional calculus as well as the standard axioms for identity; (2) *Rules*: modus ponens, substitution of terms, elimination of ∀, introduction of ∀, elimination of ∃, introduction of ∃.

3 Cf. e.g. Wójcicki (1982).

4 If A is a sentence, then the universal closure of A is equal to A itself. If A is a sentential function and $x_{i1}, \ldots, x_{in}$ (where $x_{i1} < \ldots < x_{in}$) are the all free variables of A, then the universal closure of A is of the form $\forall x_{i1} \ldots \forall x_{in} A$. The universal closure of a d-wff A will be designated by $\bar{A}$.

5 Cf. note 4.

6 The paper Wiśniewski (1987) is an abstract of the paper presented at the VIII International Congress of Logic, Methodology and Philosophy of Science, Moscow 1987. For some mysterious reasons the organizers retyped the manuscript before printing without making any proof reading. As a result this paper presumably wins the world record for misprints.

7 In Belnap's terminology a prospective presupposition is a d-wff which *expresses the presupposition* of the question. Belnap, however, defines this concept without applying the concept of mc-entailment.

8 Buszkowski, who in fact introduced this concept (cf. Buszkowski 1989), uses here the term *correct question*.

9 The definition proposed above is similar to that given by Popper (cf. Popper 1959, p. 120), who, however, uses the concept of derivability.

REFERENCES

Ajdukiewicz, K. (1974). *Pragmatic Logic*. Dordrecht: D. Reidel.

Apostel, L. (1969), A proposal in the analysis of questions. *Logique et Analyse* **12**, 376–81.

Åqvist, L. (1971), Revised foundations for imperative epistemic and interrogative logic. *Theoria* **37**, 33–73.

Åqvist, L. (1972), On the analysis and logic of questions. In: R.E. Olson, A.M. Paul (Eds.), *Contemporary Philosophy in Scandinavia*. Baltimore: John Hopkins University Press, pp. 27–39.

Åqvist, L. (1975). *A New Approach to the Logical Theory of Interrogatives. Analysis and Formalization*. Tübingen: Verlag Gunter Narr.

Åqvist, L. (1983). On the "tell me truly" approach to the analysis of interrogatives. In: F. Kiefer (Ed.), *Questions and Answers*. Dordrecht: D. Reidel, pp. 9–14.

Belnap, N.D. (1963). *An Analysis of Questions: Preliminary Report*. Technical Memorandum 7 1287 1000/00. Santa Monica, CA: System Development Corp.

Belnap, N.D. (1966). Questions, answers and presuppositions. *The Journal of Philosophy* **63**, 609–11.

Belnap, N.D. (1969a). Aqvist's corrections-accumulating question sequences. In: J.W. Davis et al. (Eds.), *Philosophical Logic*. Dordrecht: D. Reidel, pp. 122–34.
Belnap, N.D. (1969b). Questions, their presuppositions and how they can fail to arise. In: K. Lambert (Ed.), *The Logical Way of Doing Things*. New Haven: Yale University Press, pp. 23–37.
Belnap, N.D. and Steel, Th.B. (1976). *The Logic of Questions and Answers*. New Haven: Yale University Press.
Bromberger, S. (1992). *On What We Know We Don't Know: Explanation, Theory, Linguistics, and How Questions Shape Them*. Chicago: The University of Chicago Press.
Buszkowski, W. (1987). Erotetic completeness. *Abstracts of the 8th International Congress of Logic, Methodology and Philosophy of Science*, Vol. **5** Part 3, 239–41.
Buszkowski, W. (1989). Presuppositional completness. *Studia Logica* **48**, 23–44.
Carnap, R. (1943). *Formalization of Logic*. Cambridge Mass.: Harvard University Press.
Cohen, F.S. (1929). What is a question. *The Monist* **39**, 350–64.
Gentzen, G. (1934). Untersuchungen über das logische Schliessen. *Mathematische Zeitschrift* **39**, 176–210 and 405–31.
Groenendijk, J. and Stokhof, M. (1984). *Studies on the Semantics of Questions and the Pragmatics of Answers*. Amsterdam.
Hamblin, C.L. (1973). Questions in Montague English. *Foundations of Language* **10**, 41–53.
Harrah, D. (1961). A logic of questions and answers. *Philosophy of Science* **30**, 40–6.
Harrah, D. (1963). *Communication: A Logical Model*. Cambridge Mass.: MIT Press.
Harrah, D. (1975). A system for erotetic sentences. In: A.R. Anderson, Ruth Barcan Marcus, R.M. Martin (Eds.), *The Logical Enterprise*. New Haven: Yale University Press, pp. 235–45.
Harrah, D. (1984). The logic of questions. In: D. Gabbay and F. Guenthner (Eds.), *Handbook of Philosophical Logic, Volume II: Extensions of Classical Logic*. Dordrecht: D. Reidel, pp. 715–64.
Hausser, R.R. (1983). Syntax and semantics of English mood. In: F. Kiefer (Ed.), *Questions and Answers*. Dordrecht: Reidel, pp. 97–158.
Hintikka, J. (1974). *Questions about questions*. In: M.K. Munitz and P.K. Unger (Eds.), *Semantics and Philosophy*. New York: New York University Press, pp. 103–58.
Hintikka, J. (1976). *The Semantics of Questions and the Questions of Semantics*. Amsterdam: North-Holland. (*Acta Philosophica Fennica*, Vol. **28**, No 4).
Hintikka, J. (1978). Answers to questions. In: H. Hiż (Ed.), *Questions*. Dordrecht: D. Reidel, pp. 279–300.
Hintikka, J. (1983). New foundations for a theory of questions and answers. In: F. Kiefer (Ed.), *Questions and Answers*. Dordrecht: Reidel, pp. 159–90.
Hiż, H. (1962). Questions and answers. *The Journal of Philosophy* **59**, 253–65.
Hiż, H. (1978). Difficult questions. In: H. Hiż (Ed.), *Questions*. Dordrecht: D. Reidel, pp. 211–226.

Karttunen, L. (1978). Syntax and semantics of questions. In: H. Hiz (Ed.), *Questions*. Dordrecht: D. Reidel, pp. 165–210.

Keenan, E. and Hull, R. (1973). The logical presuppositions of questions and answers. In: J. Petöfi and D. Franck (Eds.), *Präsuppositionen in der Linguistik und der Philosophie*. Frankfurt/M.: Athenäum, pp. 441–6.

Koj, L. (1972). Analiza pytań II. Rozważania nad struktura pytań. *Studia Semiotyczne* III, pp. 23–39. (English translation: Inquiry into the structure of questions. In: L. Koj and A. Wiśniewski. *Inquiries into the Generating and Proper Use of Questions*. Lublin: Wydawnictwo UMCS, 1989, pp. 33–60.)

Kubiński, T. (1960). An essay in to the logic of questions. In: *Atti del XII Congresso Internazionale di Filosofia (Venezia 1958)*, Vol. 5, Firenze, pp. 315–22.

Kubiński, T. (1971). *Wstęp do logicznej teorii pytań*. (An Introduction to the Logical Theory of Questions). Warszawa: Państwowe Wydawnictwo Naukowe.

Kubiński, T. (1980). *An Outline of the Logical Theory of Questions*. Berlin: Akademie-Verlag.

Leszko, R. (1983). *Charakterystyka złożonych pytań liczbowych przez grafy i macierze* (The characteristics of compound numerical questions by graphs and mattrices). Zielona Góra: Wyższa Szkoła Pedagogiczna.

Materna, P. (1981). Question-like and non-question like imperative sentences. *Linguistics and Philosophy* **4**, 393–404.

Popper, K.R. (1959). *The Logic of Scientific Discovery*. London: Hutchinson.

Scott, D. (1974). Completeness and axiomatizability in many-valued logic. In: L. Henkin et al. (Eds.), *Proceeding of the Tarski Symposium* (Proceedings of Symposia in Pure Mathematics, vol. 25). Providence: American Mathematical Society, pp. 411–35.

Shoesmith, D.J. and Smiley, T.J. (1978). *Multiple-conclusion Logic*. Cambridge: Cambridge University Press.

Stahl, G. (1962). Fragenfolgen. In: M. Käsbauer and F. Kutschera (Eds.), *Logik und Logikkalkül*. Freiburg: Alber, pp. 149–58.

Tichy, L. (1978). Questions, answers, and logic. *American Philosophical Quarterly* **15**, 275–84.

Wiśniewski, A. (1986). *Generowanie pytań przez zbiory zdań*. Ph.D. Thesis. Poznań: Adam Mickiewicz University.

Wiśniewski, A. (1987). The generating of questions and erotetic inferences. *Abstracts of the 8th International Congress of Logic, Methodology and Philosophy of Science*, Vol. **5** Part 1, 347–50.

Wiśniewski, A. (1989). The generating of questions: a study of some erotetic aspect of rationality. In: L. Koj and A. Wiśniewski, *Inquiries into the Generating and Proper Use of Questions*, pp. 91–155.

Wiśniewski, A. (1990a). *Stawianie pytań: logika i racjonalność* (The Posing of Questions: Logic and Rationality). Lublin: Wydawnictwo UMCS.

Wiśniewski, A. (1990b). Implied questions. *Manuscrito* **13**(2), 21–38.

Wiśniewski, A. (1991). Erotetic arguments: a preliminary analysis. *Studia Logica* **50**, 261–74.

Wiśniewski, A. (1994a). Erotetic implications. *Journal of Philosophical Logic* **23**, 173–95.

Wiśniewski, A. (1994b). On the reducibility of questions. *Erkenntnis* **40**, 265–84.

Wiśniewski, A. (1995). *The Posing of Questions: Logical Foundations of Erotetic Inferences*. Dordrecht: Kluwer. ("Synthese Library", Vol. 252).

Wójcicki, R. (1982). *Wyklady z metodologii nauk* (Lectures on the Methodology of Sciences). Warszawa: Państwowe Wydawnictwo Naukowe.

Zygmunt, J. (1984). *An Essay in Matrix Semantics for Consequence Relations*. Wrocław: Wydawnictwo Uniwersytetu Wrocławskiego.

Poznań Studies in the Philosophy of the Sciences and the Humanities 1997, Vol. 51, pp. 213–221

Jan Woleński

SCIENCE AND GAMES

Although Hintikka looks at science in terms of the so called interrogative model of inquiry, his game-theoretical semantics (GTS, for brevity) also provides some interesting tools for philosophy of science. In this paper, I would like to combine together Giedymin's game-theoretical approach to science (Giedymin 1959) and GTS.

The analysis of science as a game is very natural. This game is played by two gamblers: Researcher (**R**) and Nature (**N**). **R** intends to discover something, that is, metaphorically speaking, to "extract" the mysteries from the Nature. In order to perform this task, **R** uses various hypotheses as strategies. On the other hand, **N** "tries" to stop discoveries and employs something which can be labelled as antistrategies, respectively to **R**-strategies. I hope that this antropomorfization of Nature, which is necessary for considering it as a player, is admissible. Thus, what is a cognitive success of **R** is simultaneously a loss of **N**, and reversely. This means that the **R-N** game is based on a conflict. We can consider this conflict as a two-person zero-sum game.

Jerzy Giedymin described an interesting and important aspect of science as a game. Asssume that we wish to regard the scientific game (SG, for brevity) as rational. A natural requirement here is this: SG is rational, if it is a proper game, not a *pseudo*-game. A given game **G** is an improper or *pseudo*-game, if it admits a dictatorial strategy, that is a strategy which secures the victory (or at least, non-loss, if such result is possible) for one of the players, independently of the moves of the second. At the first sight, dictatorial strategies of **R** consist in using of hypotheses which cannot be refuted by **N**, that is non-analytic propositions which are either true or coherent with arbitrary facts. However, there are serious reservations against including true hypotheses into dictatorial strategies. Firstly, **R** does not know in general which hypotheses are true and which are false. Secondly, true hypotheses can be helpful in discovering other hypotheses. Both reasons suggest that dictatorial strategies should be restricted to non-analytic propositions coherent with arbitrary facts.

Giedymin's analysis is based on Popperian ideas. In particular, Giedymin accepts Popper's criterion of demarcation: *A* is a genuine scientific proposition if and and only if *A* is falsifiable. This stipulation entails that *A* is not falsifiable, it is coherent with arbitrary facts, because the expression "*A* is coherent with arbitrary facts" means "there is no basic proposition *B* such that $\neg A$ is a logical consequence of *B*. Such propositions are, according to Popper, metaphysical, and they do not belong to science. In particular, Giedymin stated the so called condition of the critical confirmation

(1) if a hypothesis *H* is not negatively decidable (refutable), then *H* is not confirmable, that is, one cannot regard facts coherent with *H* as its confirmation-instances.

However, it seems that (1) restricts the game-theoretical analysis of science, because it links the rationality of games with a concrete approach to empirical testability, namely with falsificationism. It is known that this approach leads to some troubles which concern, above all, the propositions which begin with existential quantifiers. In fact, the propositions of the from $\exists xPx$ are not refutable (falsifiable). For example, the sentence "there are white swans" is not refutable by any finite amount of empirical data. On the other hand, this proposition is verifiable by pointing out a white swan. In order to avoid misinterpretations let me remark that speaking on the existential propositions I have in mind so called purely existential statemens, that is propostions which are not restricted by temporal or spatial coordinates. Thus, the propositions "there are white swans in Poland" is not purely existential and it may be refutable by a finite stock of observations. Now, it seems that we should accept the following principle of duality (the symbol *dA* stands for "*da* is the dual of *A*"):

(2) if *A* is a genuine proposition, *dA* is also genuine.

Thus, if a criterion of being genuine holds for a proposition, it should be also valid for its dual. However, if general propositions are genuine for their falsifiability but existential are not, the principle (2) is violated. Furthermore, on the base of (1), we should say that the purely existential propositions are not confirmable what is certainly not intuitive. This difficulty could be overcome, when falsifiability were replaced by falsifiability-or-verifiability. This solution requires abandonment of the finitistic concept of falsifiability and verifiability, because statistical hypotheses and propositions of the form $\forall x \exists y Pxy$ are neither finitary falsifiable nor finitary verifiable. Thus, the concept of partial fasifiability and partial verifiability appears as proper. In this situation, the distinction of falsifiability and verifiability becomes only secondary, because both procedures fall under a more general concept of testability, partial or complete. On the other hand, there is a room for preserving the falsifiability/verifiability distinction. In particualr, (2) leads to

(3) if a criterion **C** of being a genuine propostion holds for *A*, then the dual criterion **dC** holds for *dA*.

Thus, if the criterion of falsifiability holds for a certain type of propositions, then the criterion of verifiability, which is the dual criterion with respect to falsifiability, holds for their dual counterparts. This solves the problem of existential propositions, because they are duals of universal statements. Moreover, this suggests a generalized criterion of being a genuine scientific proposition

(4) a proposition *A* is genuine if and only if *A* is falsifiable or verifiable, that is negatively or positively testable.

The principle (4) can be also applied to non-genuine propositions:

(5) if a proposition *A* is not genuine, *dA* is also not genuine.

Assume that a hypothesis *H* is not genuine, because it is coherent with all facts. What about *dH*? According to (5), it is also non-genuine, but not because it is coherent with arbitrary facts. Our duality principles motivate that *dH* is not genuine for its incoherence with arbitrary facts. The genuiness *resp.* non-genuiness of propositions are explicitly relative to their logical form, in our case to their universal or existential character.

Assume that a proposition

(6) there is such *x*, that *x* is a ghost.

is recognized as non-genuine, because there is no proposition *Pa* such that (6) is its logical consequence. It seems that (6) is non-genuine, because it is incoherent with all facts. In general, we propose

(7) a proposition $\exists x Px$ is not (empirically) genuine if and only if it is not analytic and incoherent with arbitrary facts.

Giedymin considered dictatorial strategies for **R** only. Of course, this problem concerns **N** as well. According to (7), **N** acts by dictatorial strategies, when it plays with hypotheses incoherent with all facts. Similarly as in the case of **R**, it is admissible for **N** to play with false hypotheses. These moves are even rational for **N**, because it helps to conceal facts which **R** is looking for: disseminating of false messages is a good way for concealing truth. Adapting Einstein's well-known metaphor, both **R** and **N** can be perfidious or refined. The perfidity of **R** consists in deceiving of **N** by hypotheses which are non-falsifiable, but **N** is perfidious when it plays with non-verifiable information. The refinement of **R** consists in a reasonable use of truths, but **N** is refined when it plays with falsehoods. Thus, both **R** and **N** are refined, when rational, and irrational, when perfidious.

The methodological difference of **R** and **N** has its clear expression in logic used by both players. GTS provides a very impressive insight. Assume that **R** and **N** play a game **G** in the first order language **L** defined for a given domain **D** (see Hintikka 1987). **G** is understood here as an idealized process of verification on the part of **R**. Thus, **R** plays on truth and, dually, **N** plays on falsity. Hence, **G** is a semantic game. When **G** is on a formula A, **R** wins if A is true and loses if A is false. Dually, **N** wins, when A is false, and loses, when A is true. In general, **R** is the initial verifier, and **N** the initial falsifier.

We assume that atomic sentences of **L** have fixed logical values. When A is atomic, **R** wins if A is true. When A is a complex formula, **G** is reduced to games on subformulas of A. Let $A = B \wedge C$. **G** begins with choosing of B or C by **N**. Clearly, **R** wins **G**$(B \vee C)$ if and only if both B and C are true. If $A = B \vee C$, the choice belongs to **R** who wins, when at least one disjunct is true. The game on negation requires that **R** becomes the falsifier, and **N** the verifier. Thus, if $A = \neg B$, **G**$(\neg B)$ concerns B. **G**$(B \rightarrow C)$ can be defined as **G**$(\neg B \vee C)$, and **G**$(B \leftrightarrow C)$ as **G**$[(B \rightarrow C) \wedge (C \rightarrow B)]$. **G**$(\exists xFx)$ begins with selecting an element a from **D**. The further course of this game is on Fa. When $A = \forall xFx$, the start belongs to **N**, and the game proceeds like in the case **G**$(\exists xFx)$. The above considerations are summarized by

(8) (a) if A is atomic, then **R** wins and **N** loses, when A is true; **N** wins and **R** loses, when A is false;
(b) if $A = \neg B$, **G**(A) is the same as in the case (a), except the roles of both players are interchanged;
(c) if $A = B \wedge C$, **G**(A) begins with the choice of B or C by **N**, and it continues **G**(B) and **G**(C);

(d) if $A = B \vee C$, **G**(A) begins with the choice of B or C by **R**, and it continues as **G**(B) or **G**(C);
(e) if $A = \forall xB$, **G**(A) begins with the choice of a member of **D** (say a) by **N**, and it continues as **G**(Ba);
(f) if $A = \exists xB$, **G**(A) begins with the choice of a member of **D** (say a) by **R**, and it continues as **G**(Ba).

Since GTS is equivalent to the classical model-theoretic semantics first-order logic (the proof for the clause (8e) requires the axiom of choice; see Hodges 1983, p. 94), **R** has always winning strategy for any true formula of **L**. This result shows in what sense GTS provides the idealized model of the verification-process. However, one may ask whether this result does not generate dictatorial strategies for **R**. It does, but only provided that **R** knows the logical values of all atomic sentences of **L**. However, (8) is not intended as a description (idealized or not) of the actual verification-processes, but rather as an account of the semantic behavior of **R**.

The clauses (8a)–(8f) define semantic games from the point of view of **R**. As I already remarked, **N** is the intitial falsifier. Thus, **N** performs games which are the idealized models of falsification. These games are duals of the verification games. In order to give a formal account of such games, we assume that false atomic sentences are distinguished. Thus, we assume that **N** directly identifies falsehoods. This means that when **R** plays on conjuction, **N** plays on disjunction. Let the symbol *d***G** denote the game which is the dual of **G**. If $A = B \vee C$, *d***G**(A) begins with the choice made by **N**, and proceeds on B and C. **N** wins when both B and C are false. The full description is this:

(9) (a) if A is atomic, **dG**(A) = **G**(A), because atomic formulas are self-dual;
(b) if $A = \neg B$, **dG**(A) = **G**(A), because negation is self-dual;
(c) if $A = B \wedge C$, **dG**(A) begins with the choice of B or C by **R**, and it continues as **dG**(B) or **G**(C);
(d) if $A = B \vee C$, **dG**(A) begins with the choice of B or C by **N**, and it continues as **dG**(B) or **G**(C);
(e) if $A = \forall xB$, **dG**(A) begins with the choice of an element of **D** (say a) by **R**, and it continues as **dG**(Ba);
(f) if $A = \exists xB$, **dG**(A) begins with the choice of an element of **D** (say a) by **N**, and it continues as **dG**(Ba).

The idealized process of verification can be identified with the structure $<\mathbf{R},\mathbf{N},\mathbf{L},\mathbf{D},\mathbf{G}_1,\ldots,\mathbf{G}_n>$ with $\mathbf{G}_1,\ldots,\mathbf{G}_n$ as subgames of the global game.

Similarly, the idealized process of falsification is modelled by $<$**N**,**R**,**L**,**D**, **dG**$_1$,...,**dG**$_n$$>$. A more realistic picture is obtainable under the assumption that the number of subgames is finite, because we cannot ascribe to the Researcher and the Nature the ability of effective performing of infinitary strategies. This explains why the concept of partial verification and falsification is necessary in the philosophy of science. This observation favours those conceptions of induction which connect its origin with the fact that the human mind is effectively able to manage only finite amounts of empirical information. The approach to science *via* GTS clearly confirms non-effectiveness of verification with respect to universal hypotheses and falsification with respect to existential one. Thus, we are entitled to the conclusion that neither verification nor falsification are universal tests for hypotheses of an arbitrary kind. There are situations in which verification is suitable as well as situations in which falsification is effective. In particular, it follows from (8) and (9) that **R** is the natural verifier, but **N** the natural falsifier. However, sometimes the exchange of their roles is a reasonable step in the game.

The hitherto considerations employed the classical logic. On this approach, the verification (falsification) consists in solving the problem of logical values of formulas, relatively to **D**. Even if we disregard the problem of complete verification and falsification, the above account provides an ideal model also for purely logical reasons. It is a temptation to identify winning strategies (at least in games on logical or mathematical truths) with decidability or provability. Now, decidability fails as a general criterion just in the case of first-order logic, but provability is not sufficient for arithmetic and its oversystems. Since we have no universal recursive winning strategy for generating the set of all truths, even in mathematics, the games considered from the classical point of view are ontological in a sense, because they concern the question: how things are in **D**? However, verification and falsification can be also treated epistemically as sequences of playoffs on assertions and refutations of propositions. If this approach is adopted, intuitionistic logic provides good analytical tools (see Grzegorczyk 1964, Kripke 1964). The intutionistic semantic model is the triple $<$**I**, **S**, **F**$>$, where **I** is a non-empty set of states of information, **S** is a reflexive and transitive relation on **I**, and **F** is the forcing operator. The relation **S** represents extensions of already obtained information: symbol *i***S***j* means "the information *j* is a possible extension of the information *i*", and the symbol *i***F***A* means "the information *i* forces the formula *A*". GTS for intuitionistic games is as follows (we assume that *i* is the initial information):

(10) (a) if A is atomic, then **R** wins and **N** loses, when $i\mathbf{F}A$; **N** wins and **R** loses, when $\neg(i\mathbf{F}A)$;
(b) if $A = \neg B$, **R** wins and **N** loses,
(c) if $A = B \wedge C$, **G**(A) begins with the choice of B or C by **N**, and it continues as **G**(B) or **G**(C);
(d) if $A = B \vee C$, **G**(A) begins with the choice of B or C by **R**, and it continues as **G**(B) or **G**(C);
(e) if $A = \forall xB$, **G**(A) begins with the choice of a member of **I** (say k) by **N**, and it continues as **G**(k**F**(B)); **R** wins and **N** loses, when for every j such that $i\mathbf{S}j$, $j\mathbf{F}B$;
(f) if $A = \exists xB$, **G**(A) begins with the choice of a member of **I** (say k) by **R**, and it continues as **G**(k**F**(B); **R** wins and **N** loses, when for some j such that $i\mathbf{S}j$, $j\mathbf{F}B$.

The above clauses reflect the semantic behaviour of **R** on intutionistic logic as the base. Pragmatically speaking, "i**F**" can be read "i forces the assertion of A"; the the logic of **F** is the logic of assertion. In order to model the intuitionistic behaviour of **N**, we can introduce the concept of antiforcing. Let the symbol $i\mathbf{dF}A$ mean "the information i forces the rejection of A". We have the following conditions for intuitionistic dual games:

(11) (a) if A is atomic, **dG**(A) = **G**(A);
(b) if $A = \neg B$, **dG**(A) = **dG**(A);
(c) if $A = B \wedge C$, **G**(A) begins with the choice of B or C by **R** and it continues as **dG**(B) or **dG**(C);
(d) if $A = B \vee C$, **dG**(A) begins with the choice of B or C by **N**, and it continues as **dG**(B) and **dG**(C);
(e) if $A = \forall xB$, **dG**(A) begins with the choice of a member of **I** (say k) by **R**, and it continues as **dG**(k**F**(B); **N** wins and **R** loses, when for some j such that $i\mathbf{S}j$, $j\mathbf{dF}B$;
(f) if $A = \exists xB$, **dG**(A) begins with the choice of a member of **I** (say k) by **R**, and it continues as **dG**(k**F**B; **N** wins and **R** loses, when for every j such that $i\mathbf{S}j$, $j\mathbf{dF}B$.

The games governed by (10) and (11) are not on truth or falsity, but they proceed on assertion or rejection of propositions on the base of the initial information and its extensions. Now, we have a problem whether epistemic intuitionistic games allow us to solve every problem concerning **D**. For the intuitionists, if an answer is neither forced nor antiforced by information from **I**, the problem in question is meaningless. This means that every meaningful problem is epistemically solvable. Thus, the intuitionists say that there are not absolutely undecidable problems; the undecidability is only temporal. This view is motivated by the general metalogical thesis (see Martin-Löf 1983):

(12) ($\neg A$ is intuitionistically true) is provable if and only if (A is intuitionistically true) is not provable.

This result suggests that the intuitionistic provability is a general winning strategy in intuionistic games. Thus, we note a sharp contrast between intuititionism and classicism in looking at semantic games from both perspectives. From the point of view, there are undecidable problems, relatively to the to the initial i and its extensions. This position is based on the assumption that the extensions of information needed for solving some problems may be not effectively realizable. Hence, not every information effectively forces or antiforces A, for arbitrary A. Thus, assuming that a given problem is decidable if and only if it is algorithmically forceable (it is a version of the Church thesis), we can conclude that not every proposition is decidable, although every proposition is true or false. Looking at science (and more generally, at any cognitive activity) through the glasses of GTS enables us to see a principal diffrence between truth and their criteria and, *a fortiori*, between the classical and the criterial definition of truth. For the defenders of the classical theory, intutionistic logic (or any other which identifies conditions of truth and conditions of assertibility) is the logic of truth-criteria, but the domain of truth itself is subjected to classical logic. The criterial truth-definitions do not distinguish between the logic of truth and the logic of truth-criteria. This controversy is certainly not apparent, because each standpoints is assiociated with a different formal machinery. We also see that there is a price which must be paid by the criterial truth-definitions, namely recognizing that some classically undecidable problems are criterially meaningless. On the other hand, GTS shows that the controversy between verificationism and falsificationism is apparent. Since verification and falsification are mutually dual, the expressive power of both theories is the same. Everything that reamins is to decide when it is reasonable the epistemic roles of **R** and **N** in scientific games. The clauses in (8)–(11) answer this question.

Institute of Philosophy
Jagiellonian University
Kraków, Poland

REFERENCES

Giedymin, J. (1959). Uogólnienie postulatu rozstrzygalności hipotez (A Generalization of the Refutability Postulate). *Studia Filozoficzne* **5**(14), 139–60. Eng. summary, pp. 159–60.

Grzegorzyk, A. (1964). A Philosophically Plausible Formal Interpretation of Intuitionistic Logic. *Indagationes Mathematicae* **26**, 596–601.

Hintikka, J. (1987). Game-Theoretical Semantics as a Synthesis of Verificationist and Truth-Conditional Meaning Theories. In: E. Lepore (Ed.), *New Directions in Semantics*. New York: Academic Press, pp. 235–58.

Hodges, W. (1983). Elementary Predicate Logic. In: D. Gabbay and F. Guenther (Eds.), *Handbook of Philosophical Logic*, v. 1. Dordrecht: Kluwer, pp. 1–131.

Kripke, S. (1964). Semantical Analysis of Intuitionistic Logic. In: J.N. Crossley and M. Dummett (Eds.), *Formal Systems and Recursive Functions*. Amsterdam: North-Holland, pp. 92–130.

Martin-Löf, P. (1983). On the Meaning of the Logical Constants and the Justification of the Logical Laws. In: C. Bernardi and P. Pagli (Eds.), *Attii Degli Incontri di Logica Matematica*, v. 2. Siena: Università di Siena, pp. 203–81.

IV

GROWTH OF KNOWLEDGE: EXPLANATION AND DISCOVERY

Poznań Studies in the Philosophy of the Sciences and the Humanities
1997, Vol. 51, pp. 225–238

Matti Sintonen

EXPLANATION: THE FIFTH DECADE

1. The Four Decades

In his article "Four Decades of Scientific Explanation" Wesley Salmon (1989) divides the modern history of the notion of scientific explanation into four decades. In the beginning was Hempel and Oppenheim (1948), and then a decade of silence. This silence ended abruptly with Michael Scriven's attack on Hempel's and Oppenheim's model in 1958. The same year, Norwood Russel Hanson published his *Patterns of Discovery* which, together with the works of Stephen Toulmin, expanded the attack beyond explanation to all areas of logical-empirist rational reconstruction. Salmon locates the basic tension if not conflict to be between the formal language philosophers who pertained to the syntactic and semantic aspects of these reconstructions, and the ordinary language philosophers who advocated their abandonment. According to the ordinary language philosophers, the focus should not be in the relationships between the uninterpreted symbols or the relationships between the symbols and the world, but in the triadic relationships between the symbols, the world, the users of the symbols. This is how Salmon (1989) describes the rift:

> Although Hanson was an American by birth and Scriven an Australian, both received their advanced training in England. Ludwig Wittgenstein's *Philosophical Investigations* had been published in 1953, and his influence at Cambridge and Oxford was formidable. During that era a strong opposition developed between the Wittgensteinians, who practiced ordinary language analysis, and the logical empiricists, who might be characterized as artificial language analysts. The Hempel–Oppenheim article is an outstanding example of the use of an artificial formal language for purposes of explicating a fundamental scientific concept. After Hanson and Scriven moved to America, Stephen Toulmin, who was English by birth, was the best known philosopher of science of the ordinary language school remaining in England.

Hanson, Scriven, and Toulmin were the most conspicuous Wittgensteinian opponents of the received view of scientific explanation during the second decade.

I am here interested in this rift, as well as in attempts to cope with it during the later decades. As Salmon viewed the agenda earlier (1984), the field is badly divided into rival schools, with little hope for reconciliation. There are the followers of Hempel who subscribe to the epistemic notion in which the rationale of explanation is located in the power of a putative explanans to make an event (nomically) expected, the modalists who maintain that explanations must demonstrate that their chosen explananda are necessary, and the defenders of the ontic notion who see the power of an explanation in its capacity to show that the explanandum phenomenon fits in the causal nexus of the world.

Interestingly enough Salmon shows compassion to all three models, and he locates their common roots in Aristotle. This I think is right. Nevertheless it seems to me that his account of the four decades misses the most important ecumenical message. He says, namely, that the so-called erotetic view of explanation in which explanations are answers to why-questions is but one variant of the epistemic notion, suggesting that, on this view, answers to why-questions always have the role of making things expected.

A more balanced view, I would suggest, is that explanations sometimes have this role. But by and large the erotetic or interrogative perspective is best viewed as an umbrella notion under which the epistemic, modal and ontic notions of explanation all fit. I shall argue that this is systematically fruitful and historically justified. It is systematically fruitful because it explains why there is a unitary notion – that of a satisfactory answer to a certain form of query – behind the various intuitions of explanation, although satisfactoriness can be cashed out differently in different cases. This view not just restores the angle which somehow got lost during the four decades but also shows that Aristotle was right in thinking that explanations, different kinds of answers to why-questions, fall into distinct types (corresponding, at first approximation, to the four types of why-questions; see Brody 1972).

But this is not the end of the interrogative story. Aristotle fathered both the logic of science and the science of logic. The two were sundered apart in this century, and one spectacular example is precisely the theory of explanation in which the science of logic has failed to do justice to the logic of science. I see Hintikka's interrogative model (I-model, for short) as an attempt at a reconciliation, for it combines erotetic logic with a more broadly conceived interrogative model of knowledge acquisition. What is more, the I-model seems to be flexible in the right place, for its notion of a conclusive answer leaves open the possibility that not all conclusive answers are of a piece. Yet the question-answer relationship guarantees that there is unity behind plurality. Also the rift between the ordinary language philosophers and the formal language analysts is best understood from this perspective: artificial language and formal language analyses are not opposed as first appears. I shall suggest that returning to the roots of this insight opened the fifth decade in the theory of scientific explanation.

2. *Why why-questions?*

But first, where and why was the original interrogative insight lost? It is a striking fact that two of the major explications during the first decade set the stage for the theory of explanation along interrogative lines. The first sentence of Hempel's and Oppenheim's (1948, p. 245) classic paper reads as follows: "To explain the phenomena in the world of our experience, to answer the question "why?" rather than only the question "what" is one of the foremost objectives of empirical science." The second example is Braithwaite (1953, p. 319) who wrote that "Any proper answer to a 'Why?' question may be said to be an explanation of a sort". The contrast Braithwaite draws is not with what-questions but with how-questions, and this I think refers to the logical positivists most of whom maintained that the sciences cannot explain why the world is the way it is, although it is within their competence to describe how phenomena follow other phenomena in regular succession, or how something works.

But why was the idea not put to more systematic use? The interrogative promise in Hempel's and Oppenheim's analysandum was not redeemed in the analysans. The explanatory relationship was construed as a deductive relationship between two sets of sentences, but this relationship did not have the structure of a question-answer relationship. Why not? I have suggested two main reasons for the failure to develop a viable interrogative model of inquiry along these lines (Sintonen 1989). One was the lack of adequate logical tools at that time, the other one is a special difficulty with why-questions. I'll comment on the first one of these difficulties here, and deal with the special difficulty with why-questions in section 3.

So, one possible reason for lack of progress on the interrogative front during the first two decades is that neither erotetic nor epistemic logic were at hand. Consequently, an explicit analysis in question/answer -terms must have appeared formally intractable. Similarly, there were no adequate analytical tools to deal with belief change. This would have been an alternative strategy: one could have modelled growth of (explanatory or other) knowledge in terms of epistemic states quite independently of the questions and answers which might express those states (See Gärdenfors 1980, Sintonen 1984b). In any case, with scant tools to deal with growth of knowledge, that aspect of inquiry was to be bracketed.

This leads us back to rift identified by Salmon, for the natural interrogative alternative to using a logic of questions was to think that what counts as a question worth raising, or an answer worth accepting, is an anthropomorphic issue. As Michael Scriven (1975, p. 4) later put it, statements owe their status as explanations to subjective and psychological factors, i.e., precisely to those ingredients that the positivists shunned. Scriven's proposal was of course that what counts as an explanation – and he didn't see much of a difference between scientific and everyday explanations – depended on the persons involved, as well as on the context at large.

In this Scriven followed, in part, Braithwaite. I shall quote Braithwaite (1953), for he coupled the seeds of a relativist interrogative formulation with two important challenges. One of these is that a questioner who faces a fact requiring explanation does not necessarily know beforehand "what sort of answer will satisfy him". The other one is the idea of "partial or complete intellectual satisfaction":

> What is demanded in a 'Why?' question is intellectual satisfaction of one kind or another, and this can be provided, partially or completely, in different ways. Frequently the questioner does not know beforehand what sort of answer will satisfy him. And what gives partial or complete intellectual satisfaction to one person may give none whatever to a person at a different stage of intellectual development.

Here we meet a fundamental difference of opinion again. If explanations are requests by real people for information phrased in terms of why-questions, we must ask who raises the questions and in whose presence, in what scientific and cultural context, and, most importantly, what exactly is the question. How else could we know what a correct, good, or even admissible answer, that is, explanation, is?

This line of reasoning lead to the various pragmatic models of explanation, with a varying number of variables for persons, contexts, scientific communities, paradigms, etc. (See, e.g., Bromberger 1965 and 1992; see also Achinstein 1983). I think Hempel or any other scientific empiricist, even in the heyday of logical positivism, would have admitted that pragmatic factors effect what questions are raised. However, he would hardly have accepted that the choice of answers is similarly up to the background knowledge of the inquirer, not to speak of the extra-curricular interests of the inquirer. Hempel was a Berlin empiricist, and just like the Viennese positivists he wanted to weed out from science (and philosophy), all traces of psychologism, subjectivism and anthropocentrism. Therefore, relativizing explanations to the knowledge and interest of real or ideal inquirers was simply off limits.

The original beauty of the Hempel–Oppenheim model was precisely in its creation of a new standard of precision by showing that the tools of modern logic can be used to explicate a notion which was dear to them anyway, namely, that to explain the occurrence of an individual event, objectively, is to show that it is a logical consequence from well-established and true laws, and equally well-established and true singular statements. But of course beauty and truth are two different things, and there was a dark cloud in the horizon. Hempel's original interest was with what later became the deductive-nomological model (DN-model) of explanation, the paradigm exemplar of the covering law model. Although he was cautious enough to note that not all scientific explanations conform to the DN-model there proposed, the full articulation of the inductive-probabilistic model had to wait almost 20 years. In any case, it can hardly be doubted that despite this acknowledgment Hempel thought that statistical explanations are as objective as DN-explanations. And once he realized that they have to be relativized to the fleeting knowledge-states of the inquirer, something utterly alien to the metaphysics of the original program crept into the theory of explanation.

3. The Question-Answer Relationship: Answers and Conclusiveness Conditions

The dilemma was that the formal notions did not work, but the informal ones seemed to lead to unacceptably psychological accounts. However, it is clear that relativizing explanations to the knowledge states of the inquirer does not necessarily lead to these unpalatable consequences. One obvious remedy for coping with the dilemma was to invent new formal and semi-formal tools. The question, then, is: Now that we do have a logic of questions, as developed by Hintikka (see also Belnap and Steel 1976), does it to throw light on the notion of explanation?

I must here confine to rehearsing the main ideas and innovations. The attractive idea is to extend the well-behaved logic of propositional yes-no -questions or wh-questions to cover why-questions. It can be coupled with a view of inquiry and knowledge-seeking which develops, around the core of the logic of questions, a comprehensive view of inquiry: a logic of science in the rather traditional sense of the logical empiricists. This sounds paradoxical, because there are no logical empiricists around any more. But the allusion is nevertheless appropriate, for the view of inquiry developed around Hintikka's erotetic logic continues the tradition of logical reconstructions.

The suggestion is that the logical relationships between questions, their presuppositions and answers to these questions can restore the interrogative perspective lost during the early decades. The attraction of the suggestion is in the possibility of modelling why-questions along the logic of wh-questions. Take as an example the question: "Who shot General Governor Bobrikoff?" What the questioner wants to know is the identity of the person who shot Bobrikoff, and she presupposes that Bobrikoff in fact was shot. A direct answer gives information which is syntactically of the right type, i.e., specifies, by help of a name or a definite description, an individual person (or persons, if more than one). Here the correct direct answer is "Eugen Schauman", referring to a student involved in the resistance towards Russian rule over Finland around the turn of the century. Formally, we can represent the presupposition and the desideratum of the question as follows:

(1) $(Ex)KM(x, b)$
(2) $K(Ex)M(x, b)$

where K is the epistemic operator, $M(x,y)$ stands for "x murdered y", and b for Bobrikoff. The difference between the state desired and the presupposition is that in the state desired the questioner knows who the assassin is and not just that someone assassined Bobrikoff. In this view the relationships between the logical forms of the question, its presupposition, and its desideratum is clear. Moreover, approaching knowledge acquisition from the point of view of the conditions under which an answer is satisfactory or conclusive for an inquirer brings logic to bear on Braithwaite's challenge. An answer is conclusive for a questioner if she is after hearing it, in a position to say, truly, "I know what (who, where, etc.) X is". A correct direct answer, such as "The son of the baker" to our question is not conclusive, if the inquirer does not know who the baker is. A direct answer of form (3) requires the conclusiveness condition (4) to entail the desideratum:

(3) $KM(x, b)$
(4) $K(s=x)$

This analysis also gives an operational test for a conclusiveness: a putative candidate is conclusive only if it does not give rise to such important further questions as "But who is Schauman?"

The distinction between the presupposition and the desideratum of a question has an even more important role in the theory of explanation because it helps to distinguish between knowledge and reasoned knowledge in Aristotle's sense. One of the ways in which the previous covering law paradigm fails to illuminate actual inquiry is its concentration on the logical relationship between a sentence, the explanandum, and another sentence (or conjunct), the explanans. This gives an essentially static picture of the "finished product" as the critics of the formal-language account were eager to point out. What we need is an account which mirrors the process of inquiry.

The search for a (singular) explanation is a dynamic process from partial ignorance to partial or complete knowledge. There is, first the epistemic state in which the questioner/inquirer does not suspect that an explanadum phenomenon *E* will or is likely to take place, the state in which she does know that *E* took place but does not know why, and the happy end state in which she knows both that *E* took place and why *E* took place. As I argued (Sintonen 1984a), following and generalizing Gärdenfors (1980), this dynamic sequence of epistemic states is reflected in a sequence of questions and answers. Furthermore, I tried to show that this dynamic point of view explains why the covering law paradigm in the theory of explanation was so hard: the various formal models tried to capture essentially pragmatic constraints on the transition from (partial) ignorance to knowledge by syntactic and semantic means. The syntactic devices which where designed to ban self-explanations are good cases of point: there seemed to be no motivated way to rule out inferences in which the explanandum implied the explanans. The intuitive reason built into the I-model (see Sintonen 1989) was simple: what the inquirer already knows in the intermediate knowledge state cannot get her out from the explanatory dilemma.

In the new epistemic logic developed by Hintikka the role of new knowledge, what fulfilling the desires of the questioner amounts to apart from knowledge of the desideratum, is made more explicit still by the slash-notation of IF-logic. The slash notation allows us to represent the desideratum of the inquirer, the happy end state of the inquiry, in which the news is not knowledge *that* the explanandum is the case *but a demonstration that it is the case which is independent of this*. (See Hintikka and Halonen 1995). What needs to be added to this picture, I suggest, is the third, initial knowledge state in which the inquirer does not suspect that the explanandum *E* will take place because it is unlikely (or impossible, although handling this case would require abondoning information already accepted) given the background theory *T*. This step is needed to account for the way presuppositions of questions arise: not all pieces of information are singled out as initial reasearch questions.

4. *Braithwaite's Challenges*

I now come to the two challenges which Braithwaite presents in the quotation above. The first one was that, in explanatory contexts, "frequently the questioner does not know beforehand what sort of answer will satisfy him". This is not true of all explanatory tasks, e.g., when explaining amounts to finding suitable initial conditions, but it does seem to hold in the many cases in which explanation seeking is not search for evidence to determine which one of a set of well-defined potential answers is the actual one. (See Sintonen 1989.) In salient who-questions knowing who essentially amounts to the ability to pick out an individual person apart from others (to ability to identify this individual in the various possible worlds not ruled out by the rest of the inquirer's knowledge).

Now, sometimes a why-question has the logical structure of a finite which-question and therefore carries the force of a request to the addressee to provide information which is conclusive for singling the right disjunct in the presupposition of the question. But this is not always the case. In fact, in their own erotetic variant Belnap and Steel explicitly ruled out explanation-seeking questions in science and philosophy. Thus someone may puzzle over a why-question and fail to be in the position to delimit a range of candidates. In such cases the interrogative operator ("Why?") ranges over causes or, more generally, reasons, whose identity is contextual in the extreme, or may require conceptual enrichment. Just witness Ignaz Semmelweiss's seemingly simple query over the causes of childbed fever: there were very few unproblematic assumptions which could be used to narrow down candidates.

Such contextually open (if not ill-defined) why-questions point to another possible reason why Hempel and Braithwaite did not develope the interrogative approach further, and hence failed to develop a viable interrogative explicate for the notion of scientific explanation. The reason is of course that the logic of why-questions does not exist, or at least does not match in precision the logic of simple propositional questions and wh-questions.

I pursued this main difficulty about why-questions in (Sintonen 1984b) and later attempted to see how it could be avoided (1989, 1990, 1993), arguing that the most crucial help for a viable interrogative account of explanation comes from some broad pragmatic elements of inquiry. This help comes from two more precise corners. One of them has to do with the products of inquiry, the other one (mainly) with the process. If the above diagnosis of why-questions is correct, their logic is disappointingly thin. My suggestion was that the thin logic could be augmented by a suitably structured notion of a theory. Scientific theories characteristically specify types of potential answers, and lay down further restrictions on what count as plausible or acceptable options, and thus narrow down the semantically open or even ill-defined set of potential answers. In fact the interrogative viewpoint can illustrate both inquiry which is conducted within a reasonably structured theory, call it normal science, and "preparadigmatic" inquiry in which such theory is not available. But even in the latter case inquiry must start from a structured form of curiosity, to use Sylvain Bromberger's apt expression, where structure comes from the inquirer's initial knowledge.

The interrogative proposal also responds to Braithwaite's (and Scriven's) second challenge. Braithwaite suggested that "what gives ... intellectual satisfaction to one person may give none whatever to a person at a different stage of intellectual development".

We can now see why: different people know different things. In fact, Braithwaite's allusion is precisely to different stages in the intellectual development.[1] Since actual acceptance can vary from one person and context to another, it could therefore also be argued that the logic itself does not specify what conclusiveness is. But although erotetic logic admittedly fails to supervise the details of concrete applications it nevertheless gives a principled analysis of what counts as a satisfactory answer. This flexibility is a virtue rather than a vice, and simply draws attention to the interface between semantics and pragmatics. In any case, the game rules of interrogative inquiry and independence-fiendly epistemic logic set a normative standard: knowledge in the more demanding sense requires ability to identify the referent in the various possible worlds (chosen by nature, or the falsifier, in a game-theoretical analysis of knowledge statements, see Hintikka et al.). In the same way explanation is pragmatic, but not subjective, psychologistic, or anthropomorphic in any damaging sense. Although what counts as a satisfactory answer depends on what the inquirer happens to know, and on what questions are available in the scientific and cultural context anyway, pragmatic does not equal psychological.

5. Knowledge-Seeking Games and the Product Process Ambiguity

The second type of help to erotetic logic from the more comprehensive I-model comes from the realization that the dynamics of knowledge acquisition constrains acceptable answers. Epistemic logic helps to address issues of how questions and answers are to be analysed, and of the relationship between a question and its answers. But as such it gives a static snapshot and does not capture the process in which questions and answers are sought.

In Hintikka's I-model this feature is accommodated by looking upon inquiry as a game (against Nature) where an inquirer strives for a cognitive objective by subjecting Nature to an array of questions, and by trying to achieve the objective by help of Her answers and the Inquirer's own background knowledge. A process of querying and answering can be codified in terms of semantical tableaux employed by Beth and Hintikka, with explicit rules governing deductive rules, and further rules governing admissible interrogative moves (cf. Hintikka 1981). Again, the theme allows for variation, but in general items cannot be rushed from the right hand column to the left hand one in a tableau. It is also assumed that before an inquirer can raise a question its presupposition must occur in the left column.

I shall not pause over the details of the I-model. Suffice it to say that as a general "logic of science" it excels in philosophical virtues. It disarms some of the criticism of the "ordinary language philosophers" because it brings the logic of science closer to the way scientists themselves view their work, to actual practice (cf. the quotation from Salmon above). If there is a philosophy of a working scientist it certainly is the idea that inquiry is search for questions and answers. Furthermore, the I-model emphasizes the process in which knowledge is constructed, and therefore does not confine to the context of justification, leaving discovery and actual reasoning to psychologists and historians.

The advantages are easiest to appreciate in the theory of explanation. Here the aim can be that of finding an answer to a big initial or principal question, by subjecting nature to an array of small operational questions. This takes seriously the interrogative idea touched upon but not developed by Hempel and Braithwaite, because now the relationship between the explanandum (the presupposition of the question) and the explanans (the answer) can be made more perspicuous by the new independence friendly logic (IF-logic) which suits even better the representation of the dynamic process or explanation seeking and inquiry at large.

6. The Ontology of Explanation

The interrogative model also throws light on the ontology of explanations. Hempel's original position was that explanations are deductive (later also inductive) arguments for the chosen explanandum. Apart from the explanations which are incomplete for rather trivial pragmatic reasons, there are sketches, blueprints for complete explanations which need some filling in. But of course, on the deductive view the notion of a complete explanation is an either-or-affair: either the explanation is complete or it isn't. In the ideal case they are complete in that they contain all the premises needed for the derivation of the explanation.

This was challenged by Scriven (1962) who argued that explanations are not arguments or inferences but rather singular descriptive statements which "fill in a particular gap in the understanding of the person or people to whom the explanation is directed." That explanations are not arguments which employ general laws is testified, he argued, by the fact that sometimes mentioning of just one singular fact, a cause or the cause, is sufficient to bring about understanding. Scriven was seconded by writers who proposed that explanations in fact are displays of statistically relevant properties (Jeffrey and early Salmon), accounts of causal mechanisms (later Salmon), pairs of speech-act types and propositions (Achinstein), ideal explanatory texts (Railton), etc.

Scriven also criticised the ideal of completeness and the alleged conflation between a statement and the grounds for accepting a statement. Just as we distinguish between the content of a statement from the reasons for making or accepting it, we should separate the content of an explanatory statement from the reasons or grounds of accepting it as an explanation. In Scriven's view the explanations-as-inferences view is tied up with yet another chimera, that of a finally or completely justified explanation. There are, he thinks, no such animals: Justifications are searched as defenses against specific doubts. Although there characteristically is no "purely formal", that is, context-free way to bring a chain of justification to an end, there are, in any actual explanatory context, facts that are taken to be beyond reasonable doubt.

But of course explanations are literally answers, and separating them from their conclusiveness conditions helps to sort out these difficulties. Although the notion of a conclusive answer does not attempt to give a transcendental argument for the size and shape of all possible explanatory answers it nevertheless reinstates the original insight, and provides answers to Braithwaite's challenges. How much of the complete answer or ideal story is to be specified in the answer and what can be relegated to that part of the background knowledge which specifies the conclusiveness condition is a contextual matter. This shows that there is a clear sense in which the completeness of an answers vis-à-vis a questioner's question is a matter of degree.

This also gives a more precise account of the difference between what Peter Railton (1981) has called ideal explanatory texts and explanatory information. Railton's idea is that actual explanations never specify the full causal or statistically relevant items for the occurrence of a singular phenomenon, nor the details of the mechanisms of processes that produce a phenomenon. But although the notion of a complete story is an idealization it nevertheless serves the useful purpose of guiding the search of explanatory information: what matters is not the actual possession of a detailed story but the ability to produce arbitrary passages from it. Science is a social affair and is based on the possibility of cooperation. What interests one scientists is mere background knowledge for another, and therefore need not be brought to the forefront. Nevertheless the understanding is that there is such an ideal story whose varying passages may be chosen for a more detailed scrutiny.

In a review article on scientific explanation Noretta Koertge (1992, p. 91) notes a difficulty about Railton's account. Once we admit that actual explanation leave out portions of ideal texts, we face another question: "How much of Railton's ideal explanatory text need not be cited". Similarly, if explanation works by embedding events and processes in a causal nexus, "how much (or which parts) of the causal nexus is required?"

Koertge rightly notes that one should not dump too much of the interesting issues into the pragmatic waste-basket (formerly known as the psychological waste-basket). And she writes: "Salmon and Railton claim this as a pragmatic question about salience which a theory of scientific explanation need not answer. But although individual interests undoubtedly guide the direction which our inquiry takes, I think it unwise to suggest that they also dictate what counts as an adequate (though less than ideal) explanatory answer."

Can anything be said about these pragmatic aspects? Koertge notes (pp. 96–7) that the interrogative proposal makes progress in the right direction by taking seriously "the cliché" that explanations are answers to why-questions. And in fact the logical structure of questions does tell us something interesting about "explanation-providing answers and the modes of searching for them." She refers to the suggestion (in Sintonen 1989) that research programs or theories constrain candidates for admissible answers: "Requests for explanation arise within a knowledge system and are often prompted by objective inadequacies of the system, such as inconsistencies or gaps."

Since the interrogative approach contains the assumption that search for explanation initiates with the inquirers theoretical premise T, the vocabulary and presuppositions from T restrict possible questions and answers. It should be noted, also, that this way of looking at the search of explanations comes close to the explanation-by-specification view of Wiśniewski and Kuipers (1994). They suggest that intentional explanation (explanation of action), functional explanation (of biologial traits) and types of causal explanation have a common structure in which, however, meaning postulates derived from specific types of inquiry impose constraints on the soundness of questions and hence on possible answers. The interest of this notion of explanation is that it acknowledges an explicit logic for the way in which more specific research questions arise from the more general initial questions. The closer affinities between this and the I-model remains to be explored, however.

Department of Philosophy
P.O. Box 24
00014 University of Helsinki
Finland

NOTES

1 Braithwaite's example of intellectual development is that of a child, but Moritz Schlick (1974) had detected the phenomenon in adults: "The untutored person is scarcely conscious of this distinction between "erudition" (Wissen) and knowledge (Erkenntnis). His mind is set at rest as soon as some name or other is assigned to each thing or phenomenon. How clever the gardner fancies himself because he can tell us the Latin names of all his plants! How often do we hear people priding themselves on their storehouse of names, phrases and numbers, which they would pass off as knowledge."

REFERENCES

Achinstein, Peter (1983). *The Nature of Explanation*. New York and Oxford: Oxford University Press.

Belnap, Nuel D. and Steel, Thomas B. Jr. (1976). *The Logic of Questions and Answers*. New Haven and London: Yale University Press.

Braithwaite, R.B. (1953). *Scientific Explanation. A Study of the Function of Theory, Probability and Law in Science*. Cambridge: Cambridge University Press.

Brody, B.A. (1972). Towards an Aristotelean Theory of Scientific Explanation. *Philosophy of Science*, Vol. **39**, 20–31.

Bromberger, Sylvain (1965). An Approach to Explanation. In R. J. Butler (Ed.), *Analytical Philosophy*, 2nd Series. Oxford: Basil Blackwell, pp. 72–103.

Bromberger, Sylvain (1992). *On What We Know We Don't Know: Explanation, Theory, Linguistics, and How Questions Shape Them*. Chicago: University of Chicago Press.

Gärdenfors, Peter (1980). A Pragmatic Approach to Explanation. *Philosophy of Science* **47**, 405–23.

Hempel, Carl and Oppenheim, Paul (1948). Studies in the Logic of Explanation. Reprinted in C. Hempel, *Aspects of Scientific Explanation and Other Essays in the Philosophy of Science*. New York: The Free Press, 1965, pp. 245–95.

Hintikka, Jaakko (1981). On the Logic of an Interrogative Model of Scientific Inquiry. *Synthese* **47**, 69–83.

Hintikka, J. and Halonen I. (1995). Semantics and Pragmatics for Why-Questions. *Journal of Philosophy*, Vol. **XCII**, No 12, 636–57.

Hintikka, J., Halonen, I. and Mutanen, A. Interrogative Logic as a General Theory of Reasoning. (forthcoming)

Koertge, Noretta (1992). Explanation and its Problems. *British Journal for the Philosophy of Science* **43**, 85–98.

Kuipers, Theo and Wiśniewski, Andrzej (1994). An Erotetic Approach to Explanation by Spefication. *Erkenntnis* **40**, 377–402.

Moravcsik, Julius M.E. (1974). Aristotle on Adequate Explanations. *Synthese* **28**, 3–17.

Railton, Peter (1981). Probability, Explanation and Information. *Synthese* **48**, 201–23.

Salmon, Wesley (1984). *Scientific Explanation and the Causal Structure of the World*. Princeton, New Jersey: Princeton University Press.

Salmon, Wesley (1989). Four Decades of Scientific Explanation. In P. Kitcher and W.C. Salmon (Eds.), *Scientific Explanation*. Minnesota Studies in the Philosophy of Science, Vol. XIII. Minneapolis: University of Minnesota Press, pp. 253–82.

Schlick, Moritz (1974). *General Theory of Knowledge*. Second edition. Translated by Albert E. Blumberg. Wien, New York: Springer-Verlag. (Originally published in 1925 as *Allgemeine Erkenntnislehre*.)

Scriven, Michael (1962). Explanations, Predictions, and Laws. In. H. Feigl and G. Maxwell (Eds.), *Minnesota Studies in the Philosophy of Science*, Vol. III. Minneapolis: University of Minnesota Press, pp. 170–230.

Scriven, Michael (1975). Causation as Explanation. *Noûs* **9**, 3–15

Sintonen, Matti (1984a). *The Pragmatics of Scientific Explanation. Acta Philosophica Fennica*, Vol. **37**. Helsinki: Societas Philosophica Fennica.

Sintonen, Matti (1984b). On the Logic of Why-Questions. *PSA 1984*, Vol. One. Edited by P.D. Asquith and P. Kitcher. East Lansing, Michigan: The Philosophy of Science Association, pp. 168–76.

Sintonen, Matti (1989). Explanation: In Search of the Rationale. In P. Kitcher and W.C. Salmon (Eds.), *Scientific Explanation*. Minnesota Studies in the Philosophy of Science, Vol. XIII. Minneapolis: University of Minnesota Press, pp. 253–82.

Sintonen, Matti (1990). How to Put Questions to Nature. In D. Knowles (Ed.), *Explanation and Its Limits*. Cambridge: Cambridge University Press.

Sintonen, Matti (1993). In Search of Explanations: From Why-Questions to Shakespearean Questions. *Philosophica* **51** (1), 55–81.

Poznań Studies in the Philosophy of the Sciences and the Humanities
1997, Vol. 51, pp. 239–259

Erik Weber*

SCIENTIFIC EXPLANATION AND THE INTERROGATIVE MODEL OF INQUIRY

1. Introduction

The aim of this article is to analyze the nature of the construction process of scientific explanations. It is obvious that, to reach this aim, we need a definition of what scientific explanations are. Therefore, my article starts with a short presentation of the concept of explanation that will be used (section 2). In section 3 I apply Jaakko Hintikka's interrogative model of inquiry to the construction process of explanations, starting from the work of Matti Sintonen on this topic. I believe (and will show) that the I-model is a useful tool for analyzing the construction process of explanations: the search for scientific explanations is a type of inquiry which can be nicely codified by means of the I-model.

In several articles, Hintikka has pointed out the difference between the definitory rules of a game and the strategic rules of a game (e.g. J. Hintikka 1989, pp. 5–7). The definitory rules of a game lay down the aim of the game and the moves that are allowed. The strategic rules tell us which moves are good and which are bad. Knowing the definitory rules of chess is not sufficient for being a good chessplayer: even a mediocre chessplayer knows some strategic rules which help him to make the right moves. Likewise, knowing the rules of inference of a logical system is not sufficient for being good at producing logical proofs: we need strategic rules which help us to make good moves. In order to assess the fruitfulness and limitations of the I-model when applied to the construction process of explanations, a similar distinction must be made. On the one hand, applying the I-model results in a clear codification of the aims of the explanation seeking process and of the types of moves that occur in this process. This will be shown in section 3. On the other hand, an analysis of the construction process of explanations is incomplete until we have provided strategic rules (or strategic knowledge of another kind) which helps the inquirer to make the right moves. This strategic knowledge is not provided by the I-model. In sections 4 and 5 of this article, I will develop a method for constructing scientific explanations which is based on the I-model but also takes into account the various forms of strategic knowledge we use when we construct scientific explanations.

2. What are explanations?

Because Hempel's deductive-nomological model of explanation is very well known, I will use this model as a starting-point for developing the concept of explanation that I will use in this article. According to Hempel we have to distinguish between true explanations and explanations that are well-confirmed (see e.g. C. Hempel 1965, p. 338). Both concepts are defined by means of the auxiliary concept of *potential explanans*, which is characterized as follows (cf. C. Hempel & P. Oppenheim 1948, part 3; readers who are well acquainted with Hempel's DN model will notice that I make some simplifications):

(H) The ordered couple (L, C) constitutes a potential explanans for the singular sentence E if and only if
    (1) L is a purely universal sentence and C is a singular sentence,
    (2) E is deductively derivable from the conjunction $L\&C$, and
    (3) E is not deductively derivable from C alone.

A purely universal sentence consists of one or more universal quantifiers, followed by an expression which contains no quantifiers and no individual constants. (L, C) is a true explanans for E if and only if (L, C) is a potential explanans for E and both L and C are true. (L, C) is a well-confirmed explanans for E if and only if (L, C) is a potential explanans for E and both L and C are well-confirmed.

Let's consider an example (inspired by an example in C. Hempel & P. Oppenheim 1948, p. 246). A mercury thermometer is rapidly immersed in hot water. We observe a temporary drop of the mercury column, followed by a swift rise. According to definition (H), the following construction is a potential explanans for this phenomenon:

C_1: This thermometer consists of a glass tube which is partly filled with mercury.
C_2: This thermometer was rapidly immersed in hot water.

L: In glass tubes which are partly filled with mercury and are rapidly immersed in hot water, the mercury level first drops and then rises.

This example illustrates one of the major shortcomings of Hempel's account: though the construction satisfies the conditions in (H), it does not solve the problem it is supposed to solve. Even if we accept that C_1, C_2 and *L* are true, we do not yet understand why the explanandum *E* happened. A potential explanans as defined in (H) (henceforth called potential deductive-nomological explanans or potential DN explanans) does not have explanatory power in itself: it acquires explanatory power if it is accompanied by a derivation of the purely universal statement *L* from a general scientific theory or law. In our example, we must add a derivation of *L* which starts from three general laws: the law of thermal expansion of fluids, the law of thermal expansion of solids and the law of heat conduction. The law of thermal expansion of fluids is:

$$v = v_o + v_o\beta t$$

(v: final volume; v_o: initial volume, β: coefficient of expansion of the fluid considered; t: rise (+) or fall (−) of the temperature). The law of thermal expansion of solids is:

$$l = l_o + l_o\alpha t$$

(l: final length; l_o: initial length; α: coefficient of linear expansion of the solid considered). Finally, the law of heat conduction says that the quantity of heat dQ (joules) which, in a time period dt, flows between any two plane surfaces normal to the direction of heat flow, is given by the equation

$$dQ = -KA(\delta\theta/\delta n)dt$$

(K: thermal conductivity coefficient (joule/°C.sec.m); A: area the surfaces (m^2); $\delta\theta/\delta n$: temperature gradient (°C/m)). The law "In glass tubes which are partly filled with mercury and are rapidly immersed in hot water, the mercury level first drops and then rises" can be derived from these three laws. The derivation is as follows. The law of thermal expansion of fluids gives us the equation

$$v = v_o + v_o \times 0.00018 \times t_m$$

(v: final volume of the mercury; v_o: initial volume of the mercury; 0.00018:

coefficient of expansion of mercury; t_m: rise of the temperature of the mercury). On the other hand, the tube of the thermometer will expand in accordance to the following equation, which is an implementation of the law of thermal expansion of solids:

$$r = r_o + r_o \times 0.000008 \times t_g$$

(r: final radius of the tube; r_o: initial radius of the tube; 0.000008: coefficient of linear expansion of glass; t_g: rise of the temperature of the tube). Glass is a bad thermal conductor: its thermal conductivity is small. As a consequence, in a first phase, only the temperature of the tube will rise: $t_m=0$ and thus $v=v_o$. The relation between the level of the mercury and its volume is given by the geometrical formula

$$h = v/\pi r^2$$

(h: height of the mercury column; v: volume of the mercury; r: radius of the tube). From this formula and $v=v_o$ we derive that

$$h = v_o/\pi r^2$$

The initial mercury level is given by

$$h_o = v_o/\pi r_o^2.$$

Because the temperature of the tube rises ($t_g>0$), r is larger than r_o. The two last equations and $r>r_o$ imply that

$$h < h_o$$

In other words: the mercury level drops in a first phase. But as a result of heat conduction, the temperature of the mercury will start to rise after a few instants. The mercury expands (law of thermal expansion of fluids). Because the coefficient of expansion of mercury is much larger than the coefficient of cubic expansion of glass (approximately 3 times the linear coefficient, i.e. 0.000024) the mercury level will rise.

From now on, I will use the term "explanation" to refer to a combination of a potential DN explanans and a derivation, as illustrated in the example. The potential DN explanans has the following general form (the explanandum is "Object *a* has property β"):

C_1: $\alpha_1 a$
C_2: $\alpha_2 a$
⋮
C_n: $\alpha_n a$
L: $(\forall x)[(\alpha_1 x \& \alpha_2 x \& \ldots \& \alpha_n x) \supset \beta x]$

The derivation must show that L is derivable from a set of general laws and theories.

To conclude this section, some remarks must be made concerning my use of the term "explanation". It is widely agreed upon that one of the functions of scientific knowledge is to provide the theories and laws we need in order to understand the world. But there is much controversy about what understanding of the world consists in. I adhere to the view defended by Wesley Salmon (1993, p. 15). He claims that there are at least two irreducible kinds of understanding humans search for: on the one hand there is the desire to fit all the phenomena into a unified world picture (epistemic aspect of understanding); on the other hand there is the desire to know how things in the world work, i.e. the desire to know the mechanisms that produce the phenomena we observe (causal aspect of understanding). In general, explanations are the instruments by means of which understanding is obtained. For each aspect of understanding, we may characterize the type of scientific explanation which is the appropriate instrument for increasing understanding in that sense, and then try to provide insight into the ways in which explanations of this type are constructed on the basis of scientific knowledge. The concept of explanation which I use here is meant to be a characterization of the instruments by means of which understanding in the sense of unification is reached. It is therefore on a line with the unification account of explanation developed by Philip Kitcher (1989). The concept I use here is not adequate for discussing the causal aspect of understanding; there we need other (causal) concepts of explanation, as e.g. developed in W. Salmon (1984).

3. The I-model of inquiry applied to the construction process of explanations

3.1 An individual who wants to construct an explanation for βa is confronted with two tasks, viz. (i) deriving a statement of the form $(\forall x)[(\alpha_1 x \& \alpha_2 x \& \ldots \& \alpha_n x) \supset \beta x]$, and (ii) establishing that a possesses the characteristics $\alpha_1,\ldots,\alpha_n$. Showing that a possesses the characteristics $\alpha_1,\ldots,\alpha_n$ sometimes requires new observations or experiments. So the two types of moves an inquirer makes in the I-model (asking questions to Nature and making deductions) can be found in the construction process of scientific explanations. This means that it makes sense to apply to interrogative model of inquiry to the construction process of explanations. In two recent articles, Matti Sintonen has done this (M. Sintonen 1990 and 1993). Sintonen summarizes the result of applying the I-model to the construction process as follows:

> More specifically, applied to the notion of explanation the model has it that explanation is a process in which an inquirer *I* who does know that *E* but does not know why *E* tries to find suitable pieces of information $A_1, \ldots, A_i$ which, together with *I*'s background knowledge *T*, entail *E*. In the tableau construction this process is codified as follows. The inquirer's background knowledge, including the relevant theory *T*, is entered on the left hand column, and the explanandum *E* on the right hand column. By help of the interrogative moves the inquirer obtains whatever auxiliary facts and generalizations $A_1, \ldots, A_i$ are needed to close the tableau. Now closing the tableau shows that *T* and $A_1, \ldots, A_i$ entail *E*, a feat which parallels deductive entailment in the D-N model of explanation (M. Sintonen 1993, p. 60).

Because it makes sense to apply the I-model to explanation, my first proposal for a method for constructing explanations is based on Sintonen's account. More specifically, as a first approximation I would suggest that explanations are to be constructed by means of the following procedure:

(P_1) (1) Put the explanandum sentence (*E*) in the right hand column (the one containing the false statements).
(2) Choose a scientific theory or law *T* (or a set of laws and theories) by means of which you want to explain *E*, and put it in the left hand column.
(3) Put all background knowledge (knowledge about singular facts, mathematical or logical theorems and inference rules) which you consider relevant in the left hand column.
(4) Ask questions to Nature and put its answers (obtained by observation or experiment) in the left hand column.
(5) Try to close the tableau by applying the tableau construction rules.
(6) If it is impossible to close the tableau, ask more questions to Nature or add elements from your background knowledge.

In the terminology of the I-model, instruction 4 leads to interrogative moves, while instruction 5 leads to deductive moves. In procedure (P_1), two ideas are added to Sintonen's account in order to obtain a more realistic picture. Firstly, a clear distinction is made between *T* (the element(s) of our scientific knowledge to be used for constructing the explanation) and other background knowledge. This other background knowledge contains a mathematical part (henceforth written as B_M) and a part which contains information about singular facts (B_S). I think this distinction is necessary because an inquirer who constructs an explanation first chooses his main tool (T) and then selects parts of his background knowledge which seem to be relevant for applying T. Secondly, (P_1) admits that the inquirer adjusts the content of B_M and B_S (instruction 6). These adjustments must be permitted because it is possible that in the course of the construction process the inquirer ascertains that some part of his background knowledge which he considered irrelevant at the beginning (and therefore is not contained in B_M or B_S), is needed to construct the explanation.

3.2 I will give two examples to illustrate how procedure (P_1) works. In the first example, we consider an inquirer *I* who observes that pendulum *a* has a period in the interval [2.005 *s*, 2.015 *s*[. *I* wants an explanation of this fact and uses procedure (P_1). Then his first move will be:

(1) *I* puts the explanandum, "*a* has a period of 2.01 *s*", in the right hand column.

The predicate "has a period of 2.01 *s*" is used here as an abbreviation for "has a period in the interval [2.005*s*,2.015*s*]". In the same way, "has a period of 2.1 sec" means "has a period in the interval [2.05*s*,2.15*s*]", "has a length of 1.00 *m*" means "has a length in the interval [0.995*m*,1.005*m*]", etc. In his second move, *I* has to choose a general theory or law which will function as the main instrument for constructing the explanation. The most obvious choice is the law of the pendulum, which relates the period (P) of a pendulum to its length (L). If *I* chooses this law, the construction process continues as follows:

(2) *I* puts the law of the pendulum, $P=2\pi\sqrt{L/g}$, in the left hand column (as T).
(3) *I* puts "*a* is a pendulum" in the left hand column (as background knowledge concerning singular facts, B_S). He also includes some arithmetical theorems and/or inference rules (mathematical part of the background knowledge, B_M).
(4a) *I* asks the following question to Nature:
Q_1: What is the length of *a*?

(4b) I puts the answer of Nature in the left hand column:
A_1: a has a length of 1.000 m.

(5) I derives the law "All pendulums that are 1.000 m long have a period of 2.01 s" from T and B_M. Then he derives the proposition $A_1\&B_S \supset E$ from this law (universal instantiation). Finally, he closes the tableau by means of the propositions $A_1\&B_S \supset E$, A_1 and B_S (which are in the left hand column) and E (in the right hand column).

Closing the tableau shows that E is deductively derivable from T plus the background knowledge and the answers given by Nature. Because a specific law is derived in order to close the tableau (step 5), the tableau contains all the elements of the following potential DN explanans:

C_1: a is a pendulum.
C_2: a has a length of 1.000 m.
L: All pendulums that are 1.000 m long have a period of 2.01 s.

The tableau also contains a derivation of L from T and B_M, so we have an explanation (in the sense defined in section 2) of E.

In our second example we consider an inquirer I who wants to construct an explanation for the explanandum of the thermometer example in section 2. If I applies procedure (P_1), the following construction process may result:

(1) I puts the explanandum, "In this object the mercury column first drops and then rises", in the right hand column.

(2) I puts the law of thermal expansion of fluids, the law of thermal expansion of solids and the law of heat conduction in the left hand column (as T).

(3a) I puts "This object was rapidly immersed in hot water" in the left hand column (as B_S).

(3b) I puts some arithmetical theorems or inference rules in the left hand column (as B_M).

(4a) I asks the following questions to Nature:
Q_1: What is the structure of the thermometer?
Q_2: What is the coefficient of expansion of mercury?
Q_3: What is the coefficient of linear expansion of glass?
Q_4: Is glass a good conductor of heat?

(4b) I puts the answers of Nature in the left hand column:
A_1: This thermometer consists of a glass tube which is partly filled with mercury.

A_2: 0.00018 A_3: 0.000008 A_4: No

(5) I derives the law L ("In glass tubes which are partly filled with mercury and are rapidly immersed in hot water, the mercury level first drops and then rises") from T, B_M and the answers of Nature to the questions 2−4. Then he derives the proposition $A_1\&B_S\supset E$ from this law. Finally, he closes the tableau by means of the propositions $A_1\&B_S\supset E$, A_1 and B_S (which are in the left hand column) and E (in the right hand column).

Because the tableau can be closed, E is deductively derivable from T, B and the answers of Nature. The tableau will contain all the elements of the potential DN explanans given in section 2, and a derivation of L.

3.3 (P_1) is not a satisfactory account of the construction process of explanations. There are two problems. The first problem is caused by the fact that the tableau construction rules only allow us to add propositions to the tableau that are simpler than the ones we already have. For instance, if we have $p\&q$ in the left hand column, we may add both p and q in this column. However, adding $p\&q$ if the tableau already contains p and q on different lines of the left hand column, is not allowed. Analogously, we can eliminate universal quantifiers, but not introduce them: we can write Pa in the left hand column if we have $(\forall x)Px$ in the same column, but not conversely. This means that (contrary to what I have tacitly assumed in my examples) the covering law L we need in an explanation cannot be obtained within the tableau we are constructing. There is an obvious way out of this problem. Instead of regarding the construction process of explanations as the construction of one tableau, we regard it as a process in which a number of interrelated tableaus are constructed. Besides one central tableau, which is constructed as prescribed in (P_1), there may be several ancillary tableaus. We allow that, besides answers of Nature, conclusions of ancillary tableaus are added to the left hand column of the central tableau. This will e.g. allow us to add the laws in our examples to the central tableau: we start an ancillary tableau with T and B_M in the left hand column and L in the right hand column. By closing this tableau we show that L is derivable from T and B_M. Because we have shown this, we may add L to the left hand column of the central tableau.

An inquirer who uses procedure (P_1) has to choose questions to be ask to Nature and (in the modified version outlined in the preceding paragraph) useful derivation goals to be pursued in ancillary tableaus. The outcome of the procedure (whether or not an explanation is obtained and which one) depends largely on the choices the inquirer makes. So a complete account of the construction process of explanations has to provide strategic rules by means of which the inquirer can select relevant questions to be asked to Nature and useful derivation goals. (P_1) does not contain such rules, and this is the second problem.

In the sections 4 and 5 I will develop a method for constructing explanations which solves the second problem and also incorporates the solution to the first problem outlined above. The basic principles of this method are:

(1) When an inquirer wants to construct an explanation, he has to engage in two processes: an implementation process and a tableau construction process.
(2) The implementation process (which will be discussed in section 4) results in questions to be asked to Nature and in derivation goals.
(3) In the tableau construction process (which will be discussed in section 5) we construct tableaus in order to reach the derivation goals that have arisen during the implementation process, and in order to verify whether we really have found an explanation.

4. The implementation process

4.1 The first step of the implementation process consists in writing down the *general scheme of potential DN explanations*. As noticed in section 2.1, the general scheme of a potential DN explanans for βa is:

$$\begin{array}{ll} C_1: & \alpha_1 a \\ C_2: & \alpha_2 a \\ \vdots & \\ C_n: & \alpha_n a \\ L: & (\forall x)[(\alpha_1 x \& \alpha_2 x \& \ldots \& \alpha_n x) \supset \beta x] \end{array}$$

Descriptions of implementation processes will be called *implementation records*. An implementation record is series of numbered entries, each one describing a step in the implementation process. The general scheme constitutes the first entry of every implementation record. The last entry is always a potential DN explanans. The entries in between belong to three groups. Firstly, there are entries that state decisions made by the inquirer. Typical decisions to be made by the inquirer are the choice of T and the elements of B, and the choice of questions to be asked to Nature. Secondly, there are entries that state restrictions on the admissible instances of the scheme. We distinguish between weak and strong restrictions. A weak restriction limits the predicates we are allowed to use to replace a schematic letter of the scheme, but leaves at least two possibilities; a strong restriction excludes all possibilities but one. If an inquirer actually replaces a schematic letter with its unique instantiation prescribed by a strong restriction, we say that he has *materialised* this strong restriction. If we materialize a strong restriction, we either obtain a potential DN explanans or a schematic DN explanans (every partial implementation of the general scheme of DN explanations, in which some of the schematic letters are replaced with predicates but others are not, will be called *schematic* DN explanans). The entries of the third group are those describing a move in which a strong restriction is materialized.

The implementation process is guided by the following procedure:

(P_2) (1) Write down the general scheme of potential DN explanations.
(2) In the general scheme, replace β with the property attributed to a in the explanandum E.
(3) Choose a scientific entity T by means of which you want to explain E.
(4) Determine the content of B.
(5) State all weak and strong restrictions that have been obtained.
(6) Try to create new restrictions out of existing ones.
(7) Materialise all strong restrictions that you have. If this results in a potential DN explanans, stop; otherwise go to 8.
(8) Choose one of the following options:
(a) Formulate a question to be asked to Nature, and try to find an answer to the question by means of observation or experiment
(b) Formulate a question that expresses a decision problem and make a decision.
(c) Adjust the content of B.
Go back to instruction 5.

The first instruction results in a schematic DN explanans which materialises the strong restriction imposed by the explanandum. When we create a new restriction by combining existing ones (instruction 6), this is called a *combinative* move. The most important function of combinative moves is to create strong restrictions by combining previously obtained weak ones. The moves in which an inquirer materialises strong restrictions (instruction 7) will be called *constructive* moves.

4.2 To illustrate how (P_2) works, we reconsider the inquirer who observes that pendulum a has a period in the interval $[2.005s, 2.015s[$ (cf. the first example in section 3.2). If the inquirer uses (P_2), his first move will be to write down the general scheme of potential DN explanations. His second move will be to replace β in this scheme with "has a period in the interval $[2.005s, 2.015s[$". This move results in the following schematic DN explanans:

$$\alpha_1 a$$
$$\alpha_2 a$$
$$\vdots$$
$$\alpha_n a$$
$$(\forall x)[(\alpha_1 x \& \alpha_2 x \& \ldots \& \alpha_n) \supset P_{2.01} x]$$

$P_{2.01}$ is the formal representation of "... has a period in the interval $[2.005s, 2.015s[$". This schematic DN explanans is the second entry of the implementation record.

We assume that I makes the following decisions[1]:

(3) I decides that T is the law of the pendulum, $P = 2\pi\sqrt{L/g}$.
(4) I decides that the content of B_S is "a is a pendulum".

We do not need to know the exact content of B_M in order to construct our example; therefore it will be left unspecified. The fact that I has chosen the law of the pendulum has three consequences: (i) one of the singular sentences of the explanation must state that a is a pendulum, (ii) a second singular sentence must give information about the length of a, and (iii) no other singular sentences stating antecedent conditions are needed. If the inquirer translates these consequences into explicit restrictions on the schematic explanation obtained in move 2, the result is:
(a) the explanation will contain only two singular sentences ($\alpha_1 a$ and $\alpha_2 a$),
(b) α_1 must be replaced with "is a pendulum", and
(c) α_1 must be replaced with a region of the modality "length".
Regions are properties which are defined by associating an interval with a modality. Examples of modalities are length, age, weight, temperature, etc. A region can be regarded as an ordered couple of a modality and an interval. Examples of regions are "has an age between 35 and 50 years" and "has a length between 1 and 2 meters".

In his fifth move, the inquirer will state the restrictions (a)−(c) above. Since no combinative moves can be made, the implementation process will go on as follows:

(7) The strong restrictions (a) and (b) are materialised by writing down the following schematic DN explanans:
Sa
$\alpha_2 a$
$(\forall x)[(Sx \& \alpha_2 a) \supset P_{2.01}x]$

S is the formal representation of "is a pendulum". In order to determine what his next move will be, I must choose between three options. Because he still has weak restriction (c) at hand, a reasonable thing for I to do is to try to create a weak restriction which in combination with (c) entails a strong restriction. He can obtain such restriction by choosing a partition of the modality "length" and trying to determine to which element of this partition the object a belongs (a partition of a modality is a set of mutually exclusive and jointly exhaustive regions of this modality). So a rational way to continue the construction process is:

(8a) I formulates a question which expresses a decision problem:
Q_1: Which partition of "length" shall I use to construct an explanation of $P_{2.01}a$?
(8b) I decides to use the partition $F_1,\ldots,F_m$.

As the result of step 8 is not a potential DN explanans, I goes back to instruction 5. No combinative or constructive moves can be made, so we get back to instruction 8 almost immediately:

(5′) α_2 must be replaced with the element of the partition $F_1,\ldots,F_m$ to which a belongs.
(8′) I formulates a question to be asked to Nature:
Q_2: To which element of the partition $F_1,\ldots,F_m$ does a belong?

When I has obtained an answer to his question, he must again try to make a combinative move. Let's suppose that Nature's answer to Q_2 is "a has at t a length in the interval $[0.9995m, 1.0005m]$)". On this condition, the construction process goes on as follows:[1]

(5″) The element of the partition $F_1,\ldots,F_m$ to which a belongs is $[0.9995\,m, 1.0005\,m]$.
(6) I ascertains that the restrictions in 5′ and 5″ entail a strong restriction: α_2 must be replaced with "... has a length in the interval $[0.9995m, 1.0005m]$".

(7′) I materialises this restriction and obtains the potential DN explanans:
Sa
$L_{1.000}a$
$(\forall x)[(Sx \& L_{1.000}a) \supset P_{2.01}x]$

$L_{1.000}$ is the formal representation of "... has a length in the interval [0.9995m, 1.0005m]".

4.3 The implementation process directed by procedure (P_2) results in a potential DN explanans. The law in this potential DN explanans constitutes a derivation goal: in order to obtain an acceptable explanation with unifying effect, we have to show that this law is derivable from T and B. So procedure (P_2) results in a derivation goal which we can pursue in the tableau we are constructing (the second tyre of the construction process of the explanation). I will come back to this in section 5. Now I concentrate on another claim made in section 3.3, viz. that an implementation process guided by (P_2) tells us which questions we have to ask to Nature. If an inquirer uses (P_2), questions to Nature result from decisions he makes and from combinative and constructive moves. More specifically, the restrictions and schematic DN explanations he obtains have a heuristic value: they suggest questions to be asked to Nature and questions that express decision problems. In order to exploit the heuristic value of the restrictions and schematic DN explanations, we use *strategic background information*. This background information is situated on a meta-level: it tells us how scientific knowledge has to be applied. It must not be confused with the background knowledge B. The content of the strategic background information varies with the type of explanation we are constructing. Two large categories must be distinguished in this respect: explanations that are based on *one* phenomenological law or theory, and explanations that are based on *two or more* scientific entities. The pendulum example belongs to the first type, the thermometer example to the second kind. The second category will be discussed in section 5. I now confine myself to explanations based on one scientific entity. If, like in our example, this entity is a functional law relating simultaneous properties of a class of objects, we will call the explanation a *coexistence explanation*. From our example in 4.2 we can derive six items of strategic background information which we can use in combination with (P_2) if we have taken one functional coexistence law as T. The items are:

(a) If an inquirer who wants an explanation for "a has characteristic G" decides to use a functional coexistence law T with domain D, this decision implies that the explanation will contain only two singular sentences, $\alpha_1 a$ and $\alpha_2 a$. It also implies that α_1 must be replaced with the predicate D, and that α_2 must be a combination of regions of the explanatory modalities of T. The *explanatory modalities* of a law are the modalities of which the property that is to be explained is *not* a region. In our example the explanatory modality of $P=2\pi\sqrt{L/g}$ was "length". These three restrictions are typical for coexistence explanations.

(b) The first and second restrictions can always be materialised. So we can always obtain a schematic DN explanans of the following form:

Da

$\alpha_2 a$

$(\forall x)[(Dx \& \alpha_2 x) \supset Gx)$,

where D is the domain of the law we have chosen.

(c) The third restriction is weak and has a heuristic value: it is rational that the inquirer looks for an additional restriction which together with the third one entails a strong restriction. So the third restriction provokes the question "Which partition(s) of the explanatory modality or modalities of the law T will I use to construct an explanation for Ga?". This question expresses a decision problem.

(d) When the inquirer has made a decision in response to this question, a second question (to be asked to Nature) arises: "To which element(s) of the selected partition(s) does a belong?".

(e) Together with the restriction entailed by the answer to the first question, the answer to this second question always entails a strong restriction: α_2 must be replaced with F (I assume that the answer of Nature to the second question is "a belongs to F", where F is a conjunction of elements of the selected partitions). By materialising this restriction we obtain a potential DN explanans which has the following form:

Da

Fa

$(\forall x)[(Dx \& Fx) \supset Gx]$.

5. The tableau construction process

5.1 The tableau construction process has three functions. Firstly, it is the process in which the main derivation goal that arises at the end of the implementation process (the universal statement L) must be reached. The second function relates to a restriction on questions we are allowed to ask to Nature. In Hintikka's I-model, only questions of which the presupposition occurs in the left hand column of the tableau may be asked to Nature. This guarantees that we do not ask a question to Nature unless we have a good reason to believe that its presupposition is true. I have not mentioned this requirement till now, but I want to hold on to it. The second function of the tableau construction process is to enable us to ask the questions suggested by the implementation process without violating the requirement that no question must be asked to Nature unless we have a good reason to believe that its presupposition is true. The third function of the tableau construction process is to confirm that the result of the implementation process is a potential DN explanans. If the implementation process is coherent (if not more than one predicate is used for each schematic letter), the result is a potential DN explanans. We can check this coherence by closing the central tableau: if the implementation process is coherent, it is possible to close this tableau as soon as *L* has been derived.

The tableau construction process must be guided by the following procedure:

(P_3) (1) If you have executed the instructions 1 till 4 of (P_2), start a central tableau with *T* and *B* in the left hand column, and *E* in the right hand column.
(2) When a question to Nature is suggested during the implementation process, verify its presupposition in an ancillary tableau. If this tableau can be closed, add the presupposition in the left hand column of the central tableau.
(3) Put all the answers of Nature to questions asked during the implementation process in the left hand column of the central tableau, as soon as they are obtained. If the background knowledge *B* is adjusted, add the relevant statements to the left hand column.
(4) When the implementation process is finished, try to derive the statement *L* from *T* and *B* in an ancillary tableau. If this tableau is closed, add *L* in the left hand column of the central tableau.
(5) Try to close the central tableau.

If the central tableau is closed, we have verified that we really have an explanation. So the central tableau is the one that is important for the third function of the tableau construction process. The ancillary tableaus meant in instruction 2 relate to the second function of the tableau construction process. These ancillary tableaus will start with elements of *B* and/or previously obtained answers of Nature in the left hand column, and the presupposition to be proved in the right hand column. The ancillary tableau of instruction 4 is meant to derive the covering law we need in our explanation (first function of the tableau construction process). This tableau starts with T and B_M in the left hand column and L in the right hand column. In ancillary tableaus of both types, answers of Nature to questions asked by the inquirer may be added in the left hand column (see section 5.3 for an example).

In the remainder of this article I will reconsider the thermometer example in order to illustrate that the strategic background information to be combined with (P_2) varies with the type of explanation that is constructed (section 5.2) and in order to illustrate procedure (P_3) (section 5.3). Two preliminary remarks must be made first:

(1) In the thermometer example, the explanandum can be reformulated as "Object *a* at time *t* starts to show a temporary drop of the mercury column, followed by a rise". Because reference to time is crucial here, I will use singular sentences of "Object *a* has property *G* at time *t*", formally represented by expressions of the form $G(a,t)$.

(2) The thermometer example clearly belongs to the category of explanations in which more than one scientific entity is used. Moreover, it is an example of a *structural-interactive* explanation: the laws that are used as premises in the derivation do not tell us anything about the behaviour of the object as a whole; we start from laws that give information about the behaviour of *parts* of the object that is considered, and derive the behaviour of the whole object (the way in which it responds to the interaction) from the structure and the behaviour of the parts.

5.2 If the inquirer uses procedure (P_2), he will first write down the general scheme and then replace β with "starts to show a temporary drop of the mercury column, followed by a rise". The second move results in the following schematic DN explanans:

$\alpha_1(a,t)$
$\alpha_2(a,t)$
$\vdots$
$\alpha_n(a,t)$
In objects with properties $\alpha_1,\ldots,\alpha_n$, the mercury level first drops and then rises.

We suppose that *I* makes the following decisions:

(3) *I* decides that *T* is the conjunction of the law of thermal expansion of solids, the law of thermal expansion of fluids and the law of heat conduction.

(4) I decides that the content of B is "At time t object a is immersed in hot water".

In step 3, I has implicitly decided that he will construct a structural-interactive explanation, because the laws that are chosen do not relate to a is a whole, but to components of a. This implicit decision creates three restrictions:
(a) the explanation will contain only two singular sentences ($\alpha_1(a,t)$ and $\alpha_2(a,t)$),
(b) $\alpha_1(a,t)$ is a description of the structure of a at t, and
(c) $\alpha_2(a,t)$ is a description of an interaction of a with another object that takes place or starts at t.
In his fifth move, the inquirer will state these restrictions. Since no combinative moves can be made, the implementation process continues as follows:

(7) I materialises the strong restriction (a); this amounts to transforming the result of move 2 into the following schematic DN explanans:
$\alpha_1(a,t)$
$\alpha_2(a,t)$
In objects with properties α_1 and α_2, the mercury level first drops and then rises.

In his next move, the inquirer must decide which structural properties and which interactions he will use. So the first part of move 8 consists in formulating the following questions:

Q_1: Which structural properties of a will I use?
Q_2: Which interaction of a with another object will I use?

Before answering these questions, I may execute subordinate interrogative moves in order to obtain information about the structure of a and the interactions that took place (this possibility was not mentioned in procedure (P_2) in order to avoid it becoming too complex). We have assumed that B_M only contains the proposition "a at t has been immersed in hot water". Consequently, I has to ask the subordinate question "What is the structure of the thermometer?" to Nature before he can make a decision in response to Q_1.

Let's assume that the inquirer makes the following decisions in response to the questions:

(8) As structural properties I use the information that a is a thermometer consisting of a glass tube which is partially filled with mercury. As interaction, I take the fact that a has been rapidly immersed in hot water.

On this condition, the construction process goes on as follows:

(5′) α_1 must be replaced with "is a thermometer consisting of a glass tube which is partly filled with mercury", and α_2 with "is immersed in hot water".

(7′) The inquirer materialises the two strong restrictions obtained in step 5′; this amounts to transforming the result of move 7 into the following potential DN explanans:
At time t object a is a thermometer consisting of a glass tube which is partly filled with mercury.
At time t object a is immersed in hot water.
In glass tubes which are partly filled with mercury and are rapidly immersed in hot water, the mercury level first drops and then rises.

The implementation process described above is the result of applying (P_2) in combination with the following items of strategic background information:

(a) The implicit decision to construct a structural-interactive explanation entails three restrictions: (i) the explanation will contain only two singular sentences ($\alpha_1(a,t)$ and $\alpha_2(a,t)$, (ii) α_1 must be replaced with a predicate describing structural properties of a, and (iii) α_2 must be replaced with a predicate describing an interaction between a and another object.

(b) The first restriction can be materialised; so we can always obtain a schematic DN explanans of the following form:

$\alpha_1(a,t)$
$\alpha_2(a,t)$
$(\forall x)[(\alpha_1 x \& \alpha_2 x) \supset Gx)$,

where G is the property attributed to (a,t) in the explanandum.

(c) The two weak restrictions created by the implicit decision to construct a structural-interactive explanation have a heuristic value: they provoke two questions, each expressing a decision problem.

(d) The decisions which the inquirer makes in response to the questions meant in (c) create two strong restrictions; if we materialise them, we obtain a potential DN explanans of the following form:

$S(a,t)$
$I(a,t)$
$(\forall x)[(Sx \& Ix) \supset Gx)$,
where $S(a,t)$ describes the structural characteristics of a at t which I has chosen, and $I(a,t)$ describes the selected interaction.

5.3 When the implementation process described in 5.2 has been completed, the inquirer has obtained his main derivation goal: he has to derive the law "In glass tubes which are partly filled with mercury and are rapidly immersed in hot water, the mercury level first drops and then rises" in an ancillary tableau, as prescribed by instruction 4 of (P_3). Because *B* does not contain information about the coefficients of expansion of glass and mercury, nor about the thermal conductivity of glass, the inquirer will not succeed in deriving this law. A few subordinate interrogative moves are necessary:

Q_3: What is the coefficient of expansion of mercury?
Q_4: What is the coefficient of linear expansion of glass?
Q_5: Is glass a good conductor of heat?

These questions must be asked to Nature. When answers have been obtained, the inquirer can derive the law *L* from *T*, *B* and these answers.

6. Conclusion

In this article I have shown that a clear codification of the aims of the explanation seeking process and of the types of moves that occur in this process, can be obtained by applying Hintikka's interrogative model of inquiry (section 3). But an analysis of the construction process of explanations is not complete until we have provided strategic knowledge which helps the inquirer to make the right interrogative and deductive moves. Therefore, a method for constructing scientific explanations was developed in which the I-model was combined with two ideas: gradual implementation of a general explanation scheme, and strategic background information relating on how this scheme must be implemented (section 4 and 5).

Vakgroep Wijsbegeerte en Moraalwetenschap
Universiteit Gent

NOTES

* Postdoctoral Fellow, National Fund for Scientific Research (Belgium).

1 The numbers of the moves refer to the instruction on which they are based: (7) is the first move based on instruction 7, (7′) the second move based on instruction 7, and so on.

REFERENCES

Hempel, C. & P. Oppenheim (1948). Studies in the Logic of Explanation. *Philosophy of Science* **15**, 135–75. (Reprinted in C. Hempel, *Aspects of Scientific Explanation and other Essays in the Philosophy of Science*, 1965. New York: Free Press, pp. 245–90. Reprint pagination used for reference in this article.)

Hempel, C. (1965). Aspects of Scientific Explanations. In: *Aspects of Scientific Explanation and other Essays in the Philosophy of Science*, pp. 331–496. New York: Free Press.

Hintikka, J. (1985). A Spectrum of Logics of Questioning. *Philosophica* **35**, 135–50.

Hintikka, J. (1989). The Role of Logic in Argumentation. *The Monist* **72**, 3–24.

Kitcher, P. (1989). Explanatory Unification and the Causal Structure of the World. In: P. Kitcher & W. Salmon (Eds.), *Scientific Explanation*, Minneapolis: University of Minnesota Press, pp. 410–505.

Salmon, W. (1993). The Value of Scientific Understanding. *Philosophica* **51**, 9–19.

Salmon, W. (1984). *Scientific Explanation and the Causal Structure of the World*. Princeton: Princeton University Press.

Sintonen, M. (1990). How to Put Questions to Nature. In: D. Knowles (Ed.), *Explanation and its Limits*. Cambridge: Cambridge University Press, pp. 267–84.

Sintonen, M. (1993). In Search of Explanations. *Philosophica* **51**, 55–81.

Poznań Studies in the Philosophy of the Sciences and the Humanities 1997, Vol. 51, pp. 261–285

George Gebhard

SCIENTIFIC DISCOVERY, INDUCTION, AND THE MULTI-LEVEL CHARACTER OF SCIENTIFIC INQUIRY

1. Introduction

Recent advances in the so called independence friendly (IF) epistemic logic developed by Jaakko Hintikka and his associates now enable us to apply the interrogative model of inquiry to more complicated cases. The particular application I have in mind is the nature of experimental inquiry. My contention is that viewing scientific inquiry as a multi-level process in which small operational questions are used to deliver answers to big research questions unveils the important role that experiments play in scientific discovery. The interrogative approach also forces us to revise some generally accepted views on the nature of scientific discovery and induction. The logical problem of induction, or 'Hume's problem', as its is often called, will get its (partial) solution when we realize that its presuppositions (the atomistic postulate, defined below, being one of them) are not wholly tenable. The traditional view of induction and rationality puts a heavy burden on the certainty of individual steps of inquiry and their justification. When looking at this process from the point of view of the interrogative model, the holistic nature of inductive inquiry is exposed, and it will be seen that 'justificationism' is important only in certain less prominent contexts.

The interrogative model utilizes game-theoretical semantics (GTS) which extends a truth-value assignment to atomic sentences of a language by assigning to each sentence a two-person game, thus valuating the entire language. This puts emphasis more on a holistic point of view, which means that it is not meaningful to assign utilities to individual steps of knowledge acquisition in evaluating the success of inquiry. Whole strategies must be considered instead, and that will make rationality more a matter of strategy, where inductive rules are best seen as strategic rules (like rules of acceptance in statistics) rather than rules of logical inference. The interrogative model also sheds more light on the actual discovery process than the traditional hypothetico-deductive view, for it gives logical reasons for certain strategy choices that scientists make in their pursuit of knowledge.

In my presentation I shall first give a brief description of the interrogative model (section 2) and describe how different interrogative games can be classified in the order of their logical complexity. This has direct bearing on the types of answers the inquiry can be expected to deliver (section 3). I shall then present the success criteria of any interrogative game, where the principal premises and available answers play a major role (section 4). In section 5 the role of experiments in connection with scientific discovery will be discussed starting with Hempel's dichotomy of experimentation as a method of test vs. discovery. Then we will move to the much debated question of the logic of discovery (section 6), where reasons are given to support the suggestion that the interrogative model qualifies as *the* logic of discovery. This discussion is followed by an explication of how the model conceptualizes the logical aspects of experimental knowledge-seeking by reference to a two-level view of the scientific enterprise (section 7): a lower level of experiments where logically simpler answers are relied on, and a higher level of theory construction where logically more complex answers are used as an input. From section 8 onwards the logical problem of induction will be discussed from the point of view of the interrogative model. I shall here defend the view that, firstly, the atomistic postulate should be rejected because Nature can give in controlled experiments, answers that have at least the quantificational complexity of a $(\forall x)(\exists y)$-prenex, and secondly, that induction on the two levels of an experimental science (described in section 7) is different in nature, and no inductive or other rules of ampliative inference will be needed on either of the levels. On the higher-level inquiry induction gets on by utilizing the old deductive logic, because logically strong premises are provided by the lower-level inquiry. On the lower-level inquiry of the experimental sciences the interrogative process is initially supported by premises in the form of previously established (often, partial) generalisations and some general hypotheses concerning the uniformity of nature which can be corrected or rejected during the process; hence no ampliative rules of inference will be needed here, either.

2. *The Interrogative Model of Inquiry*

The interrogative model of inquiry,[1] developed by Jaakko Hintikka and his associates, can be described in a few words as a two-person zero-sum game in the sense of mathematical game-theory. One of the players ("the inquirer") is active and the other one ("Nature" or "Oracle") functions only as a source of answers. In the simplest type of inquiry the inquirer tries to prove a predetermined conclusion C starting from a given theoretical premise T. The inquirer has a choice of two types of moves in this interrogative game: in addition to normal deductive moves[2], which rely solely on previously available information, he may use interrogative moves[3] by posing questions to Nature. If Nature gives a conclusive answer, it is added to the inquirer's list of available premises and thus forms new information which is available to the inquirer later. These answers are either individual constants or function symbols of some language L.

The interrogative game is formalized by using a variant of Beth's method of semantic tableaux[4] where the premises in the left column are thought to be true in some L-model M, and the conclusion in the right column false in M. Before the inquirer places a question to Nature the presupposition[5] of this question must have been established by the inquirer (i.e. it must occur in the left column of a subtableau).

The interrogative process can thus be seen as an attempt to construct a counter-example to the presumed consequence relation. If the inquirer has a winning strategy, i.e. he can choose his moves so that he will win no matter what Nature does, it is said that C is *interrogatively derivable* from T (in M), expressed as

(1) $M{:}T \vdash C$

This relation may be compared with the relation of deductive consequence

(2) $T \vdash C$

and with the relation of truth-in-model

(3) $M \models C$

In a certain sense, (1) is intermediate between (2) and (3), and comprises (2) and (3) as limiting cases, where case (2) represents the simplest version of the interrogative case. In that version all the initial premises and all the Nature's answers are presumed to be true. Here the interrogative consequence relation reduces to a logical consequence relation $T \cup A \vdash C$, where A is the totality of answers given by Nature.

3. *Quantifier Hierarchy*

To specify any one interrogative game we also need to know what restrictions are put to the kinds of answers Nature is allowed to give. The restrictions that can be imposed on the logical (quantificational) complexity is probably the most important kind of restriction. The resulting system of types of inquiry is named the *AE*-hierarchy.[6]

AE-hierarchy thus concerns the maximal complexity of the quantifier prefix to an available answer. It is measured by the number of changes between different kinds of quantifier strings (strings of universal and existential quantifiers) in the prefix of a logical expression. Thus e.g. $(\forall x)(\exists y)$-prefix has an $A(2)$-complexity and $(\exists x)(\forall y)$-prefix an $E(2)$-complexity.

Restricting Nature's answers to quantifier-free negated or unnegated atomic propositions would in these terms be called an $A(0)=E(0)$ case. The assumption that Nature's answers are thus restricted may be named *the atomistic postulate*. This postulate or dogma seems to be a tacit assumption in traditional philosophy of science and it is usually manifested only through its consequences.

The next step in the hierarchy of possible answers is the A-case, which is characterized by universally quantified answers of the form $(\forall x_i),..(\forall x_k)$ $S[x_i,..x_k]$.

Further up in the hierarchy then is the double quantifier *AE*-case characterized by answers of the form $(\forall x)(\exists y)$ $S[x,y]$.

Finally, there is the unlimited case, where no restrictions are put to the degree of quantification.

Classification of interrogative games in the *AE*-hierarchy helps to distinguish different actually occurring games from each other, and to assess their logic accordingly. Examples of such games are: hearing of a witness in a court, physician's search for a diagnosis and the conduct of an empirical experiment in physics.

4. *What determines the Success of Inquiry*

One of the most important features of any type of interrogative inquiry is the range of questions that can be answered by its means, assuming that the inquirer employs an optimal strategy. This range depends mainly on two things:

i) the totality of available answers;
ii) the principal premises available.

Reduction of one of the two can often be compensated partly by extending (strengthening) the other.

4.1. Set of Available Answers

Logical restrictions on the set of available answers can be pictured in terms of the quantifier hierarchy described above. The aforementioned atomistic postulate restricts the set of available answers to one particular case only, expressed by an unquantified answer. This restriction is of the most severe kind, reflecting one of the most pervasive dogmas of traditional philosophy of science: the belief that Nature can only provide the inquirer with answers concerning particular facts.

A somewhat weaker form of restriction would be to allow answers that tell something of all objects of the inquirer's study, taken one by one, but not in relation to other objects. A typical example is the statement "all men are mortal". Logically speaking these answers would take the universally quantified form of $\forall x S[x]$, or more generally the form $\forall x_1 \forall x_2 \ldots S[x_1, x_2, \ldots]$, representing the *A*-case in the above described quantifier hierarchy.[7]

A still weaker restriction is to allow double-quantified answers of the form $\forall x\ \exists y\ S[x,y]$ (in addition to those mentioned above). This is a type of answer we need when we want to describe a functional dependency of one variable on an other in controlled experiments when using a first-order language, as will be seen later.

A further weakening of restrictions in certain types of inquiry would be to allow answers of type *AEA*, *AEAE*,...etc., finally giving up all restrictions whatsoever. The set of available answers also influences the selection of the questioning strategy.

4.2. Strength of Initial Premises

Inquiry can be conducted with equal force by using weaker initial premises along with weakening restrictions on the logical form of the answers. Scientific theories usually express dependencies between variables, and are of the *AE*-type. Therefore the inquirer usually wants to establish a universally quantified general conclusion (or a more complex *AE*-type one) as a result of the inquiry. If he, however, has received only atomistic answers (i.e. quantifier-free statements) from Nature, it would not be logically possible to arrive at a general conclusion without some strong initial premises. These initial premises then have to contain quantifiers to make up for the lack of them in the atomistic answers of the interrogative process.

4.3. Set of Available Tautological Premises

If strong initial premises are not available, the interrogative model provides for an additional strengthening of the inquiry by stipulating that we can add tautological premises of the form $(S \vee \neg S)$ to the left column if *S* has already occurred in the same subtableau. The relevance of this move is that we can thus:

a) raise simple yes-no questions that otherwise could not be raised, and

b) eliminate many wh-questions in favor of simpler propositional questions. (The tautological disjunction $(S \vee \neg S)$ functions then as a presupposition of a simpler yes-or-no question, and its introduction is admissible given that the presupposition of a *wh*-question, an existentially quantified formula, is available.)

The introduction of tautological premises has implications to the inquirer's strategic considerations in that answers to new questions can introduce new individuals, the number of which is a crucial strategic parameter in the purely logical type of argument. (As the case is, even new functions can be introduced by answers to questions, which will still further increase the opportunities to introduce new individuals.)

Another important consequence of introducing tautological premises is that we can thus avoid infinite proof trees, which in certain situations can be generated by *wh*-questions.[8]

5. The Role of Experiments in Scientific Inquiry

Let us first look at the role of experiments in science in order to see how the interrogative method of knowledge acquisition manifests itself in experimental and theoretical levels of scientific inquiry. Carl Hempel has seen experimentation in science to serve principally two different purposes:[9]

1) experimentation as a *method of test*, and

2) experimentation as a *method of discovery*.

According to the hypothetico-deductive conception of science we usually start with a problem situation which prompts us to forward a hypothesis, or a hypothetical theory, as a solution to our problem. Suitable test implications of conditional form are then deduced from our hypothetical theory. These are of the form 'if conditions of type *C* are realized, event of type *E* will occur'. To test the hypothesis an experimental setting satisfying the conditions of type *C* specified in these test implications is created. By appropriate observations we will then, it is presumed, be in a position to either falsify the hypothesis (or its auxiliary assumptions) or to find positive support for it. This view will regard *experimentation as a method of test* in an effort to either falsify or corroborate a theoretical hypothesis.

If we cannot technically bring about the required test conditions, we may perhaps wait till Nature does that. The possibility of doing so extends this method of test even to a few other, typically non-experimental, empirical sciences (e.g. astronomy and population genetics).

The standard view of the hypothetico-deductive conception assigns no rule to the process whereby our hypothesis is formed. Although this issue is much debated[10], the process of theory (or hypothesis) invention is mostly considered to belong to the domain of cognitive psychology dealing with creative intuition which defies formal description. If we, like Popper and Lakatos, do not entertain a view which ascribes a 'logic' to scientific discovery, we cannot, nevertheless, reasonably deny that experimentation, whether we conduct it only mentally or in actual practice, is one of the methods whereupon scientific discovery is made possible.

To start with, there may be no precise enough hypothesis advanced, but just an intuitive idea, or at best a rough guess which cannot be tested as such.[11] We can easily concieve of this guess as a question put to Nature. Here, as in all interrogation, finding and formulating the right question is usually the most demanding part of the enterprise. As it is now acknowledged, a scientist's ignorance of an appropriate question is not altogether unstructured because his background knowledge imposes constraints on admissible answers to inquiry. Questions so to speak, inherit their structure from background theories.[12] Once the question has been precicised the subsequent articulation of the hypotheses to a testable mode may take the form of a physical model, analogy, or a mathematical representation. A quantitative mathematical representation is often needed at least in physics, which is a quantitative science. The representation seeks to describe the phenomenon in such detail that empirical measurements can be used to test the theory. This phase of scientific inquiry has aptly been called calculation by Ian Hacking: "Speculation and experiment need not in general be closely connected, but the activity I call calculation brings them close enough to discern a quantitative fit between the two."[13]

Initial articulation of the emerging hypothesis by means of 'calculation' involves dealing with variables dependent on each other, and it may take a good deal of ingenuity and mental experimentation to discover what the precise mathematical form of the hypothized dependency is. In this context we seem to have good reasons for speaking of (mental) *experimentation as a method of discovery*.

Hacking[14] has discussed at some length, the question 'which comes first, theory or experiment?'. He quotes, among others, the famous German chemist Justus von Liebig who was strongly of the opinion that any experiment not preceded by a theory is futile; an experiment is only an aide to thought. Hacking sees, however, two readings of Liebig's views:

i) a weak version saying that you must have *some* ideas about nature and your apparatus before you conduct an experiment, and
ii) a strong version which says that your experiment is significant only if you are testing a theory relating to the phenomena under scrutiny.

He says that nobody would deny the weak version, but the strong version is simply false. Experiments can be conducted out of plain curiosity. One such example is botanist Robert Brown's observations of the irregular movement of pollen suspended in water in 1827. Another is *A-C*. Becquerel's observations in 1873 of how the electrical resistance of the metal selenium decreased when it was illuminated.

In both cases, however, the experimenters came to no conclusions about the nature of the phenomena. This seems to support the strong reading of Liebig's view, but history of science tells that the pre-theoretical observations of these men guided other scientists to solve these puzzling phenomena later. In the first case, simultaneous work of a theoretician, Einstein, and another experimenter, J. Perrin was still needed before it was resolved (almost a century later) that the pollen movement was caused by collisions with moving molecules. In the second case, the theory of the photon, subsequently developed by Einstein, explained the observed photoelectric effect. In these cases, with due regard to the time lag and contributions from theoreticians, we still seem to have grounds for speaking of (physical) experimentation[15] as a method of discovery (or at least as an aid leading to discovery).

6. *The Logic of Discovery*

When approaching the question whether there is a 'logic' of scientific discovery it is appropriate to define the meaning of the term logic in this context. In view of the fact that 'logic' is a normative term it is possible to give it a broad interpretation, something like "a set of normative rules for judging the procedures followed to attain a specific goal". Given this rather loose definition of logic, it can be claimed that a normative description of the established standards of experimentation in various domains of experimental science may deserve to be called the logic of scientific discovery (properly restricted to experimental sciences). Accepting this view would, in my opinion, result either in a plethora of 'logics' of discovery, one for each domain of science, or at best in an incoherent 'logic' of discovery with a strictly limited applicability.

Herbert Simon[16] has maintained that a normative theory ('logic') of discovery processes is possible for the formative stages of the initial idea generation, in what is often called the context of invention. He has support for his claim e.g. in present-day computer simulated pattern-recognition programs that use what he calls 'heuristic search algorithms' (HSA). These programs do, no doubt, work in some *AI*-oriented applications, but it is not obvious that HSA's properly apply to real-life scientific problem situations. Simon presents, however, some persuasive arguments for his view that *AI*-applications are not qualitatively different from real-life situations that scientists work in, implying that the HSA could properly represent the logic of theory invention. Be this as it may, HSA does not form a logic in any strict meaning of the word.

A more fertile approach is to study the discovery process in terms of Hintikka's interrogative model of inquiry which uses the familiar deductive-type logic supplemented with interrogative moves. This logic of the interrogative model, I maintain, is *the logic of scientific discovery*.[17] It is also a proper subset of the logic of questions and answers, which in turn is a proper subset of what is now called independence friendly (IF) epistemic logic. Hintikka's method of inquiry is thus grounded on logic which, nonetheless, is no reason to expect it to display a subject-matter-neutral algorithm.

Discovery in this setting will be seen as the development of a first tentative idea of a hypothesis to a full-fledged theory which will also receive its justification through the very same process (trivially, in a sense, in the simplified case where all Nature's answers are presumed to be true). We may thus rid ourselves of the old Reichenbachian dichotomy between context of discovery and context of justification along the lines "the friends of discovery" have done.[18] The interrogative model has the additional benefit of being able to show that vague and uncertain guesses have a well-founded place in the process of discovery. Risky guesses do not make the logic of inquiry any weaker or such steps fallacious because reliance on them is a matter of strategy decision and has nothing to do with the definitory rules of the model. Bold conjectures, as advocated by Popper, definitely belong to the process of scientific inquiry and discovery. Accepting them does not preclude the use of logic proper.

7. The Logic of Experimental Science

The logic of empirical natural sciences seems, at first blush, to be an atomistic one. An experimental scientist observes only particular readings of his measuring apparatus. Seen from the point of view of the interrogative process, all the inquirer receives from Nature are answers to questions concerning particular events. Nature does not give answers as to what the case is always and everywhere (i.e. universally quantified answers). This is the way the atomistic postulate appears in the philosophy of science, and it undoubtedly has *some* truth in it.

It cannot be the whole truth, however. If we distinguish experimental sciences from purely observational ones, an interesting fact will be observed: an experimental science turns out on closer examination to be an *AE*-case logically speaking, having at least an *A*(2) quantificational complexity.

A controlled experiment[19] in an empirical science forms a classical application of the interrogative model as "questions put to Nature". The scientist can usually vary, to some extent, one variable (the controlled one), and observe how another variable (the observed one) changes accordingly. Thus the outcome of such an experiment is a statement of dependence between the two variables: for each value of the controlled variable there exists a corresponding value of the observed variable related to it in a certain definitive way. Typically this relationship is one of functional dependence.

The form of an answer that is required to a question put to Nature in experimental science is therefore of *A*(2)-complexity.[20] What is being sought for is the *function* expressing the dependence between the two variables. This will most clearly be seen when formalizing in Hintikka's epistemic *IF*-logic the question "Whom (what) does everyone...?", which has a logical form equal to "What value of the observed variable corresponds to each value of the controlled variable?".[21] The *desideratum* of this question is of the form $K(\forall x)\ (\exists y/K)\ S[x,y]$, and the *presupposition* of the form

(i) $(\forall x)\ (\exists y)\ S[x,y]$.

The interrogative process will lead us from (i) to a *reply* of the form

(ii) $K(\forall x)\ S[x,f(x)]$

where Nature's answer has provided us with the function *f*, which selects for each *x* a specific value *y*, as required by the question. An additional premise, the conclusiveness condition is required for this reply to be satisfactory; that is, we have to know *what* function *f* is. This condition is formally stated as

(iii) $K(\forall x)\ (\exists y/K)\ (f(x)=y)$

and has a second order equivalent

(iv) $(\exists g)\ K(\forall x)\ (fx=gx)$.

7.1. The Two Levels of Experimental Science

In order to see where functional answers are needed and where atomistic ones will suffice, it is useful to conceive of an experimental science as a multi-level activity. This idea is made possible by the duality of the roles of questions in interrogative inquiry and the fact that the interrogative model allows varying logical complexity to the available answers. According to this view we can distinguish between 'micro-level' inquiry where the physical experiment takes place, and 'macro-level' inquiry, where theoretical conclusions are drawn.

In the micro-level game the individual measurement results of an experiment are Nature's atomistic answers[22] to operational questions posed by the inquirer. On this level scientific inference is typically a 'curve-fitting activity'.[23] The setup of the physical experiment, with its often requisite complex technology, then plays the role of a strong initial 'theoretical premise' in the interrogative process allowing simple atomistic answers from Nature as input to the inference.[24]

In macro-level inquiry, on the other hand, the whole controlled experiment of the micro-level inquiry (consisting of a sequence of operational questions) is treated as a single question put to Nature: a more complicated question requiring a double-quantified *AE*-type answer, which, as we have seen, typically provides us with a functional dependence.

In experimental sciences we can thus distinguish a lower level where experiments are conducted and a higher level where theory is developed. This idea may be pictured as follows: (*Fig. 1.*)

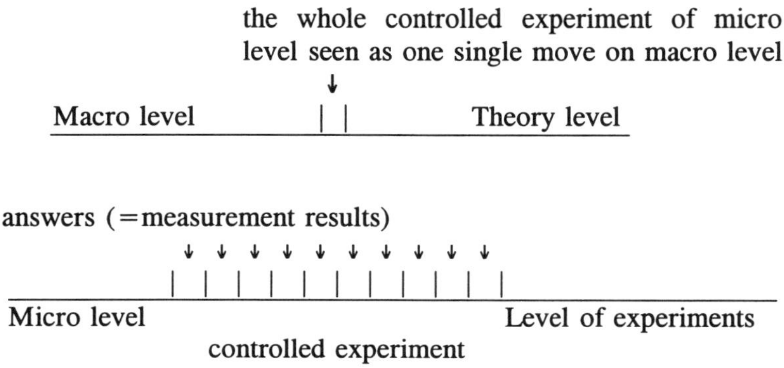

The aim of natural science is to find laws (regularities, universals) in nature. In order to establish a complex law as a conclusion of an interrogative inquiry, we need an *AE*-type answer from Nature which describes the type of functional dependence outlined earlier. Unfortunately we cannot get the required answers (of *AE*-type) directly from Nature because she does not understand logically more complex Why-questions of *AE*-type.[25] We can, however, reach our goal indirectly by using those simpler (atomistic) answers that Nature gives on the micro-level inquiry as stepping stones.

This leads us to the second distinction between different levels and two new terms. In general, we have to distinguish between *principal questions*, which are the main target of the inquiry, and *operational questions*, which are to be answered as a prerequisite to an answer to the principal question. This distinction between the two levels then amounts to the fact that what for the higher-level inquiry is an operational question is for the lower-level inquiry a principal one.

In other words: answers to principal questions are those we are after and answers to operational questions seem to be the only ones we can get from Nature directly (if we place our questions right). Given strong initial premises, Nature's atomistic answers to the micro-level operational questions will give a solution to the micro-level principal question: the form of the functional dependency. This dependency can then be used as an *AE*-type answer to an operational question on the higher-level inquiry where the principal question is of a logically more complex type.

The distinction between principal questions and operational questions can be pictured as follows (*Fig. 2.*):

The principal question of the controlled experiment as one move (=answer to an operational question) in the theory level game

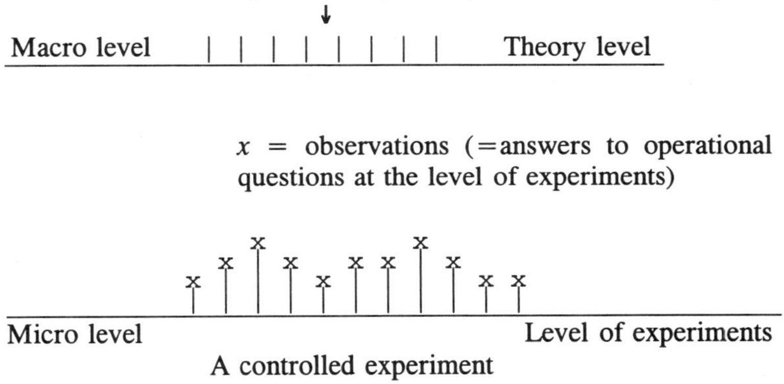

The principal question on this level is the curve fitting problem, i.e. formulating the functional dependency $y=f(x)$.

8. *Presuppositions of the Problem of Induction*

Aristotle presumed that with the aid of proper thinking our minds will get access to general truths, knowledge of "forms". Translated to the idiom of the interrogative model it means that Nature gives us, not only unquantified answers (concerning particular facts), but also more general answers of *A*-complexity, and thus we do not need initially any general presuppositions to support our inference. In the Middle Ages, nominalism effected a modification in the Aristotelian view. When, finally, empirists rejected the idea of innate knowledge and adopted the atomistic postulate, the setting for an epistemological problem was created. As is well known, the skeptical, or 'negative'[26] result of Hume to the logical problem of justifying induction brought the empiristic philosophy on the horns of a dilemma. How is scientific inference and the growth of knowledge to be accounted for if there is no logical justification for the use of ampliative inference?

As the case is, from atomistic answers alone one cannot deductively derive general laws. For that purpose, one needs either strong a priori assumptions or inductive inferences. Either alternative plunges one into difficulties, however. The former alternative violates empiricism, while the latter leads straight to Hume's problem of the logical validity of (presuppositionless) induction. For the gist of Hume's problem lies precisely in the insight that all inductive inferences need prior premises to back them up.

Now, if the interrogative model is right, Hume's problem simply does not arise, provided that the atomistic postulate is not adopted. Claiming this does not mean that we pretend having completely solved (or dissolved) the problem of induction as it was formulated by Hume. The logical validity of enumerative induction is a problem which, to the best of our knowledge, has no acceptable solution. It has been convincingly demonstrated[27] that the only way to guarantee a priori the truth of inductive generalizations is to make them analytical.[28]

What is then implied by our statement that Hume's problem does not arise? I shall here argue for the view that if scientific inquiry in experimental sciences is looked at from the point of view of the interrogative model, as outlined earlier in this paper, the presuppositions of induction are not of the kind that give rise to the problem of Hume.

9. Induction on the Macro-Level Inquiry

From what has been said earlier it should be clear that induction in the contemporary sense[29] of inference from particular cases to general truths (enumerative induction) has no role to play in the macro-level interrogative game. All inference there is deductive by nature, while *AE*-answers from the lower-level inquiry are used as premises and presuppositions to further questions. Other possible moves are definitory moves and interrogative moves which, however, are non-inferential. The resulting logical inference is interrogative ($M{:}T \vdash C$) by nature, not inductive.

An answer to a lower-level principal question in inquiries of experimental sciences was argued to normally take the form of a statement describing a function. The presupposition of a question requiring specification of a functional dependence as an answer was, as we saw earlier, of the form

(i) $(\forall x)\ (\exists y)\ S[x,y]$

The rules of the interrogative game require that before a question move is initiated, the presupposition of the question must have been established.

This requirement is evidently reasonable as regards propositional (S_1 or S_2?) or simple *wh*-questions, but appears to be less so when the question is of the more complex *wh*-type like the one above.

If the presupposition (i) of a question fails to be true for some values of x (i.e. for some x no y satisfies $S[x,y]$), it nevertheless, does not appear entirely illegitimate to ask the question concerning *the other values* of x. In these cases Nature's answer, which was earlier described as taking the form

(ii) $(\forall x)\ S[x,f(x)]$

will now take the conditional form

(iii) $(\forall x)\ ((x_1 < x < x_2) \rightarrow S[x,f(x)])$

where f is a partial function defined only for some values of x.[30] It is reasonable to suppose that this will rather be the kind of answer Nature gives to the inquirer in controlled experiments. We thus seem to be entitled to pose *AE*-questions to Nature even in cases like this, where the presupposition is not entirely established.

Having received the above type (iii) of limited generalizations as answers to his questions in the micro-level inquiry, the inquirer may use them as presuppositions on the macro-level and continue the inquiry there by:

a) trying to extend the partial generalizations to longer intervals, i.e. to further values x' of x such that $x_1' < x_1$ and $x_2 < x_2'$, (aiming perhaps to an unrestricted generalization of the type (ii) in the limit),

and

b) trying to unify the different partial generalizations into one and the same general law.

Type a) activity is what Hintikka has called the generalization task and type b) the reconciling task. The generalizing and reconciling tasks form the older sense of induction which Hintikka has identified and which is probably the most important pre-Humean sense of induction. The reconciliation of partial generalizations consists of transforming the different mathematical laws into a form where they will be seen as special cases of a more general regularity. This task usually involves conceptual development where old familiar concepts are reinterpreted or new concepts introduced.[31] New concepts are typically needed for an interpretation of the known laws as specific instances of the new generalization.

A couple of good examples of inductive reconciliations of type b) in physics are Einstein's discovery of the special theory of relativity by reconciling Newtonian mechanics with Maxwell's theory of electro-magnetism, and Planck's reconciliation of Rayleigh-Jeans law with Wien's law by introducing the statistical concept of entropy. Planck's reconciliation of two distinct laws resulted also in a new concept: quantum of energy.[32]

10. Induction on the Micro-Level Inquiry

What then is the role of induction on the micro-level of inquiry? If inquiry is constrained by the atomistic postulate, induction in the contemporary sense seems to be needed as a supplement to deductive inferences and question steps. This may be the case in a purely observational science. In an experimental science, however, the micro-level has available much stronger premises, i.e. earlier (partial) generalizations and theories which have been established in the macro-level inquiry.

Typically, the initial premises that are used in a lower-level inquiry are such well established earlier laws (or partial generalizations) which are considered unproblematic. This will be most prominent in those types of research where experiments are very expensive, for in them funding would most unlikely be available without a strong and less speculative theoretical backup for the planned experiments.

What these general initial premises help the inquirer to do in a lower-level inquiry is to carry out inductive inferences, for instance, those curve-fittings that were seen above to be typical of lower-level inquiries. These inductive inferences do not, however, need any inductive logic in the question-answer process described by the interrogative model. A curve-fitting activity of the lower-level seeks only to describe a pattern in the data that has been observed. For that, deductive logic will do fine.[33] This fact also explains Newton's claim in his Principia to have deduced the necessary general premises from experience, which evidently means controlled experiments.

Interrogatively, lower-level inquiry succeeds without stronger answers than atomic ones because the higher-level inquiry usually provides it with strong initial premises from which partial functions may be derived. In cases where we do not have the benefit of strong initial premises from above, there are other means to conduct the inquiry without resorting to inductive logic. The interrogative model uses deductive logic which implies that some sort of initial postulates are needed to get the process going.

If we look at a physical science where controlled experiments are conducted, it is evident that the inquiry process does not start completely without some presuppositions. If no macro-level partial generalizations are available, pre-theoretical ones will be used. They may ever be so implicit, but at least presumptions of continuity and differentiability seem to underlie most inquiries where we try to interpolate some previously observed functional dependencies. These presumptions are of a very general nature, and we may have obtained them from experience, just the way Newton claimed. Or else, following the view of Duns Scotus, we may maintain that our intellect, with a minimum of empirical experience, just knows them to be true (eg. the postulate of the uniformity of Nature). Anyhow, being synthetic assumptions they are in need of later empirical confirmation like any other empirical hypothesis.

These and other kind of pre-theoretical expectations now seem to form the initial premises, to replace the 'innate ideas' formerly rejected by empiricists or, as in Aristotle, well-established general opinions, endoxa, of the oracle, which we have obtained from previous interactions and interrogations with Nature. Their principal function is to form a more or less theoretical back-up, so that logically weaker answers from Nature will suffice to support the beginnings of an interrogative knowledge-seeking process. The interrogative model has the benefit of allowing one to start with rather uncertain or even speculative presumptions, because it is always possible to reject or replace them later on in the process, if they turn out to be false. The same applies naturally to answers that the Inquirer will elicit from Nature during the questioning process. The logical force of the inquiry will not suffer, provided that game rules are followed, i.e. false or inadequate presumptions and answers are later retracted or 'bracketed' and all steps relying on them likewise. This possibility enables one to engage in more risky reasoning which may produce more informative answers (cf. Popper's 'bold conjectures') and possible new discoveries at the early stages of inquiry. The bracketing operation is part of the game rules and does not indicate that anything like a 'fallacy' has been committed. The type of rules used in induction are more strategic rules of the game than ordinary rules of inference. In certain aspects they can be compared to the statisticians' decision rules of when (at what confidence level) to accept or reject a hypothesis.

It is first when we try to extrapolate or generalize partial functions of type $(\forall x)\ ((x_1 < x < x_2) \rightarrow S[x,f(x)])$ on the macro-level that rules of inductive inference seems to be needed. But, as was argued before, even in the higher-level inquiry the scientist can do without ampliative inference or strong initial premises because strong *AE*-answers to the macro-level operational questions are available from below. These answers are of such a logical complexity that general laws can be derived from them deductively. The interrogative approach has thus demonstrated that the logical problem of induction need not worry us if we correctly understand the presuppositions that lay beneath the Humean view, i.e. the atomistic postulate and the complete lack of prior (nontrivial) premises from the inquiry.

11. The Uncertainties of Induction

In his recent analysis of the experimental tradition in science Allan Franklin[34] has recognized that strong theoretical premises have an important role in the selection of experiments to be performed. His explanation of this fact is that we tend to devote our attention to what seems to be the most important aspects of nature, and theory provides one guide to what is important. This may well be so, but the reasons for directing our inquiry to domains where strong premises are available are not merely heuristic or psychological (as we are led to think) but, as was previously asserted, in a very important sense also logical. Strong initial premises allow us to proceed with weak atomistic answers.

Franklin presented also a detailed analysis of Newton's arguments concerning the derivation of his inverse-square law of gravitational force from Kepler's Laws. He concludes that the derivation was idealized and approximate, which Newton was well aware of, but regarded as harmless. To meet Popper's[35], Duhem's and Zahar's claim that Newton could not possibly have derived his (true) Law of Gravitation from false empirical generalizations (i.e. Kepler's laws) Franklin suggests that a Bayesian analysis would best explicate Newton's methods in this case (without claiming, however, that Newton was a Bayesian).

The interrogative model will nevertheless give a much more natural framework for looking at Newton's methodology in this case. The crucial judgmental component in Newtonian induction (which is not of the enumerative type[36]) on this view is not connected with approximation or observational error. It is instead connected with an assumption of independence, or near independence from further 'hidden variables' or other factors for the partial generalization which has been established. Establishing or assuming this functional independence, which is reflected in the presupposition $(\forall x)(\exists y)\ S[x,y]$ of an experimental question, is just the source of logical strength in scientific inference of this type. This assumption allows the inquirer to put *AE*-questions to Nature even when the presupposition is not wholly established, as was described before. Nature's answer will then be a restricted functional dependence. The partial generalization that Newton deduced from Kepler's laws was the two-body system consisting of the Sun and Mars in this case. By reconciling this partial generalization with other limited generalizations enabled him to arrive at his Law of Universal Gravitation along the lines discussed in section 9. Strategically speaking, it does not matter all that much how strong support we have in establishing the principal question in the first place, or how that idea is initially conceived. The derivation may be idealized and approximate. What matters is the possibility to arrive at the same conclusion via multiple paths once we have arrived there. Emphasis in discovery is therefore on a more global level, not so much on individual steps of the inquiry.

Seen from the point of view of the rules of the interrogative model it is fully acceptable, as was previously mentioned, that the inquirer suspects at any stage of inquiry some earlier answers or initial premises, and decides either to postpone or to reject them. This explains why the approximate character of Kepler's Laws was beside the point in evaluating Newton's reasoning. On the interrogative view the Newtonian inference does not depend on the approximate character of the partial generalizations on which it operates. Earlier premises that seem to lead us astray can be bracketed and later modified during the course of the inquiry. This legitimate bracketing operation has usually been interpreted among philosophers and sociologists of science as mistaken reasoning, or worse, a biased and illegitimate means to interpret one's experimental data: a questionable practice of omitting facts that do not fit the working hypothesis. The interrogative point of view helps thus to explain also the puzzling behavior of some of our most prominent scientists.

12. Reconciliation and Conceptual Development

As regards the reconciling task of induction and the connected conceptual development an excellent example is offered by the line of thought that eventually led to Einstein's discovery of the special theory of relativity by reconciling Newtonian mechanics and Maxwell's theory of electro-magnetism.[37] Instrumental in this development was Planck's quantization of energy of radiation which pointed to Einstein several conflicts between the various domains of physics. These conflicts were readily identified by Einstein because he saw physics as one unified discipline which ought to be internally consistent. He was also influenced by David Hume's and Ernst Mach's philosophy, which helped him to question the 'established truths' of Newtonian mechanics.

Einstein started with (and maintained all his life) the attitude that epistemological analysis was always required to solve foundational problems in physics. Science without such analysis is "primitive and muddled".[38] According to him foundational problems were not solved by Lorentz-type "theories of construction"; for that purpose "theories of principle" were needed. By relying on such bold "theories of principle" he presented his postulate of the invariance of the velocity of light and the postulate of the equality of reference systems for the laws of mechanics, electrodynamics and optics. His bold postulates fit very well the procedural techniques of the interrogative model presented above. The epistemological analysis which resulted in these postulates must have included experimentation with and manipulation of the relevant mathematical equations as a major component. Experimentation of this kind has very much the character of conceptual analysis of the relevant physical situation and shows clearly that mathematics is not only used as a tool for registering results of experiments, but also as a tool for making experiments on the more abstract symbolic level.

The postulate of the equality of reference systems eliminated the aether from the concept of field and put electrodynamics and mechanics on equal footing.[39] To make his postulates consistent a further analysis of time and space was then needed, which resulted in the rejection of an absolute or 'Universal' time.

Aided by the results of his analysis of simultaneity Einstein then derived new transformation rules for events in different inertial systems and presented his (special) theory of relativity in 1905. He went on to show that Maxwell's equations are invariant with respect to the new transformation rules and needed no reformulation. A *reinterpretation*, however, was called for: the separate electric and magnetic fields in Maxwell's (and Lorentz') theories should be viewed in special relativity as *one* electromagnetic field. The concept of field, however, received a totally new meaning due to the removal of the aether from electromagnetic descriptions. The electromagnetic field is no longer a state of a substance (aether) but instead something independently existing and ontologically on a par with matter; an irreducible element of description.

The shift in meaning from being a "state" to being an "independently existing entity" was *a major conceptual change* which Einstein's reconciliation of Newtonian mechanics with Maxwell's theory of electromagnetism brought about. In this connection also other, more commonly known fundamental concepts received a major change in meaning, i.e. the concepts of space and time. Still deeper conceptual changes were connected to the subsequent introduction of the general theory of relativity.[40]

The example above serves to show the difference between induction in contemporary sense and the fitting of two or more restricted laws under a more general one. In both cases the scientist moves from special cases to a more general proposition, but the reconciling process requires in addition that the restricted laws be reinterpreted. The reinterpretation precludes the use of existing inductive logics in these cases and requires a change in some of the familiar concepts in the domain of inquiry.

13. Concluding Remarks

Nature's answers in experimental situations have a sort of logical generality required by the experimental set-up where various functional dependencies prevail between the ingredients of the setting. Induction in the experimental sciences appears thus to be of the Newtonian type which advances from less general statements to more general ones. There is an increase in generality for propositions already established, which makes the 'inductive jump' rather one of degree, not of kind.

When looking at Newton's method in terms of the interrogative model of inquiry it is easier to understand, and interpret literally, his claim in Principia of having "deduced his principles from phenomena". The interrogative point of view clearly indicates that the atomistic postulate is not very well founded when we speak about the inquiry in experimental sciences. This makes it easy to see why the problem of Hume[41] need not worry us: experimental inquiry does not have to rely on enumerative induction. It relies instead on its mirror image, i.e. on the interrogative process which advances from uncertain premises using truth-preserving inference (deduction), whereas in inductive logic we advance from intersubjectively certain premises using uncertain, non-deductive inference. We have also seen how the interrogative model successfully combines the two aspects of theory formation, discovery and justification in one self-correcting process.

Department of Philosophy
University of Helsinki

NOTES

1 The interrogative model is described more in detail in Hintikka & Bachman (1991). See also Sintonen (1996b).

2 For example, the deductive rule *(L.&)* says that if a formula *P & Q* occurs in the left column, *P* and/or *Q* may be added to the same.

3 If e.g. disjunction $P \vee Q$ occurs in the left column of a tableau, the inquirer may use it as a presupposition of a question "Is *P* or *Q* true?".

4 Beth (1955). For a good introduction to the interrogative tableaux system see Harris (1994).

5 The term presupposition is explained in e.g. Hintikka (1976, p. 27), and Sintonen (1996a, section III).

6 See Hintikka (1988, p. 174).

7 This restriction is named the *A*-assumption by Hintikka and Bachman. See their (1991, p. 324).

8 For a more extended discussion see Hintikka–Harris (1991, p. 6ff).

9 Hempel (1966, pp. 19–23). Here we will disregard from such purposes of experiments that aim mainly to illustrate a phenomenon for the purposes of teaching a subject matter.

10 See e.g. T. Nickles (1980), H. Simon (1977), K. Popper (1959) and N. Hanson (1958).

11 Hacking (1983, p. 214) sees two reasons why the majority of speculative hypotheses do not mesh with the world at all:

i) one can seldom directly deduce from a speculation consequences that are even in principle testable, and
ii) a proposition which is in principle testable is often not testable simply because no one knows how to conduct the test.

12 See Sintonen (1996a, section VII) for an illuminating analysis of discovery through interrogation in a structuralistic framework.

13 Hacking (1983, p. 215).

14 Hacking (1983, pp. 152–66).

15 Systematic observations are here understood as constituting experimentation.

16 Simon (1977, pp. 326–37).

17 For a more thorough line of argument relating the interrogative model to scientific discovery see Matti Sintonen (1993).

18 See Sintonen (1996a) and (1996b) for the discovery aspect and selection of proper strategy in experimental interrogation.

19 The term 'experiment' is usually given a wider reading than the case is here. The narrower reading can be caught by the term 'controlled experiment' or 'analytical experiment'. Making this distinction and looking it from the point of view of the interrogative model may have far going consequences for our understanding of the early modern science. (See Hintikka 1988, p. 182.)

20 Complicated experimental situations can also yield answers which are more complex than the *A*(2)-variety here considered (the logical complexity of such answers are described in Hintikka 1988, pp. 177–9).

21 This is typically a *wh*-question with an outside universal quantifier.

22 Nature's answers need not in all inquiries of experimental type be atomistic type of observations. They might in some situations be eg. inquirer's intuitions of innate ideas or well-established general opinions that may be more or less certain. See further section 10.

23 This is the idea of seeing the drawing of a curve that best fits a number of points on a plane as analogous to the task of scientific inquiry.

24 The importance of the initial premises shows itself in the fact that the observational answers will have different interrogative consequences depending on the initial theoretical premise. This is a clear-cut case of Kuhnian "theory-ladenness of observations". Theory-ladenness does not, however, presuppose that the meaning of observational terms changes from theory to theory. This point is comprehensively elaborated by Hintikka in his (1992b, pp. 278–80).

25 See Sintonen (1996b, sections 3 and 6) for a more detailed description of the questioning procedure and the risks of committing petitio principii in this connection.

26 What Hume demonstrated was in short as follows: Causal laws are empirical propositions and the causal relation in them synthetical. Thus the causal relation cannot be analytically valid, it lacks logical necessity, and it therefore cannot a priori guarantee the truth of our inductive generalizations. Reasoning from cause to effect lacks logical necessity and it cannot be used to justify induction. See von Wright (1957, pp. 18–9). For Hume's 'negative solution' see Popper (1972, pp. 85–95).

27 See e.g. von Wright (1957, pp. 46–53).

28 This is what Popper attacks at as one form of a "conventionalist stratagem" in his (1959).

29 Induction in the older sense advances by unifying partial generalizations into a larger one.

30 Hintikka has also considered the possibility that Nature's answer includes a rejection of the presupposition for certain values of x. This will not, however, affect our argument here. See his (1992a).

31 See e.g. Giere (1979, p. 66).

32 These cases are described in more detail with reference to the interrogative model by Hintikka (1991c, pp. 26–7).

33 See Simon (1977, p. 330) for an argument to that effect.

34 Franklin (1986, pp. 104–23).

35 Popper (1972, chapter 5).

36 For a convincing argument to that effect see Hintikka and Garrison (fortcoming).

37 A lucid exposition of the development of Einstein's thought is presented in Nersessian (1984, pp. 121–53).

38 Nersessian (1984, p. 123).

39 What effect the Michelson-Morley experiment had on Einstein's thinking still seems to be an open question which is shortly discussed in Nersessian (1984, p. 128).

40 For example, we no longer have an 'empty space'. This concept is replaced in the general theory of relativity by the 'impulse-energy tensor'.

41 In discussing Hume's problem we have not touched what is called the pragmatic problem of induction, i.e. the question why we should rely rather on well-corroborated theories than their alternatives when planning our actions.

REFERENCES

Beth, Ewert W. (1955). Semantic Entailment and Formal Derivability. In: Hintikka, J. (Ed.), *The Philosophy of Mathematics*. Oxford University Press, 1969.

Franklin, Allan (1986). *The Neglect of Experiment*. Cambridge: Cambridge University Press.

Giere, Ronald (1979). *Understanding Scientific Reasoning*. New York: Holt, Rinehart and Winston.

Hacking, Ian (1983). *Representing and Intervening*. New York: Cambridge University Press.

Hanson, Norwood R. (1958). *Patterns of Discovery*. Cambridge: Cambridge University Press.

Harris, Stephen (1994). GTS and Interrogative Tableaux. *Synthese* **99**, 329–43.

Hempel, Carl G. (1966). *Philosophy of Natural Science*. Foundations of Philosophy Series, Prentice-Hall Inc.

Hintikka, Jaakko (1976). *The Semantics of Questions and the Questions of Semantics*. *Acta Philosophica Fennica*, vol. **28**, no. 4. Helsinki: Societas Philosophica Fennica.

Hintikka, Jaakko (1987). Concepts of Scientific Method from Aristotle to Newton. In *Knowledge and the Sciences in Medieval Philosophy*, vol. I, *Acta Philosophica Fennica*, vol. **48**. Helsinki, 1990.

Hintikka, Jaakko (1985). A Spectrum of Logics of Questioning. *Philosophica* vol. **35**, 135–50.

Hintikka, Jaakko (1988). What is the Logic of Experimental Inquiry? *Synthese* vol. **74**, 173–90.

Hintikka, Jaakko (1991a). Towards a General Theory of Identifiability. In: J. Fetzer, D. Shatz and G. Schlesinger (Eds.), *Definitions and Definability*. Dordrecht: Kluwer Academic, pp. 161–83.

Hintikka, Jaakko (1991b). Independence-Friendly Logic as a Medium of Information Representation and Reasoning About Knowledge. In: S. Ohsuga et al. (Eds.), *Information, Modelling and Databases*. Amsterdam, Washington, Tokyo: IOS Press, 1992, pp. 258–65.

Hintikka, Jaakko (1992a). The Concept of Induction in the Light of the Interrogative Approach to Inquiry. In: J. Earman (Ed.), *Inference, Explanation and Other Philosophical Frustrations*. Berkeley and Los Angeles: University of California Press.

Hintikka, Jaakko (1992b). Theory-ladenness of Observations as a Test Case of Kuhn's Approach to Scientific Inqiry. *PSA 1992*, Volume 1, 277–86.

Hintikka, Jaakko and Bachman, James (1991). *What If...? --- Toward Excellence in Reasoning*. Mountain View, California: Mayfield Publishing Co.

Hintikka, Jaakko & Garrison, James W. (forthcoming). Newton's Methodology and the Interrogative Logic of Experimental Inquiry.

Hintikka, Jaakko & Harris, Stephen (1991). Identifiability in First-Order Theories: Some Basic Results. Manuscript.

Kiikeri, Mika (1993). Kepler's Interrogative Model of Discovery. Manuscript.

Maunu, Ari (1991). Independence-Friendly Epistemic Logic and its Uses. In: O. Koistinen and J. Räikkä (Eds.), *Good Reason: Essays Dedicated to Risto Hilpinen*.

Annales Universitatis Turkuensis, Ser. B, Tom. 200. Turku: University of Turku, 1993.

Nersessian, Nancy J. (1984). *Faraday to Einstein: Constructing Meaning in Scientific Theories*. Dordrecht: Martinus Nijhoff Publishers, Kluwer.

Nickles, Thomas (Ed.) (1980). *Scientific Discovery, Logic, and Rationality. Boston Studies in the Philosophy of Science*, vol. **56**. Dordrecht: Reidel.

Nickles, Thomas (1985). Beyond Divorce: Current Status of the Discovery Debate. *Philosophy of Science* **52**.

Popper, Karl R. (1959). *The Logic of Scientific Discovery*. London: Hutchinson & Co., 1977.

Popper, Karl R. (1972). *Objective Knowledge – An Evolutionary Approach*. London: Oxford U.P., 1975.

Sintonen, Matti (1993). In Search of Explanations: From Why-questions to Shakespearean Questions. *Philosophica* **51**, 55–81.

Sintonen, Matti (1996a). Structuralism and the Interrogative Model of Inquiry. In: W. Balzer and C. Ulises-Moulines (Eds.), *Structuralist Theory of Science: Focal Issues, New Results*. Berlin: deGruyter.

Sintonen, Matti (1996b). De nouveaux instruments pour la philosophie des sciences. Paris. (forthcoming).

Simon, Herbert A. (1977). *Models of Discovery*. Dordrecht: Reidel.

von Wright, Georg H. (1957). *The Logical Problem of Induction*. Oxford: Basil Blackwell. (2nd edition)

Poznań Studies in the Philosophy of the Sciences and the Humanities
1997, Vol. 51, pp. 287–307

Mika Kiikeri

ON THE LOGICAL STRUCTURE OF LEARNING MODELS

1. Introduction

One of the most intriguing problems in behavioural and computer sciences concerns the nature of a learning process in general and first language acquisition in particular. In recent years, the problem has received much attention from scholars in these fields. This interest has brought about at least the following three approaches. First, there is the abstract learning theory which tries to formulate learning models in exact mathematical terms.[1] Secondly, there is a subfield of Artificial Intelligence in which the goal is to build working machine learning models for more or less practical purposes.[2] And thirdly, there are interesting approaches within theoretical linguistics which are related to the study of the logical problem of language acquisition. These approaches include various formal models of language acquisition formulated within modern syntactical theories.[3] These three approaches have developed relatively independently, although they put interesting constraints on each other's theoretical perspectives.

However, this list of approaches is not exhaustive.[4] It is evident that the construction of models for such a diverse and heterogenous subject as learning is not an easy matter. For one thing, it is very difficult to form any kind of overall view of the field. This fact is partly a consequence of the generality of the subject. In general, it is not even clear what is referred to by the verb "learn". Depending on the context, the term can denote many different phenomena. But this is not the whole truth. Another reason for this confusing situation is that the methodological foundations of these models have not received any systematic treatment. Since different approaches suggest different attitudes towards foundational issues, there have not been any conclusive arguments for any of them. The lack of a unifying framework has sometimes caused a feeling that the proponents of different models talk past each other.

Fortunately, it is possible to overcome at least some of these difficulties by constructing a framework within which the methodological issues can be pursued. The aim of this paper is to introduce such a new approach which is an application of the interrogative model of inquiry developed by Jaakko Hintikka and his associates (I-model, for short). I shall argue that the I-model offers promising insights into the logico-philosophical problems bothering learning models. In particular, two such important problems, namely the problem of induction and the relationship between background knowledge and data, are discussed here. Our discussion tries to call in question some of the important assumptions (i.e. the role of induction, the nature of input), and show how the alternative views can be captured by the I-model. Since the model helps us to discuss these and related problems within a general framework, it is not just another learning model. Instead, it forms, together with certain methodological views connected to it, a kind of metamodel or metatheory for various learning models, analogical to the metatheories in philosophy of science.

Before proceeding into details, let me state briefly the general idea behind Hintikka's I-model. According to it, inquiry is best reconstructed as a question-answer sequence augmented by relevant background information and logical inferences. An active agent (the inquirer) tries to establish a conclusion or to find an answer to a question by addressing further questions to an external source of answers (an oracle or Nature). She can use background information to generate relevant questions. Furthermore, she can perform logical inferences where premises come from the background information or from the previously obtained answers to questions. The inquirer's activity is the distinguishing feature of this model: she is an observer, an experimentalist or a learner who actively seeks answers to questions arising from her background information.

The rest of this article will proceed as follows. Section 2 contains a survey of the structure of various contemporary learning models as well as discuss some methodological issues. In Section 3, I shall present the fundamentals of the I-model and introduce the interrogative notion of a learning process. Methodological problems are discussed in Section 4.

2. *Structure of Learning Models*

As mentioned above, the notion of learning is so general and vague that it is very difficult to define it in any illuminating way. All the proposed definitions contain restricting idealizations which reflect one's own views. There are, however, some interesting proposals. For instance, Herbert Simon (1983, p. 28) defines learning in the following way:

> Learning denotes changes in the system that are adaptive in the sense that they enable the system to do the same task or tasks drawn from the same population more efficiently and more effectively the next time.

Simon's definition underlines the behavioral, task-oriented side of learning. It is a kind of empiristic-operationalized approach to learning, compared to other "theoretical" or mentalistic approaches which characterize learning as a process where the learner tries to construct a correct theory of her environment on the basis of a finite sample of data. According to this rival picture the learner is a scholar who tries to construct a general theory on the basis of her experiences and whatever relevant additional information she may have.[5] If we compare this to Simon's definition above, it is easy to see that the two are consistent. The construction of a theory causes "changes in the system" that can be adaptive in Simon's sense. However, it is equally easy to see that his definition is both too narrow and too general: learning a hypothesis may have no effect on the tasks the system performs, and the more efficient performance could be a result of some other factor than learning.

The formal models of learning processes have concentrated on the theory-oriented conception of learning. In fact, the existence of a formal object that is learned during the process is a presupposition of the entire approach. After these restrictions, there are possibilities that something informative can be said about the subject. Let us now try to analyze further what are the elements every learning model so restricted should contain.

One useful distinction is that between primary data and secondary data. Primary data constitute the input of a learning process. They are drawn from the learner's environment and reflect its relevant features. Secondary data, on the other hand, constitute the output state of a learning process. It describes the learner's knowledge state after the learning period. Or perhaps a better characterization is that the secondary data consist of the data that the investigator has of the learner's knowledge state after the learning period. One typical example of the secondary data are the grammatical intuitions of adult speakers. This is also the context where the distinction was first introduced. We can, however, use it more generally.

The structure of the learning situations, then, is as follows:

primary data ⟶ [????] ⟶ secondary data.

Here the problem is to find a suitable mapping which connects primary data with secondary data. Note that the primary data form usually a proper subset of the secondary data since the learner has made generalizations from the primary data.

If the situation is analyzed still further, we can isolate the following internal structure in the learning process (cf. Matthews 1989, p. 1):

primary data $\longrightarrow$ *[<T, LS, SC> → CFO]* $\longrightarrow$ secondary data.

Here *T* is the relevant background information, *LS* a learning strategy, *SC* an appropriate success criterion and *CFO* a class of formal objects that the learner can acquire in accordance with *SC*, following a strategy *LS*, on the basis of primary data and *T*. The class *CFO* is thought to generate or somehow account for the secondary data.

In the case of language aqcuisition, for instance, *CFO* is thought to consist of grammars that account for the grammatical intuitions of mature speakers. Another variant is to keep *CFO* as a class of computational devices (e.g. Turing machines) that generates all possible sequences of primary data. In both cases, the primary data form a proper subset of the secondary data, and learning is said to have been successful if the learner has picked up the correct member(s) from the class (i.e. the one(s) which generates all and only the items of the secondary data). Learning strategies are determined by suitable restrictions on the actions and the capacities of the learner. There are two types of learning strategies: general or multi-purpose strategies are domain-independent while local strategies are more or less tied to the specific features of domain.

There are many philosophical and methodological issues involved in the learning models. One problem concerns the nature and role of induction in the learning process. Are such standard conceptions of inductive inference as enumerative induction or eliminative induction operative here? The common answer to this question is affirmative, but there are also other alternatives (see Section 4.1). The challenge is to specify methods or strategies which retain the non-demonstrative character of the learning process, and diminish the role of weak, repetitive inductive procedures.

Another problem is the continuous controversy over the role and relative importance of (innate) background information and empirical experiences (e.g. Chomsky 1965, 1986, Matthews 1989b, Russell 1991). In the study of formal learning models, this controversy has many manifestations since one can use different means to incorporate background information into a learning model. Some empiristically oriented philosophers think that there either is no background information involved or if there is, it presents only general features of the learner's cognitive capacities. On the other hand, there have been influential rationalist movements in recent decades which emphasize the role of innate, genetically determined knowledge in a learning process. This innate knowledge is taken to consist of a task-specific information, so that the learning tasks and strategies in various "modules" of human cognition require different kinds of innate information to get by. This view is usually ascribed to Chomsky and his followers. They have proposed a system of innate linguistic universals as an explicate of background information (so called universal grammar-hypothesis).

Another side of this problem concerns the possible input into a learning process. For instance, is input only data which describe particular phenomena or can also stronger generalizations be included? Again, the common view is the former but I shall also examine the prospects of the latter view. The possibility of general data raises difficult epistemological questions. What is the source of strong input? What is the interplay between data, background information and induction?

3. Introduction to the Interrogative Model of Inquiry

In this section the main ingredients of the I-model will be introduced. I only describe a simple (first-order) version of the model. For a more complex version, see Hintikka et al. (1996), where more powerful conceptual resources are used. The new version also contains a general logic of questions and answers developed within the so called IF-epistemic logic (see Maunu 1993).

3.1 Logic of Questions

Epistemic logic was founded by Hintikka (1962). Subsequently, Hintikka and Åqvist[6] noticed that the logical theory of questions and answers can be given a firmer foundation in epistemic logic. The basic concepts of this epistemic-imperative erotetic logic were introduced in Hintikka (1976). In this approach, questions are conceptualized as a means of solving the tension between two knowledge states. The first of these (*presupposition*) describes the existing knowledge state which is necessary for the question to arise, the second describes the desired knowledge state (*desideratum*), the state achievable by posing a question and receiving a satisfactory answer to it. Besides these, the basic concepts are, of course, questions, answers, but also conclusiveness conditions which are epistemic criterias for the answer to be satisfactory to the inquirer. As an example, consider a typical *wh*-question such as

(1) Who hired the cook?

Now we can ask how (1) should be analyzed logically. The central notion of Hintikka–Åqvist approach is the desideratum of the question. In the case of (1), the desideratum (together with an imperative operator in front of it) is

(2) Bring it about that I know who hired the cook.

In logical symbols, this could be formulated as

(3) $!(\exists x)KH(x)$.

The presupposition and the logical formulation of this presupposition are

(4) I know that someone hired the cook,

and

(5) $K(\exists x)H(x)$.

In a special case of interrogative inquiry the inquirer answers a question by putting the desideratum as a goal or a conclusion to be established and by using the presupposition and other relevant background information as premises. The process may also involve series of *operational questions* which help to answer a *principal question*. The inquirer may, for instance, try to answer a principal wh-question by examining a series of operational yes/no-questions which query the plausibility of possible answers (provided by the relevant background information) to the principal question.

3.2 Interrogative Arguments

The interrogative model of inquiry (Hintikka 1988, 1989, 1991) could be described as a kind of a two-person game in which there are two players, the inquirer and the oracle (or Nature). The inquirer can put questions to the oracle during the course of inquiry, and, in this way, acquire information from the external source.

The inquirer has two choices at each stage of inquiry. She can make a logical move or an interrogative move, i.e., draw a deductive inference from the previously established formulas, or put a question to the oracle. If the oracle or Nature answers the question, the answer can be added to the set of established formulas.

An interrogative argument begins with an initial set of premises *T* and a conclusion *C*. The inquirer tries to establish the latter with the help of the former. The set *T* consists of the inquirer's relevant background information. They can function as premises of the deductive moves or presuppositions of the interrogative moves. Interrogative derivation differs from the usual sense of deduction since the inquirer can use the information obtained from answers as extra premises in a deduction. Consequently, interrogative derivability is generalization of the ordinary first-order notion of logical derivability or provability.

An argument is relativized to a model *M* of the languge of *T* and *C*, which could be some first-order language. The model *M* represents the domain of inquiry ("the world" or "the situation") where the interrogative game takes place. The role of model *M* is important because the answers that the inquirer receives as responses to her questions are thought to be names of individuals or functions of the domain of *M* (*dom(M)*). In this way, the answers reflect the structure of *M*.

We can capture the interrogative arguments formally by using a variant of Beth's method of semantic tableaux (Beth 1955). As usual, the table divides into the left and the right column. Formulas in the left column are thought to be true in *M*, and formulas in the right column are false in *M*. Initial premises *T* are in the left column and conclusion *C* in the right column.

If the inquirer can derive a contradiction from these assumptions (and the answers she receives), the table is closed and the alleged interrogative derivation of *C* from *T* is established. If a contradiction cannot be derived by using the usual tableau-construction rules, it is possible to construct a counterexample which shows that the supposed relation between premises and conclusion does not hold.

When the inquirer chooses a logical move, she can use the usual rules of tableau construction. For instance, the rule *(L.&)* says that if a formula *P&Q* occurs in the left column of a subtableau, *P* and *Q* may be added to the left column of the same subtableau. Accordingly, the rule *(L.v)* says that if $P \vee Q$ occurs in the left column, the entire tableaux divides into two subtableaux, where *P* is added to one of them and *Q* to the other. The rules for all the other connectives can be formulated along the same lines. In the right column, the dual counterparts of the left column rules are applied.

Finally, the interrogative moves could be dealt with by adding a rule for them:

(R.Q) If the presupposition of a question occurs in the left column of a subtableau, the inquirer may address the question to the oracle (Nature). If a conclusive answer is obtained, it could be added to the left column.

For instance, if disjunction $P \vee Q$ occurs in the left column, then the inquirer can use it as a presupposition of the propositional question

(6) *P* or *Q*?

Nature can respond by giving an answer, which is one of the two formulas *P* or *Q*. If the answer is, say, *P*, the inquirer can add it to the left column and use it as a further premise. In the same way, we can handle *wh*-questions.[7] The presuppositions of *wh*-questions are existentially quantified sentences of the form $\exists x S[x]$. If a sentence of this form occurs in the left column, the inquirer can ask the question

(7) What or who (say *x*) is such that *S[x]*?

The answers are of the form *S[b]*, where *b* is a name of the member of *dom(M)*. The answers to *wh*-questions are, then, individuals which are drawn by Nature from the domain of *M*. In fact, this task is Nature's main contribution to the interrogative process: she can give true answers or false answers, complete answers or partial answers, and there can be restrictions in the complexity of possible answers.[8]

Furthermore, in the most interesting cases, Nature is a malicious player.[9] She tries to deceive the inquirer in every possible way. What this comes up to, less metaphorically speaking, is that inquiry is unreliable because the received answers are unreliable. This unreliability can manifest in different ways: answers may be false or unrepresentative. In practice, the effect is that one play of the game is not generally enough to establish the conclusion of the interrogative argument. The answers given by malicious Nature to the very same questions can differ in the different plays of the game.

The role of possible answers is stressed when we consider more complex questions which involve both universal and existential quantifiers. The answers to these questions contain functions (so called Skolem functions) which show how the values of existentially quantified variables depend on the values of universally quantified variables.

If the inquirer manages to close the Beth tableau, viz., derive a contradiction in every subtableaux, we can say that she has interrogatively derived C from T in M. If the Inquirer has a strategy that wins (closes the tableau) whatever Nature does, we can say that C is interrogatively derivable from T in M, and express this as follows:

(8) $M: T \vdash C$

As was mentioned above, one derivation (play of the interrogative game) does not in general establish (8). The inquirer needs to have in addition a winning strategy which enables her to win in every play. Now we can also see that the notion of interrogative derivability is a generalization of the ordinary notion of logical consequence:

(9) $T \vdash C.$

If the inquirer asks no questions, then (8) reduces to (9).

3.3 An Interrogative Notion of Learning

What, then, is the relevance of the I-model to the study of the logical structures of learning models? The next section will answer this question in more details. In this subsection, we define a possible success criteria for learnability. But let us first take a closer look at the notion of learning which emerges from the I-model.

We can begin by renaming the inquirer the learner, and the oracle Nature or the environment. Now, the learner tries to seek information from the environment by putting questions to it. If the environment answers, we can say that its answers constitute the input data (or primary data) of the learning process. In addition to input data the learner can use available background information, codified as the set of initial premises T. The underlying model M could be called the domain of inquiry since it is taken to characterize the environmental situation in which the inquiry takes place.

An interesting characterization of the I-model could be achieved by using the terminology of mathematical game theory. Every questioning process could be codified as a two-person game. We can call these games *interrogative games* (*IG*, for short). Now we can associate with every alleged interrogative derivation of C from T in M an interrogative game $IG(M,T,C)$.

It is obvious that *C* is interrogatively derivable in *M* on the basis of *T* if and only if the inquirer has a winning strategy in *IG(M,T,C)*.

A general success criterion of an interrogative learning process could now be stated as follows. If the learner manages to reach her goal, viz. derive interrogatively the conclusion *C*, then we can say that she has learned *C* interrogatively or that *C* is learnable. More precisely:

(ISC) *C* is learnable (in *M* on the basis of *T*) iff *C* is interrogatively derivable from *T* in *M*, viz. $M: T \vdash C$.

The criterion can be applied to various learning situations. Conclusion *C* could be replaced by an arbitrary general hypothesis (a member of *CFO*), which would account for the learner's environment on the basis of background information *(T)*, input data (answers), and according to the domain of inquiry *(M)*. The definition of learning could be modified correspondingly.

The game-interpretation highlights the role of strategic considerations in the learning process. The success criterion for interrogative learning says that *C* is learnable just in case the Learner has a winning strategy in an interrogative game *IG(M,T,C)*. This fact is a direct consequence of the definition of interrogative derivability. The importance of strategies in learning is illuminated by our discussion of methodological problems below.

4. Methodological Problems Revisited

4.1 The Role of Induction

Learning and theory discovery are both inductive problem situations because the inquirer or the learner is trying to reach a general conclusion on the basis of a finite sample of data concerning particular phenomena. Enumerative induction is usually thought to mean just that, viz., inconclusive inferences from the set of particular cases to a general conclusion (or to another particular case).

The role of induction in learning models has always been controversial. It is often argued that simple enumerative or eliminative induction is not enough when we try to explain, for instance, language acquisition.[10] The problem is prevalent in the *AI*-systems which are thought to learn something from their environment.[11] Enumerative or eliminative induction is simply too weak: it cannot choose between vast amounts of possible hypotheses in the relatively short time available.

Despite these difficulties, some form of simple induction is usually assumed in these studies. There have not been any alternatives. As, for instance, Jerry Fodor (1980, pp. 144–5) once expressed it in the context of discussing concept acquisition:

> I take it that anybody who has ever given a theory of learning in terms of mental processes has said, in effect, that learning is a matter of inductive extrapolation, that is, of some form of nondemonstrative inference. It follows that a learning theory is some function that takes you from some sets of beliefs about your experiences into some sort of general beliefs; and secondly, that any such theory must acknowledge, among the processes involved in learning, hypothesis formation and confirmation.

One way to avoid the intrinsic limitations of these methods has been to contend that the learner has a significant amount of background information which bear on the learning task. For this reason linguists have postulated genetically endowed systems of linguistic universals, and *AI*-theorists have taken into consideration the expert knowledge (or domain theories or Bayesian priors) which can vastly reduce the number of possible hypotheses. Furthermore, even the formulation of learnable hypotheses seems to presuppose the concepts which are used in these hypotheses.

In the methodological studies of science, the focus has also shifted from the problem of induction to the role of theories, models, and other background information. Theoretical constraints are thought to account for the way scientists discover and confirm their results: empirical underdetermination can be avoided by examining the important role of prior knowledge in scientific inquiry.[12]

Similar strategies are followed in the construction of learning models. One interesting example is explanation-based learning (*EBL*, for short).[13] The motivation behind it is the observation that there are interesting situations in which the learner seems to reach general conclusions on the basis of just one example (see Russell 1991, pp. 45–6). In these situations, the learner relies on her background knowledge in order to reach an explanation of the given example. This is a kind of deductive learning process. The learner learns by deducing from her general background knowledge a more specific explanation of the example case. In this way she can generalize the information contained in the example, and relate it to the other relevant knowledge which already is on her possession.

Besides *EBL*, there are other interesting strategic methods which try to overcome the limitations of ordinary enumerative induction. These are usually discussed in the context of scientific theory-construction but they can be applied to learning, too. For example Dorling (1973) argues that there is a common deductive reasoning pattern which occurs time after time in the history of physics. He calls this reasoning pattern demonstrative induction. Demonstrative induction in Dorling's sense is a valid inference schema in which a general conclusion is deduced from a phenomenal premise (the result of experiment or low-level empirical generalization) and from general theoretical premises. Usually it is hoped that these theoretical premises are already well established. In the examples Dorling discusses they often include an indeterminate law schema which has an open empirical parameter. This parameter is then fixed by a result of an experiment and the correct form of a general law is deduced from these premises. General result is, then, achieved using only one example, as in *EBL*.

For another example, Cohen (1970) points out that Chomsky's argument for innate grammatical system, and for the impossibility of its empirical learning during a short period, also relies on the naive application of enumerative induction. He suggests an alternative method, called variational induction, by which the learner can reduce the number of possible hypotheses using only few empirical examples. The idea is the same as in many traditional methodologies (such as Whewell's consilience of inductions), namely, that the power of inductive process depends on the sufficient variability of evidence, not only on the number of instances.

By the help of the I-model these various strategies can be generalized and given a unified treatment. In interrogative arguments, deductive reasoning is combined with the examples or answers received from the external source. The relative weight of these components can vary case by case. If inquiry is data-oriented, the strategies of questioning are important since they show how data is best collected and how it helps the inquirer to reach her goal. In other cases, the focus can be on deductive reasoning which uses relevant background knowledge. This situation is prevalent in *EBL* models. It also reminds us of Plato's original solution to the Meno Paradox, so called learning-as-recollection-hypothesis. It can be interpreted as deductive learning process arising from the concepts and propositions the learner (such as Plato's slave boy) already knows (for this interpretation, see Moravcsik 1971). Furthermore, Dorling's schema of demonstrative induction is but an instance from a more general form of interrogative argument, and the demand of the variability of evidence is the common strategy of questioning which can be incorporated into the I-model.

The discussion above puts the role of induction in a learning process into a different light. In fact, it is possible, in some cases, to abandon the whole notion in its usual sense and replace it by deduction from input data and background knowledge to general conclusions. It was assumed that the results of interrogative moves (available answers) are all true. In some cases we could give up this assumption, and, instead, assume that the answers are only probable. But in this case, too, only deductive inferences are made. The premises used can be uncertain but the connection between them and the conclusion is still deductive. In this way, the role of induction could be reevaluated in the framework of the I-model.

Furthermore, these results and examples discussed seem to be generalizable. Hintikka (1992a) has suggested that there are two different concepts of induction. First, there is the usual Humean one, e.g. ordinary enumerative induction. But secondly, there is the Newtonian variant in which induction in the former sense plays no role. Instead, this variant interprets the results of experiments and other data-gathering activities to be partial generalizations, possibly established on the basis of just one example. The upshot is that these generalizations are only partial: their range of application may be severely restricted. In this case, induction means the extension of their range which may include a reconcialiation of several partial empirical generalizations into more general laws.

Hence, the Newtonian sense of induction is a form of inference in which general results are established on the basis of only one or few pieces of evidence. It is a kind of instance induction: ampliative inference in which the role of evidence is reduced to a minimum. It has not received much attention, although it seems to have an important role in the history of science as well as in various contemporary applications. Arguably, Newton used this form of induction both in *Principia* and *Opticks* (Hintikka 1992a, see also Mäenpää 1993). The strategies of learning we have discussed in this section can also be interpreted as involving instances of Newtonian induction.

In learning models, the success of instance induction was based on the deductive strategies applied on the store of background knowledge. Hintikka's original claim, however, was that general hypothesis can be established even without such strong prior knowledge. This is a strong claim which, I think, cannot be sustained in general. The apparent plausibility of this view lies in the fact that sometimes the role of background knowledge is left implicit.

But there is another logical possibility: if the acquired data is logically stronger than is usually assumed or allowed, then the role of background knowledge can, *mutatis mutandis*, be less important. This possibility will be discussed in the next section.

4.2 The Relationship Between Data and Background Knowledge

The previous section showed that the learner can utilize a rich store of background knowledge in many ways. Unfortunately, such a store is not always available: in some situations the learner has only weak initial premises, and her inquiry is more data-oriented. Another problem is to choose learning strategies which make it possible to use the available background knowledge in an efficient way. The interplay between data and background knowledge may depend on this decision.

It is usually assumed that the input data of a learning process is weak. Empirical studies of language learning, for instance, point towards a relatively small sample of sentences or utterances, and possibly of some knowledge of their syntactic structure. In order to account for these results, learning theorists have been forced to put strong restrictions on available data. In the terminology of logic, it is assumed that the input of a learning process consists of negated or unnegated atomic propositions. The learner can acquire from her environment only information that concerns particular phenomena. This assumption, which could be called the Atomistic Postulate (see Hintikka 1988), characterizes most of the views about learning and scientific inquiry.

This restrictive postulate can, nevertheless, be abandoned by allowing more logical complexity in the input data (available answers). In this respect the I-model is useful: the nature of the different types of inquiry is captured by varying the complexity of available answers. Moreover, we can define a simple logical measure which shows the logical complexity of sentences, and apply it to the study of input and background knowledge.

The measure of the logical complexity is called *AE*-hierarchy (Hintikka 1988). It measures the complexity of the quantifier prefix of a logical expression by the number of changes between different kinds of quantifier strings (e.g. strings of universal and existential quantifiers). Thus, for example, $\forall x\ \exists y$-prefix has an A^2-complexity (also called an *AE*-complexity), and $\exists x\ \forall y$-prefix has an E^2-complexity. What the Atomic Postulate then means, is that the input data of a learning process is restricted to an $A^0 = E^0$ case, i.e., to expressions without quantifiers. On the other hand, the I-model allows the answers to be restricted in any way whatsoever on the *AE*-hierarchy. Our task now is to examine whether there are interesting learning situations which do not conform to the atomistic restrictions, and what ways (if any) there are to avoid these restrictions when empirical learning situtations are modelled. Notice that even if empirical studies show that the data is atomistic, we can still try to avoid this restriction in our learning models: the model can describe and explain the level of learning process in which the data is already generalized or classified.

Can we, then, show that the Atomistic Postulate does not do justice to learning models? This is especially important for the practical applications of these models (e.g. language acquisition) because in these cases the rejection of the Atomistic Postulate has considerable methodological importance. The usual assumption of language acquisition models is the linguistic counterpart of the Atomistic Postulate, which says that the learner experiences only particular utterances during a learning period.

The most interesting way to make a case against the Atomistic Postulate is to distinguish between various levels of inquiry. The freedom to vary the strength of data is closely related to this notion of many-level process in some applications of the I-model. The interrogative inquiry could be characterized as a hierarchical structure in which the lower level inquiry seeks answers to general questions imported from the upper level. In other words, the principal question at the lower level is only an operational question at the upper level. In scientific applications, the lower level inquiry codifies the experimental and observational aspects of inquiry whereas the upper level captures the theoretical aspects of scientific enterprise (Hintikka 1992b). The difference between these two levels is manifested in the difference of the strategies employed at each level. The upper level data could also contain general statements that are obtained from the lower level.

In the case of learning processes, the many-level-idea seems to correspond to the situation in which the learner receives data at different levels of abstraction. At the lowest level a child only hears segments of sound, not words or sentences. Hence, the raw data of language acquisition models could be sound stimulus received at the neurological level. In practise, however, all learning models use as data the more refined, preanalyzed information, already classified by some lower-level device. Consequently, it is usually assumed that a child as a language learner receives as data, not sound-stimulus, but words and sentences which are syntactically categorized and structured (Wexler & Culicover 1980, pp. 60–3). This is an indispensable theoretical idealization for these models.

But is it possible to generalize these considerations? Could a child operate at the level on which the data is even more abstract than syntactically structured sentences, say, general rules of grammar? Here the analogy with the theory-construction in science may help

For a scientist, it is possible to learn empirically underdetermined general theories or hypothesis. A rich store of background knowledge is essential, but not enough. A scientist also need strategic principles which help to organize it and relate it to the evidence. These strategies vary from data-gathering and experimenting to strategies of theoretical calculation and explanation. Theory-construction process is a sum of all these strategic inquiries. In all of them the required input is different. Hence, theoretical inquiry may need strong input; perhaps empirical laws or general theoretical constraints could be possible answers for the inquirer's questions at this level. The point is, of course, that these candidate answers are already there, supplied by the other parts of the system.

The analogy should now clear up the issue in the case of learning models. The idea is to assume that same kind of division of labour can be ascribed to a learning model. Take again language acquisition as an example. The child's learning device operates on various levels. On the upper levels the device tries to construct grammatical hypothesis which account for the available data. But here the data is already generalized, consisting of the rules that the lower level inquiry has established on the basis of example sentences received from the environment.

Notice that the present view is not committed to Chomsky's strong nativistic position. The need for domain-specific prior knowledge is much smaller, and the learner only needs some theory-construction device which can implement the needed operations. Our suggestion, of course, is that the I-model is a suitable candidate. We shall soon compare it to another proposal, first though two alleged objections to our view (we can call it the construction view).

First, it could be held that the construction view does not reduce the need for strong background knowledge since the learner is still operating with the weak primary data. Even if the upper level inquiry receives strong input it merely moves the need of strong prior knowledge to the lower level. For how can a child, for instance, construct all the grammatical rules for the upper level grammar construction inquiry if she is not even hearing utterances in which all information needed is included?

But the construction view does not imply the deletion of all background knowledge, be it domain-specific or general-purpose. Intuitively, the need for strong knowledge is diminished, although it is not possible to say exactly how much without more detailed (and more empirical) research. Furthermore, the levels are not independent of each other, and the construction of a grammar and other learnable objects is an incremental, step-by-step process. As more data is received, the hypothesis becomes more accurate, and predicts more and more new instances (such as the generation of novel grammatical sentences in language acquisition models). But the initial simple hypotheses could be constructed on the basis of weak primary data alone. Thus, learning is a kind of dialectical process.

Second, it could be objected that the question whether the learning devices are, as we have argued, many-levelled is an empirical one, and it can be conclusively answered only by empirical investigations. This is indeed the case, but our suggestion is meant to be methodological: it can be rewarding to base the empirical models on the construction view rather than on the assumption of strong domain-specific prior knowledge. At least it is a viable alternative to this latter, more common view.

As mentioned, there has been at least one similar kind of proposal, discussed in the context of Chomsky's theory of language acquisition. McGinn (1981) has proposed a view he called the creativity proposal. It is reminiscient of the construction view of language acquisition in some respects but its methodological implications are closer to the hypothetico-deductive methodology. McGinn writes (*ibid.*, p. 295):

> Somewhat on the model of the creative scientist, we might conceive of the child as possessed of the capacity creatively to generate grammatical hypotheses about the language spoken around him, testing these in his own speech and by observing the speech of others. The child is said to have acquired mature linguistic competence when his grammatical hypotheses have reached maximum predictive and explanatory power. The rules of grammar, on this suggestion, would be no more innate than the propositions of quantum physics.

All this is in agreement with the construction view. But McGinn also seems to hold that hypotheses are results of intellectual creativity, produced "seemingly ex nihilo". In other words, the processes behind "creative hypotheses generation" are left without explanation. The creativity proposal is thus a variant of the hypothetico-deductive procedure applied to the domain of language acquisition. Like the H-D-view in general, it leaves the invention or discovery of hypotheses without rational explanation.

The I-model is better in this respect: it gives an account of both discovery and justification, and stresses their interdepedence. Both forms of inquiry can be reconstructed by nearly similar kind of rules (specified by the I-model), and their difference is manifested only in the strategic considerations. The inquirer's strategies of discovery are based on goals and epistemic utilities which differ from her strategies of justification.

Although I cannot argue in detail for this view here, its importance for the construction view is easy to appreciate. For its success depends essentially on the possibility to mix invention and testing procedures on the same level and between the levels, too (see Sintonen 1993). The result is the model which is based on a kind of bootstrap strategy by which the step-by-step correction of hypothesis could be captured.

In this section, we have focused on language acquisition because the role of background knowledge is most widely discussed in this context. Since the prior knowledge is thought to be innate and learning occurs on the child's subconscious level, this role is somewhat peculiar in these models. However, the discussion here could be generalized to the other learning models, too. In fact, it might be easier to argue for the virtues of the construction view and the I-model in these other cases. If the store of background knowledge and the primary data are more strictly defined than in language acquisition studies, their exact relationship could be more easily examined (see, for instance, Kelly & Glymour 1990). The same point applies to the rather vague division between levels of learning: the division could be based on the more exact logical formulation by reconstructing the models into the framework of the I-model. This formulation employs the distinction between principal and operational questions as well as the notion of *AE*-hierarchy imposed on the logical reconstructions of questions and answers.

Department of Mathematical Sciences/Philosophy
University of Tampere
P.O. Box 607
33101 Tampere
FINLAND

NOTES

1 See Gold (1967), Osherson et al. (1986), Valiant (1984), Angluin & Smith (1983, 1987).

2 See Carbonell & Langley (1987), Thagard (1990).

3 See Pinker (1979), Wexler & Culicover (1980).

4 It does not mention, for instance, the stochastic stimulus-response models associated with behaviorist psychology, and their more recent variants which are called connectionist or neural network models.

5 For example Chomsky (1959, p. 59) writes, in the context of discussing language aqcuisition, that "the young child has succeeded in carrying out what from the formal point of view, at least, seems to be a remarkable type of theory construction", and that "it is reasonable to regard the grammar of a language *L* ideally as a mechanism that provides an enumeration of the sentences of *L* in something like the way in which a deductive theory gives an enumeration of a set of theorems" (p. 58). Chomsky is well-known from his mentalistic approach to linguistic capacities such as language acquisition. Indeed, the theory-view of learning is usually associated with mentalistic approaches to psychological processes. However, it is better to keep them apart because some formal learning models which accept the theory-view do not depend on any psychological interpretation of their results.

6 See Åqvist (1975).

7 These include what-, where-, which- and who-questions. The logic of questions and semantics of different question statements is discussed in Hintikka (1976).

8 The very notion of Nature as a player does not, of course, involve any antropomorphic implications. It is a metaphor which enables us to describe different environmental situations and respective restrictions to the inquiry in a compact way.

9 In game-theoretical terms, the interrogative games in question are non-cooperative games.

10 See Cohen (1970), for a relevant discussion.

11 See Russell (1991), Bergadano (1993). This is not to say, however, that inductive inference is not studied in this context. Far from it since a great number of models based on some form of enumerative induction are proposed, and important results have been achieved. For a review of these results, see Angluin & Smith (1983, 1987). It is only that these results often function as negative ones: they may show, for instance, that it is not possible to solve a given learning problem using inductive inference.

12 For a more detailed discussion of how background knowledge narrows down both questions and admissible answers, see Sintonen (1989, 1996).

13 For *EBL*, see Mitchell *et al.* (1986).

REFERENCES

Angluin, D., and Smith, C.H. (1983). Inductive Inference: Theory and Method. *Computing Surveys* **15**, 237–69.

Angluin, D. and Smith, C.H. (1987). Inductive Inference. In: Shapiro (1987), 409–18.

Åqvist, L. (1975). *A New Approach to the Logical Theory of Interrogatives*. Tübingen: TBL Verlag Gunter Narr.

Bergadano, F. (1993). Machine Learning and the Foundations of Inductive Inference. *Minds and Machines* **3**, 31–51.

Beth, E.W. (1955). Semantical Entailment and Formal Derivability. In: Hintikka, J. (Ed.), *The Philosophy of Mathematics*. Oxford: Oxford UP, 1969.

Carbonell, J. and Langley, P. (1987). Machine Learning. In: Shapiro (1987), 464–88.

Chomsky, N. (1959). Review of B.F. Skinner's *Verbal Behavior*. *Language* **35**, 26–58.

Chomsky, N. (1965). *Aspects of the Theory of Syntax*. Cambridge, Mass.: The MIT Press.

Chomsky, N. (1986). *Knowledge of Language*. New York: Praeger.

Cohen, L.J. (1970). Some Applications of Inductive Logic to the Theory of Language. *American Philosophical Quarterly* **7**, 299–310.

Dorling, J. (1973). Demonstrative Induction: Its Significant Role in the History of Physics. *Philosophy of Science* **40**, 360–72.

Fodor, J.A. (1980). Fixation of Belief and Concept Acquisition. In: Piattelli-Palmarini (1980), 143–58.

Gold, E.M. (1967). Language Identification in the Limit. *Information and Control* **16**, 447–74.

Hintikka, J. (1962). *Knowledge and Belief.* Ithaca: Cornell University Press.
Hintikka, J. (1976). *The Semantics of Questions and Questions of Semantics. Acta Philosophica Fennica* **28**. Amsterdam: North-Holland.
Hintikka, J. (1988). What is the Logic of Experimental Inquiry? *Synthese* **74**, 173–90.
Hintikka, J. (1989). Knowledge Representation and the Interrogative Model of Inquiry. In: Clay, M., Lehrer, K. (Eds.), *Knowledge and Skepticism*. Boulder, CO: Westview Press, 155–83.
Hintikka, J. (1991). Toward a General Theory of Identiafiability. In: Fetzer, J.H., D. Schatz, and G. Schlesinger (Eds.), *Definitions and Definability: Philosophical Perspective*. Dordrecht: Kluwer Academic.
Hintikka, J. (1992a). The Concept of Induction in the Light of the Interrogative Approach to Inquiry. In: J. Earman (Ed.), *Inference, Explanation, and Other Philosophical Frustrations*. Berkeley: University of California Press.
Hintikka, J. (1992b). Theory-ladenness of Observations as a Test Case of Kuhn's Approach to Scientific Inquiry. In: *PSA 1992, vol. 1.*
Hintikka, J., Halonen, I., and Mutanen, A. (1996). *Interrogative Logic As a General Theory of Reasoning*. Manuscript.
Kelly, K.T. and Glymour, C. (1990). Theory Discovery From Data With Mixed Quantifiers. *Journal of Philosophical Logic* **19**, 1–33.
Matthews, R.J. (1989a). Introduction: Learnability and Linguistic Theory. In: Matthews and Demopoulos (1989), 1–17.
Matthews, R.J. (1989b). The Plausibility of Rationalism. In: Matthews and Demopoulos (1989), 51–75.
Matthews, R.J. and Demopoulos, W. (Eds.) (1989). *Learnability and Linguistic Theory*. Dordrecht: Kluwer Academic Publishers.
Maunu, A. (1993). Independence-Friendly Epistemic Logic and Its Uses. In: Koistinen, O. and J. Räikkä (Eds.), *Good Reason: Essays Dedicated to Risto Hilpinen*. Annales Universitatis Turkuensis, Ser. B, Tom. 200. Turku: University of Turku, 77–107.
McGinn, C. (1981). Review of *Rules and Representations* by Noam Chomsky. *Journal of Philosophy* **78**, 288–98.
Michalski, R.S., Carbonell, J.G., Mitchell, T.M. (Eds.) (1983). *Machine Learning*. Los Altos, CA: Morgan Kaufmann.
Mitchell, T., Keller, R.M., and KedarCabelli, S. (1986). Explanation-Based Generalization: a Unifying View. *Machine Learning* **1**, 47–80.
Moravcsik, J. (1971). Learning As Recollection. In: G. Vlastos (Ed.) *Plato I: Metaphysics and Epistemology*. New York: Doubleday. Reprinted in: J.M. Day (Ed.) (1994). *Plato's Meno in Focus*. London: Routledge.
Mäenpää, P. (1993). *The Art of Analysis: Logic and History of Problem Solving*. Ph.D. Dissertation. Department of Philosophy, University of Helsinki. Helsinki: Limes.
Osherson, D.N., Stob, M., and Weinstein, S. (1986). *Systems That Learn*. Cambridge Mass.: The MIT Press.
Piattelli-Palmarini, M. (Ed.) (1980). *Language and Learning: The Debate between Jean Piaget and Noam Chomsky*. Cambridge, Mass.: Harvard University Press.
Pinker, S. (1979). Formal Models of Language Learning. *Cognition* **7**, 217–83.
Russell, S. (1991). Inductive Learning by Machines. *Philosophical Studies* **64**, 37–64.

Shapiro, S.C. (Ed.) (1987). *Encyclopedia of Artificial Intelligence, vol. 1*. New York: John Wiley & Sons.

Simon, H.A. (1983). Why Should Machines Learn? In: Michalski et al. (1983), 25–38.

Sintonen, M. (1989). Explanation: In Search of the Rationale. In: Kitcher, P. and W.C. Salmon (Eds.) (1993). *Scientific Explanation*. Minnesota Studies in the Philosophy of Science, Vol. 13. Minneapolis: University of Minnesota Press, 253–82.

Sintonen, M. (1993). In Search of Explanations: From Why-Questions to Shakespearean Questions. *Philosophica* **51**, 55–81.

Sintonen, M. (1996). Structuralism and the Interrogative Model of Inquiry. In: Balzer, W., and C.U. Moulines (Eds.) (1996). *Structuralist Theory of Science: Focal Issues, New Results*. Berlin: deGruyter.

Thagard, P. (1990). Philosophy and Machine Learning. *Canadian Journal of Philosophy* **20**, 261–76.

Valiant, L.G. (1984). A Theory of the Learnable. *Communications of the ACM* **27**, 1134–42.

Wexler, K. and Culicover, P.W. (1980). *Formal Principles of Language Acquisition*. Cambridge, Mass.: The MIT Press.

V

JAAKKO HINTIKKA: REPLIES

COMMENT ON ZEV BECHLER

I am flattered and honored by Zev Bechler's acknowledgment of my influence on his thinking. Perhaps I can repay a small part of that compliment by saying that his paper has set my thoughts in motion and suggested a further generalization of the ideas that he and I have both discussed in the context of Aristotle's thought.

I agree by and large with what Zev Bechler says in his paper on Aristotle. In my 1977 monograph I discussed several ideas or perhaps conceptual models that together shaped Aristotle's theory and practice of modality. Zev Bechler focuses his attention on one of them. This is the temporary or extensional interpretation of modalities: possibility equals sometime truth and necessity omnitemporal truth. This interpretation is equivalent with what is usually referred to as the Principle of Plenitude which asserts that every genuine possibility is actualized sooner or later. As Bechler shows, this assumption pushes Aristotle to a kind of determinism, that is to say, to a position where potentialities do not do any explanatory work. The "notion of actual or real potentiality can find no place within Aristotle's conceptual scheme".

I agree by and large with what Zev Bechler says, even where he seems to think that he is criticizing me. The main case in point is my suggestion that Aristotle used the notion of *kinesis* as a way out of his "logical determinism", to use Zev Bechler's locution. I still think that Aristotle tried to do so, but I noted already in the original monograph that he was not really successful even if he could verbally reject the deterministic (Megarian) position. "Aristotle's attempted solution to the Megarian problem in terms of the *kinesis-energeia* distinction did not enable him to deal with the problem of determinism proper in a satisfactory manner" (Hintikka 1977, p. 101).

There is an aspect of Aristotle's theory of modality which I have not dealt with adequately, and I am not sure that Zev Bechler has done so, either. The really striking consequences of the Principle of Plenitude ensue only when the notion of possibility used there is taken to be what we twentieth-century philosophers would call logical or conceptual possibility. In the 1977 monograph I showed in some detail that Aristotle did not distinguish from each other logical and natural (physical) possibility and necessity. However, I did not explain why he did so. Bechler in effect relies on the same identity, but I do not see that he explains what its roots in Aristotle really were.

I have come to think that there was in Aristotle an assumption which is in a way more general that the Principle of Plenitude and which partly motivated his acceptance of both the Principle of Plenitude and the coincidence of logical and natural modalities. This assumption might be called the one-world assumption. According to it, the only reality that there is, is the actual course of events, the one and only succession of actually present *nows*. It is quite obvious that Aristotle held the one-world view. Indeed, the idea of other possible worlds came to be taken seriously only in the middle ages. (See Knuuttila 1981.)

The one-world assumption is made historically interesting by the role it has played in the history of philosophy and in contemporary philosophy. Indeed, I have recently argued that it is one of the basic assumptions of Quine's philosophy. Even though it is closely related to the so-called Principle of Plenitude, it is important for the sake of conceptual clarity to keep the two apart.

Some further remarks are needed here. Zev Bechler doubts that the Principle of Plenitude was accepted by Aristotle in a form strong enough to rule out what he calls "consistence-potentials", even unrealized ones. They are essentially unrealized logical possibilities. They will also be counter-examples to strictly taken Principle of Plenitude.

Are there such counter-examples? I am perfectly prepared to think that Aristotle was not always completely consistent. However, I am not even sure that I have to maintain that. It now seems to me that the right way of looking at the *prima facie* counter-examples is to ascertain their status in Aristotle's argumentation. The most explicit *prima facie* counter-examples are *De Int.* 9, 19a 12–14 (the coat that could be cut but wears out first) and *De Gen. et Corr.* 316a 23 (quoted by Bechler). But both these examples are from preliminary arguments pro and con which Aristotle marshals to show his problem as well as to provide *endoxa* for their solution. Such preliminary arguments do not represent Aristotle's settled view; he is in them not speaking in his own voice. Hence I do not find these two passages at all convincing.

Other *prima facie* counter-examples are more complex. I cannot deal with them here. I do not think that they are conclusive, either.

COMMENT ON MARJA-LIISA KAKKURI-KNUUTTILA

Marja-Liisa Kakkuri-Knuuttila has studied Aristotle's dialectic in an interesting and enlightening manner. It is to be hoped that her results will soon be made available to a wider scholarly audience. In the present essay, she has undertaken to suggest certain morals of Aristotle's story – meaning,

Aristotle's rhetorical and dialectical works – for our contemporary methods of argumentation in social and human studies.

I agree with Dr. Kakkuri-Knuuttila that there are interesting lessons to be learned from the Aristotelian tradition especially when it comes to what used to be called "moral sciences". However, I think that they are somewhat different from the suggestions that Kakkuri-Knuuttila makes. She offers the dialectical method as a solution to the alleged problems to which positivists' use of purely logical tools led them. This motivation turns out on a closer examination to be a mirage. Appeals to Thomas Kuhn ring hollow here. He became almost as appalled by his followers' use of such ill-defined ideas of "paradigm" as I am. Kuhn's and other philosophers' vague concept of theory-ladenness turns out to be explicable in sober consequentionalist terms, as I have shown in Hintikka (1992d). The special character of historical explanations can be given a simple explanation in information-theoretical terms; see Hintikka (1968). Even the bugbear of the "new philosophers of science", the covering law model, receives a qualified but striking *Ehrenrettung*. (See Hintikka and Halonen, forthcoming.) In all these case studies progress is made, not by using less logic, but more powerful logic, including epistemic logic. In general, it is turning out that the ideas of logical empiricists offer much better guidance in trying to reach real insights into the scientific process than Kuhn and his ilk.

Moreover, as far as Aristotle is concerned, he makes it absolutely clear that the methods of persuasion used in rhetoric are at bottom the same as the methods of scientific argumentation. This is shown by *Rhet.* I, 2, especially 1356 b 1 ff. (Cf. here my 1993a.) It would be a gross distortion of the leading ideas of Aristotle's approach to think that he has a separate rhetorical and/or dialectical methodology over and above his theory of scientific argumentation. Hence, whatever new insights Aristotle's methodology can suggest, they have to come from an analysis of his theory of argumentation in general. This is in fact what Dr. Kakkuri-Knuuttila does in her interesting discussions of notions like *topos* and *endoxa*. The main thrust of Aristotle's theory of argumentation is clear. Knowledge-seeking is a questioning process, not unlike the Socratic *elenchus* or my latter-day interrogative model of inquiry.

Such models suggest an interesting possibility of conceptualizing epistemic, moral and legal authorities that has not been utilized by twentieth-century philosophers. Take moral reasoning as an example. An interrogative inquiry uses two kinds of steps: deductive and interrogative steps. Now virtually all ethicists consider moral principles as premises for deductive steps, something like initial premises. It seems to me much more realistic to consider them instead as sources of answers ("oracles" in my technical sense). In other words, the prescriptions of the moral system in question are treated like the testimony of a witness in a court of law or the suggestions of an advisor and not like pronouncements of a judge. This makes it possible to use a moral authority selectively, subject such an authority to rational criticism (as the testimony of witness is critically weighed), consult another authority without having to reject the other one *in toto* etc. This would result in a much more realistic model of our actual moral and legal reasoning than the inference-oriented models that have dominated most of the discussion in recent decades and indeed recent centuries.

It would be a highly interesting and worthwhile task to relate this deliberative model to Aristotle's notions of *endoxa* and *topos*. This would be a promising – and challenging – application of Dr. Kakkuri-Knuuttila's ideas.

COMMENT ON MARTIN KUSCH

Coming to know the work of one's predecessors is an interesting experience. Sometimes I have found my ideas unexpectedly anticipated by earlier thinkers. This has happened to me twice in an important way in that I have found myself anticipated by C.S. Peirce. (See Hintikka 1980b; Hilpinen 1980.) In the case of the turn-of-the-century writers Kusch examines, no such surprises emerge. Rather, it seems to me we can have a glimpse of the reasons why the nature of questions and their role in human knowledge-seeking did not receive the attention they deserve. The writers Kusch deals with offer us a fair testing ground in that they not only studied questions but at least in some cases emphasized their importance in human cognition.

In reading Martin Kusch's paper, one meets all sorts of familiar ideas, such as the distinction between propositional questions and wh-questions, presuppositions of questions, etc. But I did not have any real shock of recognition, comparable with what I have experienced in studying Aristotle's interrogative dialectic, e.g. his remarks on the alleged fallacy of *petitio principii* (see Hintikka 1987; Robinson 1971).

As far as the logical and semantical properties of questions are concerned, the crucial fact should be obvious. "Question" and "answer" are concepts that cannot be handled on the level of propositions or judgments. They are discourse concepts. This does not make them any less objective notions or any less amenable to logical and semantical analysis, but the semantics must be discourse (dialogue) semantics, not propositional semantics.

One of the factors that hampered the development of a full-fledged theory of questions and answers is connected with this point. It is very much in evidence in the writers Martin Kusch discusses. It is the assimilation of questions to judgments or propositions of some sort or other. This is one of the interesting common denominators in the writers discussed in Kusch's section on "Questions and Judgments". It also explains why the writers Kusch is dealing with never developed any real logical or semantical theory of questions. For however it is that you propose to deal with judgments (and whatever they may be like), they are someone's judgments, standing conceptually on their own feet independently of the context. They belong to the theory of propositions, not to say judgments, not to discourse theory.

As far as the role of questions in knowledge-seeking is concerned, there is no trace in Kusch's paper of any attention on the part of the turn-of-the-century theorists to strategies of questioning or even sequences of questions. This is all the more remarkable as Kant had in a famous passage (*Critique of Pure Reason* B xiii) evoked precisely the role of questions as guiding inquiry. And without taking such strategic matters as the choice of the right questions into account one cannot hope to develop an adequate epistemology of questions and answers. (Cf. my 1996e.)

In brief, Kusch's paper shows indirectly that our recent theories of the logic of questions and answers as well of knowledge-seeking by questioning do go essentially beyond what the thinkers he discusses were doing.

COMMENT ON NILS-ERIC SAHLIN

The interaction between Ludwig Wittgenstein and Frank Ramsey is one of the most interesting episodes (in my biased judgment, arguably the most interesting episode) in the history of twentieth-century philosophy. It is also full of ambiguities. In his list of the most important influences on his way of thinking, Wittgenstein does not list Ramsey. Yet in the preface to the *Philosophical Investigations*, Wittgenstein acknowledges Ramsey's invaluable help in bringing him to see the mistakes he committed in his early philosophy. As Sahlin aptly tells us, in September 1923 Wittgenstein was in Ramsey's eyes a greater man than Moore, yet in March 1924 Wittgenstein was "no good for my work".

Wittgenstein acknowledges the role of "innumerable conversations" with Ramsey in shaping his thought. Yet even his 1929 notebooks show relatively few obvious signs of the effects of those conversations.

Sahlin performs an interesting thought-experiment in considering Wittgensteinian language-games as further developments of Ramsey's pragmatist ideas. This is a fascinating perspective. I agree for instance that Ramsey's decision-theoretical analysis of belief is precisely the kind of explication of the notion of belief that exemplifies a Wittgensteinian language-game for belief. (If Wittgenstein did not accept it as such, he ought to have done so.) I find this thrust of Sahlin's paper truly illuminating.

To what extent one can speak of the two thinkers influencing each other is none the less a delicate question. I agree with Sahlin that mere textual exegesis is unlikely to create genuine insights into the matter. Yet it is important to realize the overall structure of Wittgenstein's argumentation and of his development. Without taking into account such contextual matters, one cannot tell the whole story. For instance, the fascinating story of the development of Wittgenstein's and Ramsey's ideas about probability which Sahlin tells us is thrown into a sharper relief if we realize that the first main step Wittgenstein apparently took away from his position in the *Tractatus* was to give up the logical independence of elementary propositions of each other. (See here Hintikka 1988b.) This immediately destroys the kind of simple theory of logical probability Wittgenstein envisaged in the *Tractatus*. This parallels Ramsey's rejection of Keynes's views on probability, as Sahlin indeed points out. Which insight – and whose insight – was reached first can nevertheless be ascertained only through a careful study.

I have developed a partial interpretation of the main outlines of the philosophical interaction between Wittgenstein and Ramsey. Very briefly, the story runs as follows: What young Frank Ramsey thought he had found in the *Tractatus* was a completely extensional view of logic and mathematics, including what Henkin later called the standard interpretation of higher-order variables. This view was the basis of Ramsey's early work on the foundations of mathematics (e.g. his 1926). A crucial role in this view is played by the "interpretation" of quantified sentences as long conjunctions and disjunctions.

Later, Ramsey came to realize that Wittgenstein's extensional logic was merely a way of arguing for the picture theory and that Wittgenstein's true view of mathematics was an operational one. Independently Ramsey came to think of his infinitory extensional logic as being incompatible with his pragmatism. This view seems to have been self-generated (or inspired by Weyl) rather than a result of Wittgenstein's influence.

This brought Ramsey and Wittgenstein close to each other in 1929 in the philosophy of mathematics. In my view, they both failed in the same way, viz. in not realizing what the true language-games for quantifiers are, viz. the "games" of seeking and finding.

Be this as it may, the interaction of the two thinkers will repay further investigation. And I cannot think of a more inspiring starting-point for such an investigation than Sahlin's paper.

COMMENT ON THEO KUIPERS

There is little I can add to and nothing that I want to change in the details of Theo Kuipers's survey of Carnap-style approach to inductive logic. However, some general comments are in order to put this work in perspective.

(i) Theo Kuipers speaks in the very title of his paper of "the Carnap–Hintikka Programme in inductive logic". This may be appropriate if one considers merely the technical problems of constructing an inductive logic. However, it does not do justice to my philosophical position. (Cf. here Hintikka 1993b.) Carnap's project was to construct a system of purely logical probability. Even his relatively simple λ-system contains an open parameter λ. Its subjective meaning is clear: it is an index of caution. Its objective meaning is likewise clear: the optimal choice of λ depends on the order (suitably understood) in the universe. But for the very reasons that these interpretations of λ bring out, it cannot be fixed by purely logical considerations. Hence we have a reductio *ad absurdum* of Carnap's philosophical enterprise of constructing a purely logical sense of inductive probability.

One can respond to this by considering degrees of inductive probability merely as mirror images of suitable measures of information. These measures are relative to a given conceptual scheme (language system), but they can be considered as logical and hence objective. However, my work (with Niiniluoto) on more complex systems that allow us to treat inductive generalization shows that this idea does not yield a unique notion of inductive probability. There are many different types of order and disorder that cannot be weighted by purely *a priori* considerations.

(ii) Because of the presence of parameters like λ, the kind of theory of inductive inference I have envisaged is not purely Bayesian. The reason is that since there are parameters in the system whose optimal choice depends on the order in one's universe, experience may lead us to revise our ideas of this order and hence change our choice of the relevant parameters. The resulting change in our *a posteriori* probabilities is not purely Bayesian any longer.

(iii) As far as different overall approaches to scientific inference are concerned, there are at least two alternatives to straightforward inductive logic. It seems to me that there is still a great deal work to be done in and on approaches where scientific inference is thought of as maximizing expected information. One advantage of such approaches is that differences between different kinds of information can be taken into account. In hindsight, I regret that I did not develop further the ideas presented in Hintikka (1968).

Even more radically, it may be suggested that since inductive inference depends in any case on assumptions concerning the orderliness of order in one's universe of discourse, such assumptions should be expressed by explicit premises rather than choices of the value of a parameter. Then inductive logic would become simply a part of deductive logic. This would not trivialize inductive logic, for those premises themselves could be considered as lines of interrogative inquiry which can be criticized, evaluated and even rejected in the course of interrogative inquiry. This approach would turn the rules of inductive inference from definitory rules (analogous to the so-called rules of deductive inference) to strategic principles (or rules).

COMMENT ON ISAAC LEVI

In his paper, Isaac Levi adduces reasons to doubt one of the basic assumptions of recent work on rational belief change. This assumption is that the principles of ampliative reasoning are identical with the principles of belief contravening reasoning, both in the sense explained by Levi. His argument is closely argued and complex. It touches on the problems of inductive inference and inductive acceptance which I worked on with by colleagues in Finland in the sixties and seventies. I am flattered to see that Isaac Levi speaks of "the Finnish School" by reference to that work.

I cannot in this comment go back to the ideas of the Finnish School and try to see how they mesh with Levi's argument. It seems to me more useful to take the ideas of what I would like to call the Second Finnish School and relate them to Levi's argument. By the Second Finnish School I refer of course to the work that is based on the interrogative approach to inquiry. One of the many so far unattended tasks concerning this approach is to relate it to, and to bring it to bear on, theories of belief revision, non-monotonic reasoning, or even inductive inference. I cannot undertake this task here. Some things are nevertheless both obvious and related to Levi's argument.

From a questioning-theoretical viewpoint, we have to distinguish at least two different situations:

(1) The inquirer obtains a new answer *A* from nature.
(2) The inquirer anticipates an answer *B* that is made likely by the earlier answers (plus initial premises).

The former case has two subcases:

(1a) The new answer *A* is logically compatible with the earlier ones plus the initial premises.
(1b) It is incompatible with them.

Here the case (2) corresponds roughly to Levi's ampliative inference. The selection of *B* could take place by means of e.g. inductive principles or by some means of nonmonotonic reasoning. The case (1b) corresponds to a species of belief contravening reasoning, for in it the inquirer will have to give up one of the answers he or she has received (or else give up one of the initial premises).

These two main cases (1) and (2) might at first seem similar. In both cases, a step is taken beyond old evidence. Moreover, in both (1) and (2) the inquirer has to decide whether the new answer or a proposed new hypothetical answer is to be accepted.

Hence a partial analogue to Levi's question in the interrogative approach is: Should we use the same criteria in both cases (1) and (2)? We can ask this question even without specifying what principles of reasoning are being used in the two cases.

Even without such a specification we can see that the two cases (1)–(2) are not on a par. The reason is that the anticipated answer *B* (2) and its relation to the old members of the "corpus" (totality of earlier answers plus initial premises) has to be judged differently from an actual answer *A* and its impact on the corpus. An anticipated (e.g. inferred) answer *B* has to be judged only on the basis of earlier answers. Being itself inferred or conjectured, however safely, it cannot contribute anything new to the evaluation of the earlier members of the corpus. In contrast, a new actual answer *A* adds to the inquirer's store of *prima facie* information. And it is precisely on the basis of *this* corpus that that new answer itself and not only the older members of the corpus are to be judged. This is not a surprising or deep point; surely a witness can enhance the credibility of his or her earlier answers by providing a new informative and at least prima facie honest answer, quite as much as another witness can destroy his or her credibility by contradicting his or her earlier answers.

These observations support Isaac Levi's conclusion that ampliative reasoning, corresponding to (2), has to be judged by criteria different from those used to judge the impact of a new answer (1) to the corpus, especially perhaps in the case (1b) in which some changes are obviously needed in the corpus (Levi's "belief contravening reasoning").

Thus Levi is in my judgment eminently right in suggesting doubts concerning the current assumptions in the theory of belief change. This skepticism can be enhanced by noting some further consequences of the observations just made. For instance, an *AGM* belief revision has to be understood as applying in case (1) rather than (2) for anticipatory reasoning like (2) cannot induce changes in the corpus on which it is based. But if so, then it cannot be right in mandating expansion by adding as soon as the new answer *A* is consistent with the totality of earlier answers (plus initial premises). For even if *A* is logically compatible with the corpus it can make some of its old members so unlikely that they have to be rejected, at least temporarily.

For instance, if a witness in a court of law testifies both that the defendant wore a red shirt and that he, the witness, is color-blind, he or she is not contradicting himself or herself, but no jury is likely to accept both testimonies at their face value.

Even though much more work is needed here, it can already be seen that the issues Isaac Levi raises are extremely consequential and that the critical thrust of his paper is very likely to be well taken.

COMMENT ON ILKKA NIINILUOTO

Ilkka Niiniluoto defends my earlier self (Hintikka the inductive logician) against my mature *persona* (Hintikka the interrogative inquirer). Is this for me a no-win or no-loss situation? Let us see.

The right framework here is in my judgment the two-level model of inquiry which I have suggested (see Hintikka 1992d) and which is expounded in George Gebhard's paper in this volume. On the higher level (macrolevel) the input into scientific reasoning that nature's answers provide can involve general truths. This means that inductive inferences in the usual sense of the word are redundant on this level.

As Niiniluoto correctly observes, this does not exclude all "inductivist" elements from scientific inquiry. For in order to reach a viable model of the scientific procedure, we of course have to give up the simplifying assumption that all of nature's answers are true. Like any other witness nature is not totally reliable. We need guidelines for at least temporarily disregarding some of her answers and more generally speaking guidelines for testing different answers by different oracles by means of our questioning method.

But – and this is the crucial point – these principles can only be strategic. Hence they will look rather unlike the rules of conventional inductive logic, which are formulated as if they were definitory rules of scientific inquiry.

Hence it is perhaps misleading to say that inductive reasoning becomes redundant on the macrolevel according to the interrogative model. What is needed is a methodological and philosophical recasting of inductive principles in the role of strategic rather than definitory rules. I believe that the task of doing so is an interesting and important one. I also believe that it would bring inductive logic much closer to the ideas of scientific inference developed by statisticians.

On the microlevel, the Atomistic Postulate can be taken to govern nature's answers. But on this level there are also general theoretical premises available to the inquirer. Niiniluoto is quite correct in saying that inductive inference on such a mixed basis can be dealt with in a suitable inductive logic, perhaps along the lines of Niiniluoto and Tuomela (1973).

But, in principle, since we have to rely on some specific conditions in using one particular inductive logic in this lower-level enterprise e.g. such assumptions as exchangeability or some particular values of λ and α, these assumptions should be included in the initial premises of the micro-inquiry in question. But then the inductive inferences are reduced to deductive ones. Inductive logic does not become redundant, but it becomes a part of deductive logic, including deductive probability logic.

Hence inductive logic does not become irrelevant when it is considered in the context of interrogative inquiry. However it has to be integrated into a larger theoretical scheme. I suppose I can think of this as a no-loss situation for myself.

Some other remarks are in order here. Niiniluoto raises a subtle and large question in stressing the role of the human mind in the scientific enterprise. Without attempting an exhaustive discussion, I can say that it seems to me that this role is best approached in terms of the *a priori* (conceptual) knowledge that is needed to answer even purely empirical-looking questions, for instance experimental questions whose answer involves previously unknown functions. (Cf. Hintikka, forthcoming(a).)

Niiniluoto discusses at length the problem of observational errors. I do not have an exhaustive solution to this problem, in any sense of the problem. It seems to me that this entire problem has to be considered in a wider perspective, perhaps in the spirit of my comment on Isaac Levi in this volume.

COMMENT ON ARTO MUTANEN

One of the pleasant aspects of the interrogative approach to inquiry is that it yields not only an explication but a veritable theory of what in the methodology of several particular sciences, including econometrics and systems theory is called identifiability. This theory is simply a generalization of the logical theory of definability to interrogative inquiry instead of mere deductive derivability. Indeed, it can be developed as a partial analogue to the theory of definability. In his paper, Arto Mutanen offers an able summary of the basic ideas and basic results of the logical theory of identifiability.

This theory should be part of the education of every philosopher of science. It throws sharp light on several issues concerning which there has been considerable confusion in the literature. It turns out that when writers like Herbert Simon or Ernst Mach were talking about definability they strictly speaking meant identifiability. This makes a difference, for instance because all variants of identifiability do not guarantee eliminability.

Another possibility of putting the notion of identifiability to work in the philosophy of science is to clarify the observational vs. theoretical contrast. Given a theory *T(P)* containing a one-place predicate *P*, *P* can be called observable in two entirely different senses: (i) Every question of the form "*P(b)* or ~*P(b)*?" is answerable; (ii) *P* is identifiable as the basis of *T(P)*.

The theory Mutanen outlines puts us in a position to carry out such philosophically important applications of the theory of identifiability.

COMMENT ON SYLVAIN BROMBERGER

I find it hard to understand recent philosophers' interest in the idea of natural kinds. Or, more accurately speaking, I can only understand this interest as being due to certain deeper issues which have not been spelled out in the literature.

Bromberger points out that the need of the notion of natural kind first became widely recognized because of Nelson Goodman's paradoxes of induction. But what do these paradoxes amount to in the eyes of a logician? The laws of inductive logic have a certain second-order generality. They are formulated in terms of predicate variables and hence are presumably supposed to apply to any predicates. What Goodman in effect pointed out is that "any predicate" cannot be taken purely extensionally, in a sense which any extensionally possible class is the extension of some *bona fide* predicate.

Now there is in the rich conceptual arsenal of logicians a distinction which applies to questions of the kind Goodman was dealing with. In fact, as soon as one is using second-order or higher-order variables, one faces an important choice. Do these variables range over the set of all extensionally possible entities of the appropriate logical type or only over some designated subset of that set? – for instance, over the entire power set of the domain of one's individuals or only over some proper subset of the power set. The two alternatives give rise to two different kinds of interpretations of higher-order quantifiers and more generally speaking higher-order logic. The former answer yields what Henkin (1950) called the standard interpretation, the latter a generalization of what Henkin called the nonstandard interpretation. I will generalize here Henkin's terminology, as I have done before (Hintikka 1980a, 1982, 1995h; Hintikka and Sandu 1992i).

The tacit role of the standard vs. nonstandard distinction in logic, mathematics and philosophy is one of the most neglected topics in recent philosophical discussion. Some philosophers, e.g. Nelson Goodman, reject the idea that higher-order entities enjoy any real existence apart from their extensions, and hence can countenance only the standard interpretation. Others think that higher-order entities can only be dealt with in terms of first-order axiomatic set theory, which is inevitably a nonstandard theory. Philosophers of neither sort seem to be willing to come out of the closet and make a clean breast of their set-theoretical orientation.

Goodman in effect showed that we need a nonstandard higher-order logic for the purpose of dealing with the problem of induction. This puts an onus on philosophers of induction to provide a specification of the ranges of their nonstandard higher-order variables. The notion of natural kind is best understood as a candidate for that role.

Contrary to a widespread prejudice, one can choose a nonstandard interpretation in a variety of ways. The choice is not fixed by some eternal and immutable principles, but in a way which captures the range of predicates for which inductive principles can be formulated. Instead of "inductive principles" Bromberger speaks (possibly more accurately) of "descriptive and explanatory laws".

It is in this way that I can make best sense of Bromberger's project. Frankly, I cannot judge his results. There are nevertheless a couple of questions that must be raised here.

First, I wonder whether it is truly illuminating to try to characterize run-of-the-mill natural kinds in terms of the questions one can ask about them. Questions are possible to ask if and only if their presuppositions are true. In all interesting cases those presuppositions mark laws that hold of the entities in question. For instance, one can ask "What is the freezing point of *m*?"

of a class of values of m if and only if each of them always changes from liquid state to solid state in one and the same temperature. Hence it seems to be as instructive – and more so – to examine directly the laws that hold of the members of a natural kind than to try to specify the class of questions that can be asked about them.

The main exception is offered by the largest natural kinds of categories. Aristotle's categories correspond in fact to the different question words (or in some question phrases) of the ancient Greek language, and I have shown (see Hintikka 1986b) that this correspondence is rooted in the semantical nature of our language.

Second, in practically all feasible nonstandard interpretations the ranges of variables of different logical types are closed not only with respect to Boolean operations but also with respect to projective ones. Hence a great many one-place predicates, to focus on Bromberger's subject matter, that are values of nonstandard quantifiers are complex predicates involving relations (and possibly still other higher-order entities). Bromberger acknowledges this possibility in speaking of the ancestrals of relations but it deserves in my judgment more attention. There are even theories according to which such ordinary predicates as "tall" are historically speaking derived from the corresponding comparatives, in this case "taller".

I suspect that some philosophers have tended to overlook the relational character of many natural kinds because they hope to cast natural kinds in the role of the targets of direct reference à la Kripke. The entire theory of direct reference is fallacious, however, as shown in Hintikka and Sandu (1995c).

COMMENT ON SCOTT A. KLEINER

I was very pleased to read Professor Kleiner's paper, not only because he is working along the same lines as I am in approaching scientific explanation in interrogative terms. My pleasure was not prompted exclusively by the interesting and illuminating case study which he presents in his paper, either. I was also pleased because what he is doing fits in extremely well with what I have found out (together with Ilpo Halonen) about the logic of different kinds of questions and about the logic of scientific explanation.

How does a why-question operate? What is its semantics and its pragmatics? These questions are answered in Hintikka and Halonen (1995b). A why-question is answered by deriving the explanandum E from a background explanatory theory T plus nature's answers to the inquirer's specific questions. Let their conjunction be A. In scientific practice, they are typically results of observations and experiment, arrived at by some amount of curve-smoothing and other experimental reasoning. But that derivation of E from T and A together is not yet an explanation in the same sense as a covering law explanation. We have to make some further assumptions in order to find one. Assume, for instance, that E is of the form $P(b)$ and that P does not occur in A or b in T. Then it can be shown (ruling out only degenerate cases) that there exists a covering law $(\forall x)(H[x]) \supset P(x))$ such that

$$A \vdash H[b] \quad \text{and}$$
$$T \vdash (\forall x)(H[x] \supset P(x)).$$

Moreover, and this is one of the most interesting things here, $H[x]$ can be thought of as a summary of the entire chain of interconnections (as brought out by the interrogative derivation of E from T). It is this "abstract" (in the English composition sense) character of the covering law that lends it its character as an explanation.

But what happens if the requirements for the existence of a covering law are not satisfied? An answer is simple: We have an answer to a how-question but not to a why-question.

Scott Kleiner's paper is in this perspective a study of the role of such how-questions in science. It has brought home to me several important points about how-questions. First, many answers to how-questions are bona fide scientific explanations, even when there is no covering law in the offing. This might seem to violate what I said earlier. I seemed to imply that answers to why-questions, but not answers to how-questions, are "really" explanations. In reality, I have no logical or epistemological objection to calling answers to how-questions, too, explanations. This is not merely a matter of terminological tolerance, but a logical one. Even in the case of answers to why-questions, the meat of the explanation is in the derivation of E from T and A. It is merely a matter of a fortunate division of labor between the nonlogical constants involved that in the case of answers to why-questions we can in effect comprise the entire explanation into a single covering law. The fact that this is not possible to do in the case of answers to how-questions does not mean that how-explanations are any less explanatory than answers to why-questions.

All this is illustrated vividly by Professor Kleiner's paper. For instance, the holistic character of answers to how-questions is in evidence in the way in which more comprehensive how-questions are first split up in a series of smaller how-questions which then in the end are replaced by normal wh-questions.

It might be objected here that experimental questions are often expressed as how-questions: "How does the observed variable depend on the control variable?" (Cf. my 1988c.) However, this observation does not affect what I have said. On the one hand the same question could have been formulated by asking, "What is the functional dependence of the observed variable on the control variable?" On the other hand, Nature's apparent answer to an experimental question is in reality normally a result of a lower-level inquiry. (Cf. George Gebhard's paper in this volume.) In any case this observation strengthens the case for an important role of how-questions in scientific explanation.

Kleiner is also right when he points out that typical answers to how-questions do not involve a covering law in Hempel's sense. Indeed, in answers to scientific how-questions we have a large class of explanations that do not conform to the Hempelian canon. More interesting than this rather trivial point is that such explanations nevertheless are subject to explicit logical and epistemological analysis involving a fully worked out logic of questions and answers quite as much as covering-law explanations or for that matter answers to any other kinds of questions.

In a general perspective, Professor Kleiner's paper illustrates the value of specific case studies even for the proper understanding of the logic of scientific explanation.

COMMENT ON ANDRZEJ WIŚNIEWSKI

The most constructive comment I can make on Wiśniewski's able and lucid paper is to explain in what ways my theory of the logic of questions and answers goes beyond what he does.

Wiśniewski takes for granted the notion of a *direct answer* to a given question (see section 2, para. 2 of his paper). Such an approach precludes all questions as to how the precise syntactical or logical form of the question determines the set of its answers – in any sense of the notion of answer. Yet without knowing how that determination works out it is difficult to know how the resulting theory can be applied to any actual language, formal or natural.

It is nevertheless easy to see why an analyst might want to proceed in the way Wiśniewski does. For if we actually want to formulate the question-answer relationship for different kinds of questions, the result seems to be different in different cases and different in a way that does not lend itself easily to any kind of unified treatment.

This is worth spelling out by reference to examples. Consider the following sample questions:

(1) Does Jan live Poland, France, or England?
(2) Who killed Roger Ackroyd?
(3) Whom does each person admire most?

The replies to these might be said to satisfy one of the following formulas:

(4) (a) Jan lives in Poland.
(b) Jan lives in France.
(c) Jan lives in England.
(5) *b* killed Roger Ackroyd.
(6) Each person, call him/her *x*, admires most *f(x)*.

Earlier, I tried to reach a uniform treatment by introducing an explicit operator K_d, *d knows that*, and to formulate the logical properties of direct questions by reference to its *desideratum*, which is a specification of the epistemic state of affairs that will satisfy the questioner. In the case of (1)–(3) their desiderata can be said to be of the form

(7) $KS_1 \vee KS_2 \vee KS_3$
(8) $(\exists x)KS[x]$
(9) $(\exists f)K(\forall x)S[x,f(x)]$

There is still no uniformity in (7)–(9). An especially striking oddity is the need of a second-order quantifier in specifying the desideratum of a question like (3) which is logically speaking obviously a first-order one.

In my earlier work (e.g. in Hintikka 1976), no significant unification of the basic question-theoretical notions like desideratum, presupposition, (conclusive) answer etc. was achieved. Wiśniewski thus had a good *prima facie* reason to try another tack in his theory of questions and answers.

The situation has been changed radically by the introduction of the notion of informational independence (cf. Hintikka 1992f, 1995a; Hintikka and Sandu 1989). I cannot explain the notion here. Suffice it to say that in a game-theoretical semantics it means informational independence precisely in the usual game-theoretical sense. The independence of a constant Q_2 (logical or nonlogical) for which a game rule has been defined of another one, say Q_1, will be expressed by writing it (Q_2/Q_1).

Then (7)–(9) can be expressed uniformly as

(10) $K(S_1\ (\vee/K)\ S_2\ (\vee/K)\ S_3)$
(11) $K(\exists x/K)\ S[x]$
(12) $K(\forall x)(\exists y/K)\ S[x,y]$

In terms of such a slash notation, we can easily formulate a uniform general treatment of all different kinds of questions, their desiderata presuppositions, replies, answers and criteria of answerhood (conclusiveness conditions).

Indeed, $(\exists x/K)$ and $(\vee/K)$ (where K is sentence-initial) are the logical counterpart to the question constituent of natural languages. For instance, the presupposition of a question is obtained from its desideratum by omitting the slashes.

This uniform treatment shows that we can go further than Wiśniewski in that we need not postulate as given in our analysis the set of answers to each question. Among other things, we can spell out precisely when a sentence is an answer to a question with a given desideratum.

COMMENT ON JAN WOLEŃSKI

There are games and there are games. I have recently emphasized the crucial importance of distinguishing from each other three fundamentally different kinds of games. They are (i) the semantical games which lend our logical constants their meaning including their truth-conditions; (ii) the interrogative games through which we come to know what is true and false; and (iii) games of formal proof, through which logical truths are established – insofar as they can be so established. (For the distinction, see Hintikka 1995g, 1996a, chapters 2, 10–11.) Of these games, (i) is studied in game-theoretical semantics (GTS), (ii) by means of the interrogative model of inquiry, and (iii) in proof theory. Unfortunately, there is no end of confusion between the three in recent philosophical tradition. For instance, in view of the distinction it is a lost cause to try to change the truth-conditions of sentences by manipulating the rules of formal proof which are essentially rules for games of the kind (iii).

Needless to say, there are interesting interrelations between (i)–(iii). However, semantical games (i) are the fundamental ones. Changes in the other games (ii)–(iii) typically are induced by changes of some kind or other in the games (i). Of the three kinds of processes, the interrogative games (ii) are the ones most directly relevant to epistemology and philosophy of science.

Some mileage in epistemology and philosophy of science could be obtained also by studying the semantical games (i) and perhaps modifying their rules. This is a reasonable and in my judgment highly interesting direction of work. Indeed, in Hintikka (1996a, chapters 4 and 11) I have extended the game-theoretical approach from different first-order logics to epistemic logic and used the result to interpret intuitionistic ideas in the foundations of logic and mathematics. The results are quite gratifying. One can for instance see that intuitionistic mathematicians are dealing with our knowledge of mathematical objects whereas classical mathematicians are dealing with our knowledge of mathematical truths.

What Woleński is doing is closely similar. Instead of semantical games for epistemic logic, he is discussing certain kinds of semantical games that can be associated with intuitionistic logic. Since it is well known that intuitionistic logic and epistemic logic are closely related, what he is doing is not drastically different from my line of thought. The issues Woleński raises are eminently worth a closer examination.

The crucial question is what philosophical clarification or understanding can be generated here. Although I am very much in sympathy with Woleński's overall approach, I am distinctly skeptical about many of the details.

First, Giedymin's interesting ideas about science as a game apply much more happily to epistemic games (ii) than to semantic games (i). What Giedymin and Woleński call somewhat tendentiously "dictatorial strategies" are of course merely what a game theorist would call winning strategies. Ruling out such winning strategies in knowledge-seeking games can be seen as merely toeing Popperian partly line, to which I do not want to object here. It means ruling out epistemic strategies that cannot be possibly refuted. However, Giedymin-like prohibition against winning strategies in semantical games is in my view not cricket. It misconstrues the very idea of semantical games as an implementation that truth means in-principle verifiability, for the most natural explication of the in-principle verification is arguably the very existence of a winning verificatory strategy in the relevant semantical game. Hence Woleński's starting-point differs much more radically from mine than perhaps first meets the eye.

Woleński's intuitionistic games are rather close to semantical games for a suitable epistemic logic. A comparison is made difficult among other things by the fact that his rules are incomplete. There are no rules for games of the form **G**(*k***F***b*).

Instead of the notion of knowledge, Woleński uses the notion of forcing. This notion may be more familiar to a technical logician, but it is not an illuminating one for philosophical purposes, being in my judgment itself in need of a semantical interpretation. Moreover, the philosophical ideas of Per

Martin-Löf are so confused that there is little hope of elucidating anything by their means.

Moreover, one cannot speak simply of intuitionistic logic as some known, unproblematic entity. I have for instance shown that Heyting's rules for intuitionistic logic are not correct unless one wants to restrict their applicability so much as to make them philosophically irrelevant (see 1996a, ch. 11).

Finally, a smaller point. Woleński says that there are winning ("dictatorial") strategies in semantical games for ordinary first-order logic only if the verifier "knows the logical values of and atomic sentences of" the underlying language. Since the notion of knowledge plays no role in semantical games for first-order logic, I find this statement puzzling. Moreover, in any play of a semantical game, only a finite number of atomic sentences are considered. In what sense, then, does such a game presuppose knowledge – presumable *a priori* knowledge – of the truth values of all atomic sentences?

COMMENT ON MATTI SINTONEN

In his useful paper, Matti Sintonen discusses the mottled history of theories and discussions of explanation since the classical paper by Hempel and Oppenheim in 1948. He relates expertly the interrogative approach to explanation which he has himself developed in cooperation with myself to the earlier history of the problem. Sintonen's account is in many ways extremely clarifying, for instance in showing what role the development of the logic of questions and answers has played in the theory of explanation. I do not have much to object to in Sintonen's lucid and instructive account. However, it seems to me that some of the developments he examines can be put into an even sharper relief. For one thing, in the light of hindsight the initial criticisms of Hempel's and Oppenheim's covering law model were a comedy of errors. The early critics thought that they were criticizing the use of uninterpreted formal systems in the philosophy of science, and even Salmon asserts that the "Hempel–Oppenheim article is an outstanding example of the use of an artificial formal language for purposes of explicating a fundamental scientific concept". Admittedly at one stage of the development of logical empiricism some of the Vienna Circle members thought that philosophy of science could be construed as the study of the syntax of the language of science. But this preference, which was no longer prevalent in 1948, is completely irrelevant for the discussion of the covering law or of explanation in general. Indeed, the problem must be formulated by reference to interpreted languages of science. The gist of the covering law model is the claim that in typical cases of scientific explanation there exists a covering law and that this covering law is what mediates the explanation in question. In fact, in a less sharp form this view of explanation was common among logical empiricists long before the Hempel–Oppenheim paper.[1] (In view of this interest in explanation, I find it strange of Sintonen to say that logical positivism "maintains that the sciences cannot explain why the world is the way it is".) The question as to whether the explanandum and the covering law are formulated in the language of formal logic, of mathematical language of science, or in ordinary language, is beside the point. On the contrary, formulations like Salmon's hide the facts that the logic used by Hempel and Oppenheim is elementary and minimal.

What the critics were probably aiming at, however confusedly, is that explanations involve pragmatic factors. However, here they committed another fallacy, maybe because they were misled by the usual misinterpretation of the syntax-semantics-pragmatics trichotomy. They thought that this pragmatic element puts the theory of explanation hopelessly beyond the scope of logical tools and techniques. Once again, this is flatly contradictory to facts. There is no reason why we cannot develop an explicit, objective theory of many different kinds of uses of language, in other words, of different branches of pragmatics, quite independently of the psychological and sociological idiosyncracies of language users. There is no reason, either, why such a theory cannot employ logical concepts and rely on logical techniques and results.

What is true in the critics' confused views is that explanation is an epistemic notion whose application to particular inquirers is relative to their epistemic state. However, that epistemic state is no less amenable to explicit logical treatment than the database of a computer. What has to be done in any adequate analysis of explanation is merely that this epistemic element must be made explicit. These general conceptual truths are illustrated by the analysis of (nonstatistical) explanation which I am in the process of developing jointly with Ilpo Halonen. It is closely related to our theory of why-questions. (See Hintikka and Halonen 1995.) This theory is in turn a particular application of my new logical theory of questions and answers, whose distinctive element is the explicit use of the epistemic operator "knows that", more particularly the logical dependence and independence of other notions on it. Moreover, the crucial idea in vindicating the covering

---

1 See Rudolf Carnap (1966, pp. 12–6). *Philosophical Foundations of Physics*. New York: Basic Books.

law model as codifying a genuine explanation is to show that it can be interpreted as a summary of the entire explanatory argument for the explanandum. This idea is based on results from contemporary logic which are of a higher degree of logical sophistication by an order of magnitude than the Hempel–Oppenheim paper.

In our model, to find an explanation is to find an interrogative derivation of the explanandum from a background theory. This task is an application of the strategic principles of interrogative inquiry. It may be that this strategic element is one of the things that the critics of Hempel and Oppenheim had dimly in mind. It is very much in keeping with Sintonen's emphasis on the dynamic character of explanation. However, it goes much further than, say, the dynamics, of belief change. Current theories of belief change still deal with what happens (or ought to happen) to one's beliefs as a consequence of the reception of one item of new information. Strategic questions go much further and raise questions about the entire sequence of questions the inquirer is asking, especially about the choice of questions to be asked.

Thus Matti Sintonen's paper helps in several different – and important – ways to put my work (joint work with Ilpo Halonen) in an interesting historical and topical perspective.

COMMENT ON ERIK WEBER

Erik Weber's paper is very much in the same direction as the work I have done together with Ilpo Halonen (see our 1995b). Some of our results were published in our 1995 paper. Other results are unpublished.

We agree fully with Weber's first-approximation account according to which in order to explain an explanandum *E* we have to construct an interrogative derivation of *E* from a suitable scientific theory (plus other background information). The further questions that remain concern what has to be added to this recipe for finding genuine explanations. What is the role of Hempelian covering law model in this enterprise? Weber casts the Hempelian model in the role of a strategic guideline in the search of explanations. Together with Halonen, I have shown that a simple account is possible. Under relatively weak assumptions, we show that as soon as an interrogative derivation of the kind indicated is found, a covering law automatically exists as a by-product of the derivation. This covering law receives its explanatory character from the fact that (assuming that the derivation is in a suitable form) it can be thought of as a kind of summary of the entire argument that led from the initial assumptions to the explanandum. (These initial assumptions are called by us "background theory"; Weber distinguishes in them a scientific theory from other background assumptions.) Thus the covering law does not occur as an intermediate stage of the explanatory derivation.

These results do not necessarily contradict Weber's account. A closer analysis is needed to see the precise relationship of the two. There is in fact an obvious connection between the two. If the covering law can be thought of as a summary of the entire explanatory derivation, then to look for the crucial derivation is ipso facto to look for the covering law – and vice versa. In what way, if any, this relationship can be of strategic help in a search for explanations remains to be investigated further.

COMMENT ON GEORGE GEBHARD

Gebhard's paper presents a summary of the basic ideas of the interrogative approach to scientific inference and of their application to a two-tier model of the scientific process. There is nothing that I can object to in Gebhard's able exposition. Needless to say, many of the details deserve – and in some cases need – further discussion. An example is offered by the question of the ultimate role, if any, of induction within the interrogative approach.

One of the merits of Gebhard's paper is that it illustrates vividly the potentialities of the interrogative approach. It is apparently thought by some philosophers (cf. Ilkka Niiniluoto's contribution to this volume) that the epistemologically simplest form of the interrogative model is somehow the normative one. By the epistemologically simplest version I mean the one in which all of nature's answers are true, and known by the inquirer to be true. The interest of this simple version is not due to its representativeness but on the contrary to its being a special case. It is the case in which the confirmation aspect of scientific inquiry disappears. What is left (if the ultimate conclusion is not known ahead of time) is the problem of discovery. The salient fact here is that there exists an interesting theory even for this special case. This means that a genuine, rich logic of pure discovery is forthcoming as a special case of the interrogative approach. The richness of this logic is illustrated by results like the Strategy Theorem of Hintikka, Halonen and Mutanen (forthcoming).

Some of the most important uses of the interrogative model concern not the model as such, for the further developments it prompts, even though purely logically speaking these developments are no longer mere aspects of the interrogative model. Gebhard's paper offers two major examples. One of the most salient features of the use of questions and answers in knowledge-seeking is that questions play a double role in it. On the one hand, the aim of the entire interrogative game is to find an answer to a question. This can be implemented by making the desideratum of the question the ultimate conclusion of the inquiry. To be able to do so is in fact the most important reason for taking into account the epistemic element in the interrogative model. Without making this element explicit, one simply cannot formulate adequately the desiderata of question.

The double role of questions in inquiry is so fundamental that it is found already in the earliest uses of questioning as a method of argumentation. In a Socratic *elenchus*, the interlocutors are examining an answer to a overarching question ("What is knowledge?", "What is piety?") by means of a number of "small" questions. And the Aristotelian fallacy of *petitio principle* was originally nothing but an injunction against asking ("petitioning") the principal question as if it were one of the "small" ones. (See Robinson 1971, Hintikka 1987.)

But once this double role of questions and answers is acknowledged, the possibility of a two-tiered treatment of scientific inference expounded by Gebhard stares you in the face. Hence, although the two-level model of the scientific process is not part and parcel of the original interrogative model, it is forcefully suggested by it.

I strongly believe that Gebhard is right and that it is not possible to do justice to the role of induction as scientific inference except in terms of the two-level model. On the higher (theoretical) level, induction plays a subordinate role because nature's answers can already be general thus obviating the need of inferences from particular to general propositions. On the lower (experimental) level nature's answers are particular observations, but they are backed up by general initial premises concerning the experimental setup. In neither case do we have any Humean problem to deal with.

Likewise, natural, not to say unavoidable extensions of the basic interrogative throw light on another, historically important sense of induction. This is the Newtonian sense in which induction means extending partial generalizations further and of synthesizing different partial generalizations. Such "inductive" inferences are once again not simply parts of the original interrogative model but they are such an obvious extension of the basic problems of interrogative inquiry that a logician could have seen that they are bound to play and have played an important role in actual science. In my paper (1992c) I have shown that this expectation is in fact dramatically fulfilled.

COMMENT ON MIKA KIIKERI

Mika Kiikeri opens an interesting new perspective on the interrogative approach to inquiry. He suggests in effect that the interrogative model can be used as a learning model. This is indeed a highly interesting but so far unexploited possibility. Kiikeri explains carefully the wherewithal of such an application of the interrogative approach. He follows the example of such thinkers as Chomsky and Quine in considering the learning process as being largely analogous with the scientific process. This viewpoint enables him to identify the most important factors in the learning process, such as the role of initial premises (background information), and the "answers" that constitute the input into the learning process.

One of the crucial questions here is the nature of the input. Is this input to be thought of as being codifiable by particular statements, or does it also comprise logically speaking general truths? If so, how complex? And to what extent do we have to postulate strong tacit initial premises in order to understand the learning process? These questions constitute the conceptual landscape in which learning theories have to move. This landscape is outlined by Kiikeri in a clear and instructive manner. For instance, he points out that a weakening of the logical strength of available input can be compensated by strengthening the initial premises of the inquiry. An especially urgent question is whether the input can be restricted to information conceptualized by particular (quantifier-free) propositions. This assumption is known as the Atomistic Postulate. It is in effect made in almost all current formal learning theory. Yet its counterpart in the philosophy of science is far too restrictive to be realistic. Should we perhaps give up the Atomistic Postulate in learning theory, too? As Kiikeri points out, this would lead us to a new kind of formal learning theory.

By raising these questions, Kiikeri has posed extremely interesting and important questions for future research.

REFERENCES

Hilpinen, Risto (1980). On C.S. Peirce's Theory of the Proposition. *The Monist*, vol. **63**, 182–8.

Hintikka, Jaakko (1968). The Varieties of Information and Scientific Explanation. In: B. van Rootselaar and J.F. Staal (Eds.), *Logic, Methodology, and Philosophy of Science* III, *Proceedings of the 1967 International Congress*. Amsterdam: North-Holland Publishing Co., pp. 151–71.

Hintikka, Jaakko (1976). *The Semantics of Questions and the Questions of Semantics. Acta Philosophica Fennica* vol. **28**, no. 4. Helsinki: Societas Philosophica Fennica.

Hintikka, Jaakko (1977). *Aristotle on Modality and Determinism. Acta Philosophica Fennica* vol. **29**, no. 1. Helsinki: Societas Philosophica Fennica.

Hintikka, Jaakko (1980a). Standard vs. Nonstandard Logic: Higher-order, Modal, and First Order Logics. In: Evandro Agazzi (Ed.), *Modern Logic*. Dordrecht: D. Reidel Publishing Co., pp. 283–296.

Hintikka, Jaakko (1980b). C.S. Peirce's 'First Real Discovery' and its Contemporary Relevance. *The Monist*, vol. **63**, 304–15.

Hintikka, Jaakko (1982). Is Alethic Modal Logic Possible? In: Ilkka Niiniluoto and Esa Saarinen (Eds.), *Intensional Logic: Theory and Applications. Acta Philosophica Fennica*, vol. **35**. Helsinki: Societas Philosophic Fennica.

Hintikka, Jaakko (1985). A Spectrum of Logics of Questioning. *Philosophica* vol. **35**, 135–50.

Hintikka, Jaakko (1986). The Varieties of Being in Aristotle. In: Simo Knuuttila and Jaakko Hintikka (Eds.), *The logic of Being: Historical Studies*. Dordrecht: D. Reidel, pp. 81–114.

Hintikka, Jaakko (1987). The Fallacy of Fallacies. *Argumantation*, vol. **1**, 221–38.

Hintikka, Jaakko (1988a). On the Incommensurability of Theories. *Philosophy of Science*, vol. **55**, 25–38.

Hintikka, Jaakko (1988b). 'Die Wende der Philosophie': Wittgenstein's New Logic of 1928. *Philosophy of Law, Politics and Society – Proceedings of the 12th International Wittgenstein Symposium*. Vienna: Hölder-Pichler-Tempsky, pp. 380–96.

Hintikka, Jaakko (1988c). What is the Logic of Experimental Inquiry? *Synthese,* vol. **74**, 173–90.

Hintikka, Jaakko and Gabriel Sandu (1989). Informational Independence as a Semantical Phenomenon. In: F.E. Fenstad et al. (Eds.), *Logic, Methodology and Philosophy of Science VIII*. Elsevier Science Publishers B.V., pp. 571–89.

Hintikka, Jaakko (1992a). Different Constructions in Terms of 'Knows'. In: Jonathan Dancy and Ernest Sosa (Eds.), *A Companion to Epistemology*. Oxford: Basil Blackwell, 99B–104B.

Hintikka, Jaakko (1992b). Knowledge-Seeking by Questioning. In: Jonathan Dancy and Ernest Sosa (Eds.), *A Companion to Epistemology*. Oxford: Basil Blackwell, 241A–244A.

Hintikka, Jaakko (1992c). The Concept of Induction in the Light of the Interrogative Approach to Inquiry. In: John Earman (Ed.), *Inference, Explanation and Other Frustrations: Essays in the Philosophy of Science*. University of California Press, pp. 23–43.

Hintikka, Jaakko (1992d). Theory-Ladenness of Observations as a Test Case of Kuhn's Approach to Scientific Inquiry. In: David Hull et al. (Eds.), *PSA 1992: Proceedings of the 1992 Biennial Meeting of the Philosophy of Science Association,* **1**. East Lansing, MI: PSA, pp. 277–86.

Hintikka, Jaakko (1992e). The Interrogative Model of Knowledge Acquisition as a Framework for Concept Identification. In: S. Ohsuga et al. (Eds.), *Information, Modelling and Databases*. Amsterdam, Washington, Tokyo: IOS Press, pp. 174–81.

Hintikka, Jaakko (1992f). Independence-Friendly Logic as a Medium of Knowledge Representation and Reasoning about Knowledge. In: S. Ohsuga et al. (Eds.),

Information, Modelling and Databases. Amsterdam, Washington, Tokyo: IOS Press, pp. 258–65.

Hintikka, Jaakko (1992g). The Interrogative Model of Inquiry as a General Theory of Argumentation. *Communication and Cognition*, vol. **25**, 221–42.

Hintikka, Jaakko (1992h). Carnap's Work in the Foundations of Logic and Mathematics in a Historical Perspective. *Synthese*, vol. **93**, 167–89.

Hintikka, Jaakko and Gabriel Sandu (1992i). The Skeleton in Frege's Cupboard: The Standard vs. Nonstandard Distinction. *Journal of Philosophy*, vol. **89**, 290–315.

Hintikka, Jaakko (1993a). Socratic Questioning, Logic, and Rhetoric. *Revue Internationale de Philosophie*, vol. **47**, 5–30.

Hintikka, Jaakko (1993b). On Proper (Popper?) and Improper Uses of Information is Epistemology. *Theoria*, vol. **59**, 158–65.

Hintikka, Jaakko and Gabriel Sandu (1994). Why Parallel Processing? In: Robert Casati, Barry Smith and Graham White (Eds.), *Philosophy and Cognitive Sciences: Proceedings of the Sixteenth International Wittgenstein Symposium*. Vienna: Hölder-Pichler-Tempsky, pp. 265–72.

Hintikka, Jaakko (1995a). What Is Elementary Logic? Independence-friendly Logic as the True Core Area of Logic. In: Kostas Gavroglu, John Stachel and Marx W. Wartofsky (Eds.), *Physics, Philosophy, and the Scientific Community*. Dordrecht: Kluwer Academic, pp. 301–26.

Hintikka, Jaakko and Ilpo Halonen (1995b). Semantics and Pragmatics for Why-Questions. *Journal of Philosophy*, vol. **92**, 636–57.

Hintikka, Jaakko and Gabriel Sandu (1995c). The Fallacies of the New Theory of Reference. *Synthese*, vol. **104**, 245–83.

Hintikka, Jaakko (1995d). The Phenomenological Dimension. In: Barry Smith and David W. Smith (Eds.), *The Cambridge Companion to Husserl*. Cambridge U.P., pp. 78–105.

Hintikka, Jaakko (1995e). Constructivism *aufgehoben*. In: T. Childers and O. Majer (Eds.), *Logica '94*. Praha: Filosofia, pp. 1–15.

Hintikka, Jaakko and Gabriel Sandu (1995f). What Is the Logic of Parallel Processing?. *International Journal of Foundations of Computer Science*, vol. **6**, 27–49.

Hintikka, Jaakko (1995g). The Games of Logic and the Games of Inquiry. *Dialectica*, vol. **49**, 229–49.

Hintikka, Jaakko (1995h). Standard vs. Nonstandard Distinction: A Watershed in the Foundations of Mathematics. In: Jaakko Hintikka (Ed.), *From Dedekind to Gödel: Essays on the Development of the Foundations of Mathematics*. Dordrecht: Kluwer Academic Publishers, pp. 21–44.

Hintikka, Jaakko (1996a). *The Principles of Mathematics Revisited*. Cambridge U.P.

Hintikka, Jaakko and Gabriel Sandu (1996b). Game-Theoretical Semantics. In: Johan van Benthem and Alice ter Meulen (Eds.), *Handbook of Logic and Language*. Amsterdam: Elsevier.

Hintikka, Jaakko (1996c). Knowledge Acknowledged: Knowledge of Propositions vs. Knowledge of Objects. *Philosophy and Phenomenological Research*, vol. **56**, 251–75.

Hintikka, Jaakko (1996d). Cogito, ergo quis est? *Revue Internationale de Philosophie*, vol. **50**, 5–21.

Hintikka, Jaakko (1996e). Strategic Thinking in Argumentation and Argumentation Theory. *Revue Internationale de Philosophie,* vol. **50**, 307–24.

Hintikka, Jaakko (1996f). On the Development of Aristotle's Ideas of Scientific Method and the Structure of Science. In: William Wians (Ed.), *Aristotle's Philosophical Development: Problems and Prospects*. Savage, Maryland: Rowman & Littlefield, pp. 83–104.

Hintikka, Jaakko. Contemporary Philosophy and the Problem of Truth. In the proceedings of the 1995 meeting of the IIP.

Hintikka, Jaakko. Knowledge of Functions in the Growth of Mathematical Knowledge, forthcoming(a).

Hintikka, Jaakko, Ilpo Halonen and Arto Mutanen. Interrogative Logic as a General Theory of Reasoning, forthcoming(b).

Knuuttila, Simo (Ed.) (1981). *Reforging the Great Chain of Being*. Dordrecht: D. Reidel.

Niiniluoto, Ilkka and Raimo Tuomela (1973). *Theoretical Concepts and Hypothetico-Inductive Inference*. VII + 264 pp.

Ramsey, F.P. (1925). The Foundations of Mathematics. *Proceedings of the London Mathematical Society*. 2nd. Ser. vol. 25, pt. 5.

Robinson, Richard (1971). Begging the Question, 1971. *Analysis* vol. **31**, 113–7.

VI
ABSTRACTS

Zev Bechler
The Cohn Institute for the History & Philosophy
of Science & Ideas
Tel-Aviv University

HINTIKKA ON PLENITUDE IN ARISTOTLE

Jaakko Hintikka was the first to present a detailed argument for the claim that Aristotle subscribed to the so-called Principle of Plenitude according to which all possibilities are eventually actualized. This paper argues that Aristotle's "actualism", the view that only actualities are real, determined his peculiar type of logical determininism as well as his doctrine of plenitude. The paper also argues, in part contrary to what Hintikka seems to have thought, that Aristotle's definition of *kinēsis* does not provide an escape from the Principle of Plenitude or from his form of logical determinism. It is also suggested that, as a result, potentialities in Aristotle's scheme cannot have an explanatory role.

Marja-Liisa Kakkuri-Knuuttila
Helsinki School of Economics

WHAT CAN THE SCIENCES OF MAN LEARN FROM ARISTOTLE?

This paper argues that both Aristotle's notion of rhetoric and dialectic, as well as his epistemological and methodological views have a greater contribution to offer to modern research in the sciences of man than recognized so far. Having presented some basic concepts of dialectic and rhetoric, the author shows with the help of a paradigmatic example the relevance of the Aristotelian model of rhetorical speech to traditional narrative historical research. She also elaborates in which ways Kuhn's conception of interparadigmatic choice corresponds to the Aristotelian view of decision-making in rhetorical situations. In the last section, she claims that the Aristotelian concept of reputable opinions (*endoxa*) and the method of *saving the phenomena* provide fundamental epistemological solutions to problems raised by the recent criticism of positivistic philosophy of science.

Martin Kusch
Science Studies Unit
University of Edinburgh

THEORIES OF QUESTIONS IN GERMAN-SPEAKING PHILOSOPHY AROUND THE TURN OF THE CENTURY

Around the turn of the century, German-speaking philosophers displayed a lively interest in the nature of questions. Theorising concentrated on the relation between questions and judgements, the distinction between propositional and wh-questions, various further taxonomies, the problem of answerhood, and attempts to describe knowledge-acquisition as an interrogative process. Interest in question theory ran high because logical inquiry was non-formal and close to psychology, and because both applied and theoretical (experimental) psychology had independent reasons for focusing on the study of questions. Kant's use of question-theoretical terms was an important resource for Neokantian philosophers.

Nils-Eric Sahlin
Gothenburg University and
Lund University

'HE IS NO GOOD FOR MY WORK': On the Philosophical Relations between Ramsey and Wittgenstein

This is a paper on the philosophical relations between F. P. Ramsey and Ludwig Wittgenstein. The key question of the paper is to what extent Frank Ramsey influenced the thinking of the later Wittgenstein. It is well known that Wittgenstein strongly influenced Frank Ramsey's philosophical thinking. But, what is less well known is that Ramsey came to change his view of philosophy and in doing so became a major force in Wittgenstein's upheaval of the Tractarian view. From 1929 and on one can clearly spot a remoulded Ramsey in the work of Wittgenstein. Thus, it is argued that Wittgenstein probably was more influenced by Ramsey than Ramsey was influenced by Wittgenstein. The arguments are presented in three parts: (1) a few biographical glimpses are given; (2) Ramsey's British pragmatism is outlined and discussed; and (3) it is shown where this British pragmatism can be spotted in Wittgenstein's work after 1930.

Theo A.F. Kuipers
Faculty of Philosophy
University of Groeningen

THE CARNAP–HINTIKKA PROGRAMME IN INDUCTIVE LOGIC

Rudolf Carnap initiated a research programme in quantitative confirmation theory, also called inductive probability theory, or inductive logic, by designing a continuum of systems for individual hypotheses. It has guided the search for optimum systems and for systems that take analogy into account.

Carnapian systems, however, assign zero probability to universal hypotheses. Jaakko Hintikka was the first to reconsider their confirmation. His way of using Carnap's continuum for this purpose has set the stage for a whole spectrum of inductive systems of this type.

The Carnap–Hintikka programme clearly has much internal dynamics; as will be indicated, its results can also be applied in several directions.

Isaac Levi
Columbia University

CAUTION AND NONMONOTONIC INFERENCE

Default reasoning is that species of nonmonotonic reasoning that is *ampliative*. The other type of nonmonotonic reasoning is *belief contravening* as in reasoning from suppositions. It is commonly held that these two kinds of nonmonotonic reasoning share the same formal properties. According to the decision theoretic approach to ampliative reasoning shared by Hintikka, Hilpinen, Niiniluoto, Pietarinen and myself, formal similarity holds only when ampliative inferences are drawn with maximum boldness and when iterating the use of rules for inductive inference can yield no new information. Such "bookkeeping" with maximum boldness is hard to defend raising doubts about an important assumption made by the majority of students of nonmonotonic reasoning.

Ilkka Niiniluoto
Department of Philosophy
University of Helsinki

INDUCTIVE LOGIC, ATOMISM, AND OBSERVATIONAL ERROR

Jaakko Hintikka has recently claimed that the traditional problem of inductive generalization occupies only a modest place in the interrogative logic of scientific inquiry. It is argued in this paper that the Atomistic Postulate, i.e., the restriction of Nature's answers to atomic propositions and their negations, can be avoided only under very strong regularity conditions. In typical curve-fitting problems the desired function is not given by Nature but discovered and supplied by the inquirer. On the other hand, Hintikka's own system of inductive logic is not tied to the Atomistic Postulate, since it can be (and has been) applied to situations where the evidence contains a universal theory. The second part of the paper outlines a system of inductive logic which allows for observational error. In analogy to Bayesian treatments of parameter estimation, the system is based upon a probabilistic error distribution for observations.

Arto Mutanen
Department of Philosophy
University of Helsinki

THEORY OF IDENTIFIABILITY

In this paper I will introduce the notion of identifiability. The most basic properties of the notion will be stated. In my presentation I will follow, very carefully, Hintikka's ideas. The theory of identifiability is introduced in the framework of interrogative model of inquiry. We 'stabilize' interrogative process by using Abraham Robinson's diagram method. This helps us in presenting properties of the notion, and in seeing the interconnection between the notions of identifiability and definability.

Sylvain Bromberger
Massachusetts Institute
of Technology

NATURAL KINDS AND QUESTIONS

Science would be impossible without the notion of natural kinds. But that notion is replete with difficulties that have long been the subject of intense philosophic inquiry. In this paper the author begins by reviewing some of these difficulties. He then proposes an analysis of his own based on two tenets. The first tenet is that belief in the kinds needed for science consists of a range of presumptions whose more complex ones entail (but are not entailed by) simpler ones. The second tenet is that natural kinds are characterizable in terms of certain questions pertaining to their members.

Scott A. Kleiner
College of Arts & Sciences
Philosophy Department
The University of Georgia

THE STRUCTURE OF INQUIRY IN DEVELOPMENTAL BIOLOGY

What are judged to be the most important questions from a survey of researchers in developmental biology are analyzed in the context of a cytological model of development and in the context of a molecular regulative model for gene expression. These models provide interpretations and a rational ordering of these questions that supersede the intuitively based opinions surveyed. The rational support behind molecular developmental genetics comes from the epistemic import of explaining cell differentiation in embryology, discovering interfield and interlevel relations, and explaining variation, physiological stability and gene conservation in evolution.

Andrzej Wiśniewski
Tadeusz Kotarbiński Pedagogical University
Institute of Philosophy

SOME FOUNDATIONAL CONCEPTS OF EROTETIC SEMANTICS

Some concepts which may be useful in the analysis of questions and questioning are defined in terms of model-theoretical semantics and multiple-conclusion entailment. In particular, the following concepts are defined: soundness of a question in an interpretation of a language, partial answerhood, normal question, regular question, proper question, relative soundness of a question, relative informativeness of a question, self-rhetoricity of a question, and various kinds of presuppositions. The connections between these concepts are examined in detail. In addition, a general overview of the existing logical theories of questions and answers is presented.

Jan Woleński
Institute of Philosophy
Jagiellonian University

SCIENCE AND GAMES

This paper tries to combine ideas of Jaakko Hintikka and Jerzy Giedymin in order to achieve a general game-theoretical approach to science. The two players, Myself and Nature, taking part in scientific game are regarded as mutually dual. The same concerns their strategies. Following Giedymin, rational scientific games are defined as games without dictatorial strategies. Following Hintikka, game-theoretical semantics is used for describing the strategies of both players. In particular, strategies of Nature are described by dual logic, that is, logic preserving falsity.

Matti Sintonen
Department of Philosophy
University of Helsinki

EXPLANATION: THE FIFTH DECADE

In his "Four Decades of Scientific Explanation" Wesley Salmon locates the roots of all three modern notions of explanation in Aristotle. Yet his account in which the so-called erotetic or question-answer -view is but one variant of the epistemic notion of explanation misses an important ecumenical message. This paper suggests that the erotetic or interrogative notion is best viewed as an umbrella notion under which all three rivals, the epistemic, modal and ontic notions of explanation, fit.

This explains why there is a unitary notion – that of a satisfactory answer to a certain form of query – behind the various intuitions of explanation, although satisfactoriness can be cashed out differently in different cases. This view not just restores the angle which somehow got lost during the four decades but also shows that Aristotle was right in thinking that explanations, different kinds of answers to why-questions, fall into distinct types. This view also builds a bridge from the science of logic to the logic of science, and hence does justice to the long interrogative tradition of inquiry.

Erik Weber
Vakgroep Wijsbegeerte en Moraalwetenschap
Universiteit Gent

SCIENTIFIC EXPLANATION AND THE INTERROGATIVE MODEL OF INQUIRY

The aim of the article is to analyze the nature of the construction process of scientific explanations. The search for scientific explanations is a type of inquiry which can be codified by means of Jaakko Hintikka's interrogative model of inquiry. But there is more to be said; the method for constructing scientific explanations which is developed is based on the interrogative model, but also takes into account the various forms of strategic knowledge we use when constructing scientific explanations.

George Gebhard
Department of Philosophy
University of Helsinki

SCIENTIFIC DISCOVERY, INDUCTION, AND THE MULTI-LEVEL CHARACTER OF SCIENTIFIC INQUIRY

The object of this investigation is the nature of experimental inquiry and its logic, seen from the point of view of Hintikka's interrogative model of inquiry. In this framework scientific inquiry is best seen as a multi-level process in which small operational questions are used to deliver answers to big research questions.

I will start with a brief description of the interrogative model and ways of classifying interrogative games in the order of their logical (quantificational) complexity. This illuminates the type of answers we can expect from Nature on different levels of an experimental inquiry. The success of any interrogative game depend largely on the strength of the principal premises and available answers. Lack of strength in initial premises may be compensated by (logically) stronger answers and vice versa. This will be expounded by reference to an experimental knowledge-seeking situation where a two-level conception of the scientific enterprise is sketched: a lower level of experiments where logically simpler answers are relied on, and a higher level of theory construction where logically more complex answers are used as an input. Induction on these two levels is different in nature, and no inductive or other rules of ampliative inference will be needed on either of them. In this connection we will also see that the old distinction between 'the context of discovery' and 'the context of justification' loses much of its plausibility.

Mika Kiikeri
Department of Mathematical Sciences/Philosophy
University of Tampere

ON THE LOGICAL STRUCTURE OF LEARNING MODELS

This paper studies methodological issues in various learning models. The framework we adopt is the interrogative model of inquiry (I-model, for short), originally proposed and developed by Jaakko Hintikka. We shall first offer a survey of the structure of various contemporary learning models as well as discuss some methodological issues. Subsequently, the fundamentals of the I-model are presented and the interrogative notion of a learning process is introduced. Finally, we shall discuss some methodological problems which are prevalent in this context. For instance, it is interesting to study the problem of induction within this framework, and some of the most usual methodological restrictions of learning models could be critically assessed.

POZNAŃ STUDIES IN THE PHILOSOPHY OF THE SCIENCES AND THE HUMANITIES

Contents of back issues

VOLUME 1 (1975)

No. 1 (Sold out)

The Method of Humanistic Interpretation – J. Kmita, *Humanistic Interpretation*; W. Ławniczak, *On a Systematized Interpretation of Works of Fine Arts*; J. Topolski, *Rational Explanation in History*. **The Method of Idealization** – L. Nowak, *Idealization: A Reconstruction of Marx's Ideas*; J. Brzeziński, *Interaction, Essential Structure, Experiment*; W. Patryas, *An Analysis of the "Caeteris Paribus" Clause*; I. Nowakowa, *Idealization and the Problem of Correspondence*. **The Application: The Reconstruction of Some Marxist Theories** – J. Topolski, *Lenin's Theory of History*; J. Kmita, *Marx's Way of Explanation of Social Processes*; A. Jasińska, L. Nowak, *Foundations of Marx's Theory of Class: A Reconstruction*.

No. 2 (Sold out)

Articles – J. Kmita, L. Nowak, J. Topolski, *Against the False Alternatives*; É. Balibar, *Dialectique, Contradiction, Lutte des Classes*; J. Burbelka, *Historical Materialism: Its Essence and Forms*; S. Magala, *Dialectics and Praxis*. **Discussions** – A. Feenberg, *Notes on Methodology and Ontology in Lenin's Thought*; J. Kmita, *The Controversy About the Principles of the Development of Science*; I. Nowakowa, *On the Notion of Correspondence*; A. Klawiter, *Kuhn and Feyerabend on a Scientific Fact and Its Interpretation*.

No. 3 (Sold out)

Articles – K. Zamiara, *Pavlov's Theory and Marx's Directive of the Functional-Genetic Explanation in the Humanities*; J. Hilpelä, *Factors Determining Theory Formation and Methods in Sociology*; A. Klawiter, *The Category of Generation in Historical Materialism*; L. Nowak, *The Anatomy of Anti-Marxist Criticism*. **Discussions** – W. Mejbaum, *Standard and Non-Standard Concept of Social Praxis*.

No. 4 (Sold out)

Articles – J. Agassi, *Between Metaphysics and Methodology*; T. Czeżowski, *Definitions in Science*; P. Materna, *Frege Revised (on a New Concept of Semantics)*; W. Marciszewski, *The Cartesian Concept of Conviction*; M. Bunge, *Meaning in Science*. **Discussions** – J. Misiek, *On the Concept of Matter*; D.A. Rohatyn, *Why "Internal Relations" Cannot be True*; J. Kmita, *Methodological Proposals of Claude Lévi-Strauss*; W. Patryas, *Idealization and Approximation*.

VOLUME 2 (1976)

No. 1 (Sold out)

Articles – W. Mejbaum, *Value and Conceptualization*; J. Kmita, *On Axiological Heterogeneity of Evaluating Predicates*; P. Buczkowski, *The Marxian Category of Bourgeois Scientist*; L. Nowak, *Evaluation and Cognition*; T.M. Jaroszewski, *The Peculiarity of Man*. **Discussions** – M. Franklin, *Professor Topolski on Lenin's Theory of History*; F.J. Fleron, *Comments on "Lenin's Theory of History" by Jerzy Topolski.*

No. 2 (Sold out)

Articles – M. Przełęcki, *On Possibility and Possible Worlds*; Z. Augustynek, *Relational Becoming*; T. Kubiński, *On Foundations of Deontic Logic*; Z. Ziembiński, *Factual Assumptions of Normative Utterances*; J. Agassi, *On Spontaneity in the Arts*; B. Tuchańska, *A Phenomenon, the Essence of the Phenomenon, a Theory*; T. Pawłowski, *Indicators in Comparative Research. The Problems. Introductory Remarks.* **Discussions** – F.J. Vandamme, *Theory Change, Incompatibility and Non-Deducibility*; D.H. DeGrood, *Dialectical Method and "Idealizations"*; I. Supek, *Boscovich's Philosophy of Nature.*

No. 3 (Sold out)

Idealizational Concept of Science – L. Nowak, *Essence – Idealization – Praxis. An Attempt at a Certain Interpretation of the Marxist Concept of Science*; B. Tuchańska, *Factor versus Magnitude*; J. Brzeziński, *Empirical Essentialist Procedures in Behavioral Inquiry*; J. Brzeziński, J. Burbelka, A. Klawiter, K. Łastowski, S. Magala, L. Nowak, *Law and Theory. A Contribution to the Idealizational Interpretation of Marxist Methodology*; P. Chwalisz, P. Kowalik, L. Nowak, W. Patryas, M. Stefański, *The Peculiarities of Practical Research.* **Discussions** – T. Batóg, *Concretization and Generalization*; R. Zielińska, *On Inter-Functional Concretization*; L. Nowak, *A Note on Simplicity*; L. Witkowski, *A Note on Implicational Concept of Correspondence.*

No. 4 (Sold out)

Categorial Interpretation of Dialectics – L. Nowak, *On the Categorial Interpretation of History*; I. Nowakowa, *The Principle of Universal Nexus in the Categorial Interpretation of Dialectics*; S. Magala, *The Phenomenon of Complementarity in the Categorial Dialectics*; I. Nowakowa, *Partial Truth – Relative Truth – Absolute Truth. An Attempt at a Construction of the Ordering Concept of Essential Truth.* **Discussions** – J. Witt-Hansen, *Reflections on Marxian Dialectics*; A. Pałubicka, J. Kmita, *Some Remarks on the Dialectical Method of Karl Marx*, D.W. Felder, *Contradiction and Conflict*; M.A. Amer, *Contradiction: Dialectical and Logical*; O.A. Wojtasiewicz, *The Concept of Effect in Autodynamic World*; A. Miś, *Remarks on a Certain Interpretation of the Marxist Methodology.*

VOLUME 3 (1977)

Nos. 1-4 (Sold out)

Aspects of the Production of Scientific Knowledge (Edited by J. Witt-Hansen) – J. Witt-Hansen, *Marx's Method in Social Science, and Its Relationship to Classical and Modern Physics and Mathematics*; S.A. Pedersen, *Logic and Ontology in Study of Theory Change*;

H.S. Jensen, *On the Production of Scientific Knowledge.* **Articles** – R. Arthur, *The Empiricist Account of Scientific Knowledge – A Polemic Evaluation*; S. Shibata, *Die Moderne Wissenschaft und marxistische Philosophie. Zur Problematik der Methodologie der Wissenschaft*; W. Patryas, *The Sense of Empirical Testing*; K. Łastowski, *The Method of Idealization in the Populational Genetics*; B. Tuchańska, *An Idealizational View on Measurement and Indicator-Based Reasoning*; J. Topolski, *On the Class Approach to History.* **Discussions** – W. Patryas, *When do Two Theorems Correspond?*; E. Paszkiewicz, *Context of Discovery and Context of Justification – Opposition or Complementarity?*; J. Woleński, *Jörgensen's Dilemma and the Problem of the Logic of Norms.*

VOLUME 4 (1978)

Nos. 1-4 (Sold out)

Aspects of the Growth of Science (Edited by W. Krajewski) – W. Krajewski, *Preface*; I. Szumilewicz, *The Postulate of Simplicity*; E. Pietruska-Madej, *Anomalies and the Dynamics of Scientific Theories*; A. Motycka, *Kuhn's Sociological Principle of Demarcation*; A. Lewenstam, *Models of Atom. Remarks on the Development of Science.* **Articles** – J. Agassi, *Logic and Logic of*; E. Köhler, *Observation Established by Artificial Intelligence and the Solution to Goodman's Paradox*; M. Przełęcki, *Some Approach to Inexact Measurement*; S. Bartlett, *Lower Bounds of Ambiguity and Redundancy*; J. Such, *Idealization and Concretization in Natural Sciences*; J. Pogonowski, J. Wiśniewski, *A Formal Approach to Causality*; S. Richmond, *Polanyi's Critique of Methodology and Liberalism*; M. Nowakowska, *Towards a Formal Theory of Group Action*; K. Zamiara, *The Problem of Psychologistic Reduction*; M.A. Finocchiaro, *Theory and Practice in the Philosophy of Science. Comments on Galileo and Clavelin*; M.J. Siemek, *Lukács' Dialectical Epistemology*; I. Nowakowa, L. Nowak, *Marxism and Positivism: The Idea of a Scientific Philosophy.* **Discussions** – E. Kronthaler, *Induction und Gesetzsartigkeit. Zu einer Theorie von Nelson Goodman*; E. Kronthaler, *Marginalie zu G.H. von Wright's* The Logical Problem of Induction.

VOLUME 5 (1979)

Nos. 1-4 (Sold out)

Methodological Problems of Historical Research (Edited by J. Topolski) – J. Topolski, *The Basic Problems of the Methodology of History*; G. Castellan, *Histoire et Psychologie*; G.G. Iggers, *History as an Historical Science*; L. Canfora, *Epic and Historical Claim to Totality*; A. Pałubicka, *The Positivist and Instrumentalist Concepts of the So-called Archeological Culture*; T. Kostyrko, *On the Concept of Historical Sources.* **Articles** – W. Marciszewski, *Epistemological Foundations of Democratism in Cartesian Philosophy*; G. di Bernardo, *Thetic and Prohairetic Normative Systems*; J. Wróblewski, *The Theory of Law – Multilevel, Empirical or Sociological?*; T. Maruszewski, *On Humanist Interpretation in Psychology: The Case of Investigations of the Defence Mechanisms*; K. Nielsen, *On Marx and Moralizing Social Science*; H. Skolimowski, *Space in Architecture: A Phenomenological Analysis*; B. Dziemidok, *On Cathartic-Compensatory Function of Art*; W. Ławniczak, *On Different Concepts of a Model*; K. Łastowski, *On Possibility of Adaptive Interpretation of Dialectics*; I.S. Narski, *The Functions and Structure of Dialectical Logic.* **Discussions** – K. Berka, *Critical Remarks on the Axiomatized Expected Utility Theory*; J. Ławniczak, *Anti-Individualism, Scientific Discovery and Third World*; R. McCleary, *Remarks on Historical Materialism and Dialectics.*

VOLUME 6 (1982)

SOCIAL CLASSES ACTION & HISTORICAL MATERIALISM

On Classes – I. Fetscher, *Some Theories on the Role of Intellectuals in the Revolutionary Labour-Movement. Brecht and Sartre*; R.P. Sieferle, *Class Consciousness in Marx's Theory of Revolution: Some Remarks*; M. Franklin, *On the Magic of Meritocracy*; O.A. Wojtasiewicz, *Status and Social Structures.* **On Action** – J. Ritsert, *"Wertbeziehung" to the Societal Basis – Two Lines of Argument in Max Weber's Theory of Action?*; A. Klawiter, *Structure and Praxis*; W. Patryas, *The Form of the Principle of Rationality.* **The Adaptive Interpretation of Historical Materialism** – L. Nowak, *The Theory of Socio-Economic Formations as a Theory of Adaptation Processes*; K. Łastowski, *The Theory of Development of Species and the Theory of Motion of Socio-Economic Formation*; P. Buczkowski, *Toward a Theory of Economic Society. An Attempt at the Adaptive Interpretation*; J. Burbelka, *Historical Materialism: General Theory and Forms*; P. Buczkowski, A. Klawiter, L. Nowak, *Historical Materialism as a Theory of Social Whole*; A. Klawiter, *The Theory of Social Formation in Historical Materialism*; P. Buczkowski, *International Relations in the Adaptive Interpretation of Historical Materialism.* **Contributions to Historical Materialism** – J. Ritsert, *Totality, Theory and Historical Analysis. Remarks on Critical Sociology and Empirical Research*; L. Nowak, *Adaptation and Revolution. The Problem of Motion of Socio-Economic Formations in the Adaptive Interpretation of Historical Materialism*; J. Ritsert, *On Horkheimer's and Adorno's Concept of Ideology*; P. Buczkowski, *Imperialism: On the Possibility of Reinterpretation of Lenin's Concept.*

VOLUME 7 (1982)

DIALECTICAL LOGICS FOR THE POLITICAL SCIENCE
(Edited by Hayward R. Alker, Jr.)

L. Nowak, *Marxian Methodology Leads to the Generalization of Historical Materialism*; D. Mefford, *A Comparison of Dialectical and Boolean-Algebraic Models of the Genesis of Interpersonal Relations*; C. Roig, *Leninist Dialectic Seen From the Point of View of Kenneth Burke's Dialectic: Theory, Analysis and Ideology in Lenin's Speech*; H.R. Alker Jr., *Logic, Dialectics, Politics: Some Recent Controversies*; F. Loeser, *What is Dialectical Logic?*

VOLUME 8 (1985)

CONSCIOUSNESS: METHODOLOGICAL AND PSYCHOLOGICAL APPROACHES
(Edited by Jerzy Brzeziński)

Editor's Introduction; A. Klawiter, *A Scientific Theory versus Theoretical Program. A Contribution to the Problem of a Social Character of Cognition*; S. Magala, L. Nowak, *The Problem of Historicity of Cognition in the Idealizational Concept of Science*; J. Brzeziński, *The Protoidealizational Model of the Investigative Process in Psychology*; M. Gaul, *The Problem of Interaction in the Protoidealizational Model of Investigative Process. Modifications and Developments*; J. Kmita, *A Humanistic Coefficient of Activity and the Psychology-Humanities Relation*; K. Zamiara, *In Support of Psycho-Physical Parallelism*; M. Ziółkowski, *Some Remarks on the Notion of Social Consciousness*; S. Kowalik, *The Psycho-*

Social Problem and Modern Psychology; J. Topolski, *On the Concept and Structure of Methodological Consciousness of Historians*; T. Maruszewski, *Are the Idealizational Procedures Used within the Scope of Common Sense Knowledge?*; T. Zgółka, *Stylistic and Sociolinguistics*. **Discussions** – T.A.F. Kuipers, *The Paradigm of Concretization: The Law of van der Waals*; N. Ziv, *The Paradoxical Nature of Bloor's Strong Program in the Sociology of Knowledge.*

VOLUME 9 (1986)

THEORIES OF IDEOLOGY AND IDEOLOGY OF THEORIES
(Edited by Piotr Buczkowski and Andrzej Klawiter)

I. Theories of Ideology – P. Buczkowski, *The Levels of Consciousness: Some Remarks on Sociology of Knowledge*; L. Nowak, *Ideology versus Utopia. A Contribution to the Analysis of the Role of Social Consciousness in the Movement of Socio-Economic Formation*; J. Kmita, *The Antagonism of Art and Science as a Worldview Component.* **II. Ideology of Theories** – J. Kmita, *Scientism and Anti-Scientism*; L. Nowak, *Science, that is, Domination through Truth*; J. Sensat, *Recasting Marxism: Habermas's Proposals.* **Discussions** – J. Brzeziński, *A Statistical Model of Data Analysis in Interactional Psychology. Comments on the Quantitative Analysis of the Scores of the "S-R" Inventory of Anxiousness.*

VOLUME 10 (1987)

WHAT IS CLOSER-TO-THE-TRUTH?
A PARADE OF APPROACHES TO TRUTHLIKENESS
(Edited by Theo A.F. Kuipers)

Introduction. **Part I: Four Approaches to Truthlikeness** – I. Niiniluoto, *How to Define Verisimilitude*; G. Oddie, *The Picture Theory of Truthlikeness*; G. Schurz, P. Weingartner, *Verisimilitude Defined by Relevant Consequence-Elements*; T.A.F. Kuipers, *A Structuralist Approach to Truthlikeness.* **Part II: Eight Aspects of Truthlikeness** – J. van Benthem, *Verisimilitude and Conditionals*; J. Cohen, *Verisimilitude and Legisimilitude*; R. Festa, *Theory of Similarity, Similarity of Theories, and Verisimilitude*; T.A.F. Kuipers, *Truthlikeness of Stratified Theories*; I. Niiniluoto, *Verisimilitude with Indefinite Truth*; G. Oddie, *Truthlikeness and the Convexity of Propositions*; G. de Vries, *Explaining 'Truth' in a Relativist Way*; H. Zandvoort, *Verisimilitude and Novel Facts.*

VOLUME 11 (1988)

NORMATIVE STRUCTURES OF THE SOCIAL WORLD
(Edited by Giuliano di Bernardo)

Editor's Introduction; H.-N. Castañeda, *Ought, Reasons, Motivation, and the Unity of the Social Sciences*; G.H. von Wright, *Action Logic as a Basis for Deontic Logic*; S. Galvan, *Über den Begriff von möglicher Welt in den Anwendungen der Modallogik*; G. di Bernardo, *The Formal Model M of the Game of Chess as a Type of Social Context*; R. Guastini, *Constitutive Rules and the Is-Ought Dichotomy*; H.J. Heringer, *Not by Nature nor by Intention. The Normative Power of the Language Signs*; A. Kasher, *Justification of Speech, Acts and Speech Acts*; Z. Ziembiński, *Le contenu et la structure des normes concédant les compétences*; K. Opałek, *Directives, Norms, and Performatives*; G. Kalinowski, *Les performatifs en droit. Sur la distinction entre le langage prescriptif et le langage performatif*; J. Wróblewski, *Cognition of Norms and Cognition through Norms*; A.G. Conte, *Eidos. An Essay on Constitutive Rules.* **Discussions** – A. Wiśniewski, *On Some Consequences of the Postulate of an Empirical Falsifiability of Logical Theorems*; H. Gipper, *Die Sprachphilosophie Adam Schaffs.*

VOLUME 12 (1987)

POLISH CONTRIBUTIONS TO THE THEORY AND PHILOSOPHY OF LAW
(Edited by Zygmunt Ziembiński)

Editor's Introduction. **Part I** – S. Czepita, *Leon Petrażycki and Czesław Znamierowski – Founders of the Polish Theory of Law*; K. Opałek, *Normativism against the Background of the Methodological Inquiries in Polish Legal Theory*; C. Znamierowski, *The Basic Concepts of Theory of Law. Introductory Remarks.* **Part II** – Z. Ziembiński, *The Methodological Problems of Theory and Philosophy of Law: a Survey*; J. Wróblewski, *Paradigm of Legal Dogmatics and the Legal Sciences*; J. Woleński, *Jurisprudence, Science, Philosophy*; T. Gizbert-Studnicki, *Is an Empirical Theory of Language of Law Possible?* **Part III** – L. Nowak, S. Wronkowska, M. Zieliński, Z. Ziembiński, *Conventional Acts in Law*; L. Nowak, *A Concept of Rational Legislator*; S. Wronkowska, *The Rational Legislator as a Model for the Real Lawmaker*; M. Zieliński, *Decoding Legal Text.* **Part IV** – S. Wronkowska, M. Zieliński, Z. Ziembiński, *The Methodological Problems Concerning the "Principles of Law"*; S. Pałczyński, *The Concept of Human Act in the Legal Sciences.* **Discussions** – Fr. Castellani, *Some Remarks on the Semantic Interpretation of Intensional Logic.*

VOLUME 13 (1988)

DIMENSIONS OF THE HISTORICAL PROCESS
(Edited by Leszek Nowak)

Editor's Introduction. **I. Analyzing the Dimensions of the Historical Process** – A. Klawiter, *Historical Materialism and the Visions of Social Development. A Study of Transformations of Marxian Periodization Formula*; L. Kriesberg, *The Role of Consensus in Social Conflicts*; I.C. Jarvie, *Objective versus Mentalist Conceptions of Social Class.* **II. Individual/Civilizational/Transformational Approach** – A. Honneth, *Work and Instrumental Action*; P. Buczkowski, *Rationality and the Levels of Organization of Society*; A. Bartoszek, E. Bogalska-Czajkowska, W. Czajkowski, A. Małkiewicz, *The Social Division of Labor.* **III. Globalistic/Civilizational/Adaptive Approach** – A. Klawiter, K. Łastowski, L. Nowak, W. Patryas, *Adaptation, Learning, Praxis. Some Applications of the Adaptive Conceptual Apparatus*; A. Klawiter, *Adaptation and Competition. A Contribution to the Classification of Adaptive Relationships*; M. Witkowski, *On Adaptive and Functional Dependencies. An Attempt at Categorial Approach*; K. Niedźwiadek, *The Structure and Development of the Society's Modes of Spiritual Production.* **IV. Globalistic/Conflict/Transformational Approach** – T. Söderqvist, *The Macro- and Micro-Conflict Structure of the Information- and Knowledge Society*; D. Pels, *Towards a Non-Hegelian Conception of the "New Class"*; L. Nowak, *An Idealizational Model of Capitalist Society*; G. Tomczak, *The Economic Collapse in Two Models of Socio-Economic Formation*; M. Niewiadomski, *Toward a Model of Economic Institutions*; A. Falkiewicz, *The Anatomy of Utopia*; K. Paprzycka, L. Nowak, *On the Social Nature of Colonization.*

VOLUME 14 (1990)

ACTION AND PERFORMANCE: MODELS AND TESTS. CONTRIBUTIONS TO THE QUANTITATIVE PSYCHOLOGY AND ITS METHODOLOGY
(Edited by Jerzy Brzeziński and Tadeusz Marek)

VOLUME 15 (1989)

VISIONS OF CULTURE AND THE MODELS OF CULTURAL SCIENCES
(Edited by Jerzy Kmita and Krystyna Zamiara)

Notion of Pedagogical Culture; P. Ozdowski, *Logical Semantics and Psychology of Cognitive Processes*; T. Zgółka, *Study of Language as a Form of Symbolic Culture: Methodological Foundations*; G. Banaszak, *Contemporary Musical Consciousness in the Perspective of the Socio-pragmatic Theory of Culture*; A. Zeidler, *Valorization of Tradition as a Means of Preserving the Continuity of Culture.*

VOLUME 16 (1990)

IDEALIZATION I: GENERAL PROBLEMS
(Edited by Jerzy Brzeziński, Francesco Coniglione, Theo A.F. Kuipers and Leszek Nowak)

Introduction – I. Niiniluoto, *Theories, Aproximations, and Idealizations.* **Historical Studies** – F. Coniglione, *Abstraction and Idealization in Hegel and Marx*; B. Hamminga, *The Structure of Six Transformations in Marx's* Capital; A.G. de la Sienra, *Marx's Dialectical Method*; J. Birner, *Idealization and the Development of Capital Theory.* **Approaches to Idealization** – L.J. Cohen, *Idealization as a Form of Inductive Reasoning*; C. Dilworth, *Idealization and the Abstractive-Theoretical Model of Explanation*; R. Harré, *Idealization in Scientific Practice*; L. Nowak, *Abstracts Are Not Our Constructs. The Mental Constructs Are Abstracts.* **Idealization and problems of the philosophy of science** – M. Gaul, *Models of Cognition or Models of Reality?*; P.P. Kirschenmann, *Heuristic Strategies: Another Look at Idealization and Concretization*; T.A.F. Kuipers, *Reduction of Laws and Theories*; K. Paprzycka, *Reduction and Correspondence in the Idealizational Approach to Science.*

VOLUME 17 (1990)

IDEALIZATION II: FORMS AND APPLICATIONS
(Edited by Jerzy Brzeziński, Francesco Coniglione, Theo A.F. Kuipers and Leszek Nowak)

Forms of Idealization – R. Zielińska, *A Contribution to the Characteristic of Abstraction*; A. Machowski, *Significance: An Attempt at a Variational Interpretation*; K. Łastowski, *On Multi-Level Scientific Theories*; A. Kupracz, *Concretization and the Correction of Data*; E. Hornowska, *A Certain Approach to Operationalization*; I. Nowakowa, *External and Internal Determinants of the Development of Science: Some Methodological Remarks.* **Idealization in Science** – H. Rott, *Approximation versus Idealization: The Kepler-Newton Case*; J. Such, *The Idealizational Conception of Science and the Law of Universal Gravitation*; G. Boscarino, *Absolute Space and Idealization in Newton*; M. Sachs, *Space, Time and Motion in Einstein's Theory of Relativity*; J. Brzeziński, *On Experimental Discovery of Essential Factors in Psychological Research*; T. Maruszewski, *On Some Elements of Science in Everyday Knowledge.*

VOLUME 18 (1990)

STUDIES ON MARIO BUNGE'S *TREATISE*
(Edited by Paul Weingartner and Georg J.W. Dorn)

Section I: Semantics – M. Dillinger, *On the Concept of 'a Language'*; J. Ferrater-Mora, *On Mario Bunge's Semantical Realism*; A. Bartels, *Mario Bunge's Semantical Realism. An*

Antidote against Incommensurability?; J.-P. Marquis, *Partial Truths about Partial Truth*; I. Nowakowa and L. Nowak, *Approximation and the Two Ideas of Truth.* **Section II: Ontology** – J. M. Bocheński, *On the System*; J. Agassi, *Ontology and Its Discontent*; Y. Wand and R. Weber, *Mario Bunge's Ontology as a Formal Foundation for Information Systems Concepts*; D. Blitz, *Emergent Evolution and the Level Structure of Reality*; M. Espinoza, *The Four Causes*; B. Dubrovsky, *A Comment on Three Topics in Volume 7 of the* Treatise*: Teleology, the Mind-Body Problem, and Health and Disease*; W.R. Shea, *Tackling the Mind*; H. Kurosaki, *Mario Bunge on the Mind-Body Problem and Ontology. Critical Examinations.* **Section III: Epistemology** – R. Boudon, *On Relativism*; G. Vollmer, *Against Instrumentalism*; M. Kary, *Information Theory and the* Treatise*: Towards a New Understanding.* **Section IV: Philosophy of Science and Technology** – F. Miró-Quesada-Canturias, *Mario Bunge's Philosophy of Logic and Mathematics*; M. Paty, *Reality and Probability in Bunge's* Treatise; M. García-Sucre, *On the Relationship Between Mathematics and Physics*; B. Kanitscheider, *Does Physical Cosmology Transcend the Limits of Naturalistic Reasoning?*; M. Stöckler, *Realism and Classicism, or Something More? Some Comments on Mario Bunge's Philosophy of Quantum Mechanics*; M. E. Burgos, *Projections Are a Law of Nature*; A. Barceló, *Are there Economic Laws?*; R. Mattessich, *Mario Bunge's Influence on the Administrative and Systems Sciences*; Y. Gingras and J. Niosi, *Technology and Society: A View from Sociology*; D. A. Seni, *The Sociotechnology of Sociotechnical Systems: Elements of a Theory of Plans*; P. Weingartner, *The Non-Statement View. A Dialogue between Socrates and Theaetetus.* **Section V: Ethics** – E. Garzón-Valdes, *Basic Needs, Legitimate Wants and Political Legitimacy in Mario Bunge's Conception of Ethics*; F. Forman, *Virtue, The Missing Link to the Last Volume of the* Treatise; G. Zecha, *Which Values Are Conductive to Human Survival? A Bunge-Test of Bunge-Ethics.* **Section VI: The Treatise as a Philosophy** – A. Cupani, *The Significance of the* Treatise *in the Light of the Western Philosophical Tradition*; E. Rosenthal, *Scientific Philosophical and Sociopolitical Aspects of the Work of Mario Bunge.* **Section VII: Mario Bunge Replies. Section VIII: Mario Bunge's Life and Work** – M. Bunge, *Instant Autobiography*; D. Blitz, *Bibliography of Mario Bunge's Publications 1939-1989.*

VOLUME 19 (1990)

NARRATION AND EXPLANATION.
CONTRIBUTION TO THE METHODOLOGY OF THE HISTORICAL RESEARCH

(Edited by Jerzy Topolski)

G. Iggers, *Historical Studies in the 1980's. Some Observations*; I. Nowakowa, *Historical Narration and Idealization*; J. Pomorski, *On Historical Narration. A Contribution to the Methodology of a Research Programme*; G. Zalejko, *On Cognitive and Extra-Cognitive Components of the Historical Narrative*; J. Topolski, *Historical Explanation in Historical Materialism*; A. Zybertowicz, *Technological Determinism and Historical Materialism Revisited. Some Preliminary Notes*; J. Burbelka, *Types of Historical Processes in Ethnology*; W. Wrzosek, *In Search of Historical Time. An Essay on Time, Culture and History*; A. Ponzio, *Humanism, Philosophy of Language and Theory of Knowledge in Adam Schaff*; J. Rüsen, *Was ist historische Methode?*; A. Constanzo, *Several Remarks on Principles of Law.*

VOLUME 20 (1990)

Jürgen Ritsert
MODELS AND CONCEPTS OF IDEOLOGY

Chapter 1: *Preface – Enterprise and Experiment*; Chapter 2: *Some Theoretical Problems of Theories of Ideology*; Chapter 3: *Base and Superstructure – On the Generality of Some Marxian Models*; Chapter 4: *Action, Interest and Reflection – Weber and Weberian Models of Scientific Knowledge*; Chapter 5: *Analysis of Social Content and the Debate about Internalism and Externalism*; Chapter 6: *French Connections in Models of Ideology – Or: Des idées Franco-allemandes*; Chapter 7: *Critical Theory of Ideology – On Adorno's Dialectical Model of Appearance and Ideology.*

VOLUME 21 (1991)

PROBABILITY AND RATIONALITY
STUDIES ON L. JONATHAN COHEN'S PHILOSOPHY OF SCIENCE
(Edited by Ellery Eells and Tomasz Maruszewski)

Part I: Setting the Stage – L. J. Cohen, *From a Historical Point of View*; **Part II: The Method of Relevant Variables and Other Modes of Scientific Reasoning** – L. Nowak, *The Method of Relevant Variables and Idealization*; M. Fisch, *Learning from Experience*; M. A. Finocchiaro, *Induction and Intuition in the Normative Study of Reasoning*; **Part III: Subjective Probability: Pascalian and Baconian Conceptions** – D.A. Schum, *Jonathan Cohen and Thomas Bayes on the Analysis of Chains of Reasoning*; J. Logue, *Weight of Evidence, Resiliency and Second-Order Probabilities*; C.S. Nosal, *Neurobiology of Subjective Probability*; R. Stachowski, *On Standard and Non-Standard Models in Theories of Psychological Measurment*; **Part IV: Rationality and Methodological Pluralism** – L.L. Lopes and G.C. Oden, *The Rationality of Intelligence*; G. Gigerenzer, *On Cognitive Illusions and Rationality*; J.E. Adler, *An Optimist's Pessimism: Conversation and Conjuction*; T. Maruszewski, *Human Rationality – Fact or Idealizational Assumption*; I.T. Ścigała, *Are People Programmed to be Normal? Psychological Determinants of Human Normality.* **Part V: Conclusion** – L. Jonathan Cohen, *Some Comments by L.J.C.*

VOLUME 22 (1991)

THE SOCIAL HORIZON OF KNOWLEDGE
(Edited by Piotr Buczkowski)

J. Ritsert, *The Wittgenstein-Problem in Sociology or: The "Linguistic Turn" as a Pirouette*; D. Jary, *Beyond Objectivity and Relativism: Feyerabend's "Two Argumentative Chains" and Sociology*; L. Nowak, *The Defence of a Social System Against its Ideology. A Case Study*; P. Buczkowski, *Remarks on the Structure of Social Consciousness*; A. Falkiewicz, *The Individual's Horizon and Valuation*; F. Lövenich, *Heiligsprechung des Imaginären. Das Imaginäre in Cornelius Castoriadis' Gesellschaftstheorie*; T. Maruszewski, *Everyday Knowledge as Representation of Reality.*

VOLUME 23 (1991)

ETHICAL DIMENSIONS OF LEGAL THEORY
(Edited by Wojciech Sadurski)

W. Sadurski, *Introduction*; J. Wróblewski, *Moral Values and Legal Reasoning: Some Aspects of Their Mutual Relations*; H. Tapani-Klami, J. Sorvettula and M. Hatakka, *Moral Reasoning and Evidence in Legal Decision-Making*; T. Gizbert-Studnicki, *Conflict of Values in Adjudication*; M. Krygier, *Thinking Like a Lawyer*; A. Peczenik, *Prima-facie Values and the Law*; R. Wacks, *Judges and Moral Responsibility*; Z. Ziembiński, *The Concept of Morality in Philosophy of Law*; G. Haarscher, *Law, Reason and Ethics in the Philosophy of Human Rights*; T.D. Campbell, *Unlawful Discrimination*; H. Ph. Visser' Hoft, *Intergenerational Justice: Some Reflections on Methodology*; Ch. Sampford and D. Wood, *Tax, Justice and Priority of Property.*

VOLUME 24 (1991)

ADVANCES IN SCIENTIFIC PHILOSOPHY
ESSAYS IN HONOUR OF PAUL WEINGARTNER ON THE OCCASION OF THE 60TH ANNIVERSARY OF HIS BIRTHDAY
(Edited by Gerhard Schurz and Georg J.W. Dorn)

Section I: Advances in Philosophical Logic – R. Sylvan and R. Nola, *Confirmation without Paradoxes*; B. Smith, *Relevance, Relatedness and Restricted Set Theory*; G. Schurz, *Relevant Deductive Inference: Criteria and Logics*; M.L. Dalla Chiara, *Epistemic Logic without Logical Omniscience*; P. Gochet and E. Gillet, *On Professor Weingartner's Contribution to Epistemic Logic*; H. Festini, *An Application of Weingartner's Logical Proposal for Rational Belief, Knowledge and Assumption*; L. Åqvist, *Deontic Tense Logic: Restricted Equivalence of Certain Forms of Conditional Obligation and a Solution to Chisholm's Paradox*; E. Orłowska, *Relational Formalization of Temporal Logics*; W. Lenzen, *What Is (or At Least Appears to Be) Wrong with Intuitionistic Logic?* **Section II: Current Challenges in Philosophy of Science** – Ch.P. Enz, *Quantum Theory in the Light of Modern Experiments*; P. Mittelstaedt, *An Inconsistency between Quantum Mechanics and Its Interpretation: The "Disaster" of Objectification*; E. Scheibe, *Substances, Physical Systems, and Quantum Mechanics*; M. Stöckler, *Reductionism and the New Theories of Self-Organization*; B. Kanitscheider, *Unification as an Epistemological Problem*; R. Haller, *Atomism and Holism in the Vienna Circle*; M. Przełęcki, *Is the Notion of Truth Applicable to Scientific Theories?*; T.A.F. Kuipers, *Structuralist Explications of Dialectics*; C.U. Moulines, *Pragmatics in the Structuralist View of Science*; K.R. Popper, *A World of Propensities. Two New Views of Causality*; G.J.W. Dorn, *Inductive Support*. **Section III: Recent Debates in Semantics and Ontology** – F. von Kutschera, *Kripke's Doubts about Meaning*; I. Bellert, *A Linguistic Approach to Frege's Puzzles*; P.M. Simons, *Inadequacies of Intension and Extension*; E. Morscher, *Inadequacies of Peter and Paul*; J. Brandl, *Some Remarks on the "Slingshot" Argument*; A. Hieke, *Real Facts.* **Section IV: Epistemological and Ethical Problems with Society** – E. Klevakina, *The Case against Value-Free Belief*; R. Tuomela, *Mutual Beliefs and Social Characteristics*; H. Lenk nad M. Maring, *A Pie-Model of Moral Responsibility? Remarks Concerning Ethical Dilutionism*; O. Neumaier, *Are Collectives Morally Responsible?*; E. Bencivenga, *The Electronic Self.* **Section V: Analytical Philosophy of Religion** – J.M. Bocheński, *Faith and Science. A Logical Commentary on the First Question of the*

Summa; E. Nieznański, *The Beginnings of Formalization in Theology*; H. Ganthaler, *Some Comments on Paul Weingartner's Concept of Scientific Theology*. **Section VI: Methods of Philosophy** – G. Kreisel, *Suitable Descriptions for Suitable Categories*; M. Bunge, *Why We Cherish Exactness?*; L. Koj, *Exactness and Philosophy*. **Section VII: Life and Work of Paul Weingartner** – G. Zecha, *Paul Weingartner: Philosophy at Work. A Biographical Sketch*; E. Stieringer, *Bibliography of Paul Weingartner's Publications 1961-1991.*

VOLUME 25 (1992)

IDEALIZATION III: APPROXIMATION AND TRUTH
(Edited by Jerzy Brzeziński and Leszek Nowak)

Introduction – L. Nowak, *The Idealizational Approach to Science: A Survey*. **On the Nature of Idealization** – M. Kuokkanen and T. Tuomivaara, *On the Structure of Idealizations*; B. Hamminga, *Idealization in the Practice and Methodology of Classical Economics: The Logical Struggle with Lemma's and Undesired Theorems*; R. Zielińska, *The Threshold Generalization of the Idealizational Laws*; A. Kupracz, *Testing and Correspondence*; K. Paprzycka, *Why Do Idealizational Statements Apply to Reality?* **Idealization, Approximation, and Truth** – T.A.F. Kuipers, *Truth Approximation by Concretization*; I. Nowakowa, *A Notion of Truth for Idealization*; I. Nowakowa, L. Nowak, *"Truth is a System": An Explication*; I. Nowakowa, *The Idea of "Truth as a Process". An Explication*; L. Nowak, *On the Concept of Adequacy of Laws. An Idealizational Explication*; M. Paprzycki, K. Paprzycka, *Accuracy, Essentiality and Idealization*. **Discussions** – J. Sójka, *On the Origins of Idealization in the Social Experience*; M. Paprzycki, K. Paprzycka, *A Note on the Unitarian Explication of Idealization*; I. Hanzel, *The Pure Idealizational Law – The Inherent Law – The Inherent Idealizational Law.*

VOLUME 26 (1992)

IDEALIZATION IV: INTELLIGIBILITY IN SCIENCE
(Edited by Craig Dilworth)

C. Dilworth, *Introduction: Idealization and Intelligibility in Science*; E. Agazzi, *Intelligibility, Understanding and Explanation in Science*; H. Lauener, *Transcendental Arguments Pragmatically Relativized: Accepted Norms (Conventions) as an A Priori Condition for any Form of Intelligibility*; M. Paty, *L'Endoréférence d'une Science Formalisée de la Nature*; B. d'Espagnat, *De l'Intelligibilité du Monde Physique*; M. Artigas, *Three Levels of Interaction between Science and Philosophy*; J. Crompton, *The Unity of Knowledge and Understanding in Science*; G. Del Re, *The Case for Finalism in Science*; A. Cordero, *Intelligibility and Quantum Theory*; O. Costa de Beauregard, *De l'Intelligibilité en Physique. Example: Relativité, Quanta, Correlations EPR*; L. Fleischhacker, *Mathematical Abstraction, Idealization and Intelligibility in Science*; B. Ellis, *Idealization in Science*; P.T. Manicas, *Intelligibility and Idealization: Marx and Weber*; H. Lind, *Intelligibility and Formal Models in Economics*; U. Mäki, *On the Method of Isolation in Economics*; C. Dilworth, R. Pyddoke, *Principles, Facts and Theories in Economics*; J.C. Graves, *Intelligibility in Psychotherapy*; R. Thom, *The True, the False and the Insignificant or Landscaping the Logos.*

VOLUME 27 (1992)

Ryszard Stachowski

THE MATHEMATICAL SOUL.
AN ANTIQUE PROTOTYPE OF THE MODERN MATEMATISATION OF PSYCHOLOGY

Chapter 1: *The Pythagorean "Number of Man"*; Chapter 2: *The Soul as a (Self-Moving) Number*; Chapter 3: *The Soul as Harmony*; Chapter 4: *The Soul as Originative of Relational Concepts*; Chapter 5: *The Faculty of the Soul as Intensive Psychological Quantity*; Chapter 6: *Soul Equals Mathematicals*; Chapter 7: *Soul to a Greater or Lesser Degree*; Epilogue: *Why Did Aristotle Call Psychology an Exact Science?*; Conclusion

VOLUME 28 (1993)

POLISH SCIENTIFIC PHILOSOPHY:
THE LVOV-WARSAW SCHOOL
(Edited by Francesco Coniglione, Roberto Poli and Jan Woleński)

Preface. **Background and Influence** – L. Albertazzi, *Brentano, Twardowski, and Polish Scientific Philosophy*; K. Schuhmann, *Husserl and Twardowski*; G. Küng, *Phenomenology and Polish Scientific Philosophy*; F. Coniglione, *Scientific Philosophy and Marxism in Poland*. **History and Systematics** – T. Kwiatkowski, *Classification of Reasonings in Contemporary Polish Philosophy*; A. Dylus, *The Problems of Ethics of Science in the Lvov-Warsaw School*; J.J. Jadacki, *Kazimierz Twardowski's Descriptive Semiotics*; P.M. Simons, *Nominalism in Poland*; T. Rzepa, R. Stachowski, *Roots of the Methodology of Polish Psychology*; K. Trzęsicki, *Łukasiewicz on Philosophy and Determinism*; A. Jedynak, *Conventionalism in Ajdukiewicz*; A. Siemianowski, *On Certain Consequences of the Radical Empiricism Hypothesis*; J. Woleński, *Tarski as a Philosopher*; R. Poli, *The Dispute over Reism: Kotarbiński – Ajdukiewicz – Brentano*.

VOLUME 29 (1993)

Zdzisław Augustynek and Jacek J. Jadacki

POSSIBLE ONTOLOGIES

Possible Ontologies – J.J. Jadacki, *Preface*; Z. Augustynek, *Point Eventism. An Outline of a Certain Ontology*; J.J. Jadacki, *Ontological Minimum*. **Discussions** – K. Paprzycka, *Carnap and Leibniz on the Problem of Being*; P. Przybysz, *Polish Discussions about Reism*.

VOLUME 30 (1993)

GOVERNMENT: SERVANT OR MASTER?
(Edited by Gerard Radnitzky and Hardy Bouillon)

G. Radnitzky, *Introduction: The Ominous Growth of the Monstrous Leviathan*. **Part I: Theory** – A. Seldon, *Politicians for or against the People*; G. Radnitzky, *Private Rights against Public Power: The Contemporary Conflict*; A. de Jasay, *Is Limited Government*

Possible?; H. Bouillon, *Mastering the Growth of Government: A Muzzle for Leviathan*; A. de Jasay, *Ownership, Agency, Socialism.* **Part II: Case Studies** – P. Bernholz, *Necessary and Sufficient Conditions to End Hyperinflations*; G. Schwarz, *The Deterioration of Switzerland's Economic System Policy*; H.O. Lenel, *Three Turns in West Germany since 1948 – Considered from the Viewpoint of the Theory of Economic Systems*; A. Flew, *Educational Services: Abuses of State Monopoly*; J.W.F. Sundberg, *Revenue-Only Taxes vs. Multipurpose Taxes: Philosophy and Implementation in Swedish High Tax Society*; S. Pejovich, *Political and Economic Reforms in Eastern Europe: The Case of Yugoslavia.* **Part III: The Limiting Case. The Totalitarian Order** – P. Bernholz, *Necessary Conditions for Totalitarianism: Supreme Values, Power and Personal Interest.*

VOLUME 31 (1993)

CREATIVITY AND CONSCIOUSNESS. PHILOSOPHICAL AND PSYCHOLOGICAL DIMENSIONS

(Edited by Jerzy Brzeziński, Santo Di Nuovo, Tadeusz Marek and Tomasz Maruszewski)

Introduction. **Part I: What Do Philosophers Say about Creativity and Consciousness?** – K. Zamiara, *Psychological Approach to Creativity. A Critical Approach*; R.L. Franklin, *Creativity and Depth in Understanding*; Z. Piątek, *Creating of Life and F. Nietzsche's Idea of Superman*; J. Janousek, *Dialogue and Joint Activity: A Psychological Approach*; K. Zamiara, *Some Remarks on Piagetian Notion of "Consciousness" and Its Importance for the Science of Culture*; A. Gałdowa, A. Nelicki, *Attitudes Toward Values as a Factor Determining Creativity.* **Part II: The Role of Creativity in Theory-Building** – L. Nowak, *Creativity in Theory-Bulding*; I. Nowak, *Discovery and Correspondence*; J. Brzeziński, *Factors Disturbing Validity of Experimental and Non-Experimental Psychological Research – Context's Analysis of Researcher's Psychologists Methodological Consciousness*; A. Falkowski, *Cognitive Similarity in Scientific Discovery: An Ecological Approach.* **Part III: Consciousness in Historical Perspective** – K.V. Wilkes, *Inside Insight*; F. di Maria, G. Lavanco, *Conscious/Unconscious and Group-Analysis.* **Part IV: Between Expression and Projection** – M. Stasiakiewicz, *Creativity and Projection: Paradigm Opposition and Implicit Correspondence*; A. Brzezińska, *Creative Expression versus Projection.* **Part V: The Role of Psychophysiological Components in Explanation of Phenomena of Consciousness and Creativity** – M. Bunge, *Explaining Creativity*; P. Wolski, *Hemispheric Asymmetry and Consciousness. Is there Any Relationship?*; A. Kokoszka, *A Rationale for Psychology of Consciousness*; M. Fąfrowicz, T. Marek, *Creativity and Attention.* **Part VI: Psychological Explanations of Creativity and Consciousness** – S. Di Nuovo, *Consciousness and Attention*; T. Maruszewski, *Some Remarks about Consciousness*; M. Kowalczyk, *On the Question of Functions of Consciousness*; D.K. Simonton, *From Childhood Giftedness to Creativity Genius*; M. Fąfrowicz, T. Marek, Cz. Noworol, K. Żarczyński, *Creativity and Organization*; M.A. Runco and J. R. Gaynor, *Creativity and Optimal Development.*

VOLUME 32 (1993)

FROM ONE-PARTY-SYSTEM TO DEMOCRACY

(Edited by Janina Frentzel-Zagórska)

J. Frentzel-Zagórska, *Introduction.* **Part I: Theoretical Approaches** – Z. Bauman, *A Postmodern Revolution*; L. Holmes, *On Communism, Post-communism, Modernity and Post-*

modernity; L. Nowak, *The Totalitarian Approach and the History of Socialism*; J. Pakulski, *East European Revolutions and 'Legitimacy Crisis'.* **Part II: The Transitional Period** – A. Czarnota, M. Krygier, *From State to Legal Traditions? Prospects for the Rule of Law after Communism*; M. Szabó, *Social Protest in a Post-communist Democracy: Taxi Drivers' Demonstration in Hungary*; Z. Bauman, *Dismantling Patronage State*; E. Mokrzycki, *Between Reform and Revolution: Eastern Europe Two Years after the Fall of Communism*; J. Frentzel-Zagórska, *The Road to Democratic Political System in Post-communist Eastern Europe.* **Part III: The Case of Yugoslavia** – R.F. Miller, *Yugoslavia: The End of the Experiment.*

VOLUME 33 (1993)

SOCIAL SYSTEM, RATIONALITY, AND REVOLUTION
(Edited by Marcin Paprzycki and Leszek Nowak)

Introduction. **On the Nature of Social System** – U. Preuss, *Political Order and Democracy. Carl Schmitt and his Influence*; K. Paprzycka, *A Paradox in Hobbes' Philosophy of Law*; S. Esquith, *Democratic Political Dialogue*; E. Jeliński, *Democracy in Polish Reformist Socialist Thought*; K. Paprzycka, *The Master and Slave Configuration in Hegel's System*; M. Godelier, *Lévi-Strauss, Marx and After. A Reappraisal of Structuralist and Marxist Tools for Social Logics*; K. Niedźwiadek, *On the Structure of Social System*; W. Czajkowski, *Social Being and Its Reproduction.* **On Rationality and Captivity** – M. Ziółkowski, *Power and Knowledge*, L. Nowak, *Two Inter-Human Limits to the Rationality of Man*; M. Paprzycki, *The Non-Christian Model of Man: An Attempt at Psychological Explanation*; R. Egiert, *Toward the Sophisticated Rationalistic Model of Man.* **On Social Revolution** – L. Nowak, *Revolution is an Opaque Progress but a Progress Nonetheless*; K. Paprzycka, M. Paprzycki, *How do Enslaved People Make Revolutions?*; G. Tomczak, *Is it Worth Winning a Revolution?*; K. Brzechczyn, *Civil Loops and the Absorption of Elites*; R. McCleary, *What Makes Marxist Historical Materialism Objective?*; G. Kotlarski, *Classes and Masses in Social Philosophy of Rosa Luxemburg.* **On Real Socialism** – E. Gellner, *The Civil and the Sacred*; W. Marciszewski, *Economics and the Idea of Information. Why Socialism must have Collapsed?*; L. Nowak, K. Paprzycka, M. Paprzycki, *On Multilinearity of Socialism*; A. Siegel, *The Overrepression Cycle in the Soviet Union. An Operationalization of a Theoretical Model*; K. Brzechczyn, *The State of the Teutonic Order as a Socialist Society.* **Discussions** – R. McCleary, *Socioanalysis and Philosophy*; W. Heller, *Methodological Remarks on the Public and the Private in Hannah Arendt's Political Philosophy*; K. Brzechczyn, *On Unsuccessful Conquest and Successful Subordination.*

VOLUME 34 (1994)

Izabella Nowakowa

IDEALIZATION V: THE DYNAMICS OF IDEALIZATIONS

Introduction; Chapter I: *Idealization and Theories of Correspondence*; Chapter II: *Dialectical Correspondence of Scientific Laws*; Chapter III: *Dialectical Correspondence in Science: Some Examples*; Chapter IV: *Dialectical Correspondence of Scientific Theories*; Chapter V: *Generalizations of the Rule of Correspondence*; Chapter VI: *Extensions of the Rule of Correspondence*; Chapter VII: *Correspondence and the Empirical Environment of a Theory*; Chapter VIII: *Some Methodological Problems of Dialectical Correspondence.*

VOLUME 35 (1993)

EMPIRICAL LOGIC AND PUBLIC DEBATE.
ESSAYS IN HONOUR OF ELSE M. BARTH
(Edited by E.C.W. Krabbe, R.J. Dalitz, and P.A. Smit)

Part I: Interpersonal Reasoning: Conflicts and Fallacies – T. Govier, *Needing Each Other for Knowledge: Reflections on Trust and Testimony*; J. Woods, *'Secundum quid' as a Research Programme*; G. Nuchelmans, *On the Fourfold Root of the 'argumentum ad hominem'*; F.H. van Eemeren, R. Grootendorst, *The History of the 'argumentum ad hominem' Since the Seventeenth Century*; L.S. van Epenhuysen, *Debate in a Bermuda Triangle of Medical Ethics*; E.C.W. Krabbe, *Reasonable Argument and Fallacies in the Kok-Stekelenburg Debate*; **Part II: Linguistic and Conceptual Tools** – J.D. North, *Some Weak Links in the Great Chain of Being*; A. Næss, *'You assert this?' An Empirical Study of Weight Expressions*; R. Wiche, *Gerrit Mannoury on the Communicative Functions of Negation in Ordinary Language*; J. Hoepelman, T. van Hoof, *Default and Dogma*; Ch. Goossens, *On the Logic of Nonmoral Commitment*; W. Marciszewski, *Arguments Founded on Creative Definitions*; R. Jorna, *Cognitive Science and Connectivism: Friend and Enemy or Move and Counter-Move, an Application of Empirical Logic*; **Part III: Dialectical Climates and Tempests** – R.H. Johnson, *Dialectical Fields and Manifest Rationality*; P. du Preez, *Reason Which Cannot Be Reasoned With: What Is Public Debate and How Does it Change?*; M.A. Finocchiaro, *Logic, Democracy, and Mosca*; P.A. Smit, *The Logic of Virtue and Terror*; G. van Benthem van den Bergh, *On Obstacles to Public Debate*; J.P. van Bendegem, *Real-Life Mathematics versus Ideal Mathematics: The Ugly Truth*; **Part IV: The Disempowernment of Woman: Strategies and Counter-Moves** – V. Songe-Møller, *The Road of Being and the Exclusion of the Feminine. An Analysis of the Poem of Parmenides*; R.J. Dalitz, *The Subjection of Women in the Contractual Society. An Analysis of Thomas Hobbes' Theory of Agreement*; H. Schröder, *Anti-Semitism and anti-Feminism Again: The Dissemination of Otto Weininger's* Sex and Character *in the Seventies and Eighties*; J.R. Richards, *Traditional Spheres and Traditional Logic.*

VOLUME 36 (1994)

MARXISM AND COMMUNISM: POSTHUMOUS REFLECTIONS ON
POLITICS, SOCIETY, AND LAW
(Edited by Martin Krygier)

M. Krygier, *Introduction*; A. Flis, *From Marx to Real Socialism: The History of a Utopia*; P. Marciniak, *The Collapse of Communism: Defeat or Opportunity for Marxism in Eastern Europe*; J. Clark, A. Wildavsky, *Chronicle of a Collapse Foretold: How Marx Predicted the Demise of Communism (Although He Called It "Capitalism")*; L. Nowak, *Political Theory and Socialism. On the Main Paradigms of Political Power and Their Methodological and Historical Legitimation*; E. Mokrzycki, *Marxism, Sociology, and "Real Socialism"*; R. Bäcker, *The Collapse of Communism and Theoretical Models*; A. Zybertowicz, *Three Deaths an Ideology: The Withering Away of Marxism and the Collapse of Communism. The Case of Poland*; M. Krygier, *Marxism, Communism, and the Rule of Law*; A. Czarnota, *Marxism, Ideology, and Law*; G. Skąpska, *The Legacy of Anti-Legalism*; A. Sajo, *Law and the Legal Scholarship in the Happiest Barrack and Among the Hungry Liberated: Personal Recollections.*

VOLUME 37 (1994)

THE SOCIAL PHILOSOPHY OF AGNES HELLER
(Edited by John Burnheim)

J. Burnheim, *Introduction*; M. Vajda, *A Lover of Philosophy – A Lover of Europe*; P. Despoix, *On the Possibility of a Philosophy of Values. A Dialogue within the Budapest School*; M. Jay, *Women in Dark Times: Agnes Heller and Hannah Arendt*; J.P. Arnason, *The Human Condition and the Modern Predicament*; R.J. Bernstein, *Agnes Heller: Philosophy, Rational Utopia and Praxis*; Z. Bauman, *Narrating Modernity*; P. Beilharz, *Theories of History – Agnes Heller and R.G. Collingwood*; P. Wolin, *Heller's Theory of Everyday Life*; P. Harrison, *Radical Philosophy and the Theory of Modernity*; A.J. Jacobson, *The Limits of Formal Justice*; P. Murphy, *Civility and Radicalism*; P. Murphy, *Pluralism and Politics*; V. Camps, *The Good Life: A Moral Gesture*; L. Boella, *Philosophy Beyond the Baseless and Tragic Character of Action*; G. Márkus, *The Politics of Morals*; A. Heller, *A Reply to My Critics*; *The Bibliography of Agnes Heller.*

VOLUME 38 (1994)

IDEALIZATION VI: IDEALIZATION IN ECONOMICS
(Edited by Bert Hamminga and Neil B. De Marchi)

Introduction – B. Hamminga, N. De Marchi, *Preface*; B. Hamminga, N. De Marchi, *Idealization and the Defence of Economics: Notes Toward a History*. **Part I: General Observations on Idealization in Economics** – K.D. Hoover, *Six Queries about Idealization in an Empirical Context*; B. Walliser, *Three Generalization Processes for Economic Models*; S. Cook, D. Hendry, *The Theory of Reduction in Econometrics*; M.C.W. Janssen, *Economic Models and Their Applications*; A.G. de la Sienra, *Idealization and Empirical Adequacy in Economic Theory*; I. Nowakowa, L. Nowak, *On Correspondence between Economic Theories*; U. Mäki, *Isolation, Idealization and Truth in Economics*. **Part II: Case Studies of Idealization in Economics** – N. Cartwright, *Mill and Menger: Ideal Elements and Stable Tendencies*, W. Balzer, *Exchange Versus Influence: A Case of Idealization*; K. Cools, B. Hamminga, T.A.F. Kuipers, *Truth Approximation by Concretization in Capital Structure Theory*; D.M. Hausman, *Paul Samuelson as Dr. Frankenstein: When an Idealization Runs Amuck*; H.A. Keuzenkamp, *What if an Idealization is Problematic? The Case of the Homogeneity Condition in Consumer Demand*; W. Diederich, *Nowak on Explanation and Idealization in Marx's* Capital; G. Jorland, *Idealization and Transformation*; J. Birner, *Idealizations and Theory Development in Economics. Some History and Logic of the Logic Discovery*. **Discussions** – L. Nowak, *The Idealizational Methodology and Economics. Replies to Diederich, Hoover, Janssen, Jorland and Mäki.*

VOLUME 39 (1994)

PROBABILITY IN THEORY-BUILDING. EXPERIMENTAL AND NON-EXPERIMENTAL APPROACHES TO SCIENTIFIC RESEARCH IN PSYCHOLOGY

(Edited by Jerzy Brzeziński)

Part I: Probability and the Idealizational Theory of Science – M. Gaul, *Statistical Dependencies, Statements and the Idealizational Theory of Science.* **Part II: Probability–Theoretical Concepts in Psychology–Measurement** – D. Wahlstein, *Probability and the Understanding of Individual Differences*; B. Krause, *Modeling Cognitive Learning Steps*; D. Heyer, R. Mausfeld, *A Theoretical and Experimental Inquiry into the Relation of Theoretical Concepts and Probabilistic Measurement Scales in Experimental Psychology.* **Part III: Methods of Data Analysis** – T.B. Iwiński, *Rough Set Methods in Psychology*; W. Koutstaal, R. Rosenthal, *Contrast Analysis in Behavioral Research.* **Part IV: Artifacts in Psychological Research and Diagnostic Assessment** – D.B. Strohmetz, R.L. Rosnow, *A Mediational Model of Research Artifacts*; J. Brzeziński, *Dimensions of Diagnostic Space.*

VOLUME 40 (1995)

THE HERITAGE OF KAZIMIERZ AJDUKIEWICZ

(Edited by Vito Sinisi and Jan Woleński)

Preface; J. Giedymin, *Ajdukiewicz's Life and Personality*; K. Ajdukiewicz, *My Philosophical Ideas*; L. Albertazzi, *Some Elements of Transcendentalism in Ajdukiewicz*; T. Batóg, *Ajdukiewicz and the Development of Formal Logic*; A. Church, *A Theory of the Meaning of Names*; M. Czarniawska, *The Way from Concept to Thought. Does it Exist in Ajdukiewicz's Semantical Theory?*; E. Dölling, *Real Objects and Existence*; J. Giedymin, *Radical Conventionalism, Its Background and Evolution: Poincaré, Le Roy, Ajdukiewicz*; J.J. Jadacki, *Definition, Explication and Paraphrase in the Ajdukiewiczian Tradition*; G. Küng, *Ajdukiewicz's Contribution to the Realism/Idealism Debate*; J. Maciaszek, *Problems of Proper Names: Ajdukiewicz and Some Contemporary Results*; W. Marciszewski, *Real Definitions and Creativity*; K. Misiuna, *Categorial Grammar and Ontological Commitment*; L. Nowak, *Ajdukiewicz and the Status of the Logical Theory of Natural Language*; A. Olech, *Several Remarks on Ajdukiewicz's and Husserl's Approaches to Meaning*; A. Orenstein, *Existence Sentences*; J. Pasek, *Ajdukiewicz on Indicative Conditionals*; R. Poli, *The Problem of Position: Ajdukiewicz and Leibniz on Intensional Expressions*; M. Przełęcki, *The Law of Excluded Middle and the Problem of Idealism*; H. Skolimowski, *Ajdukiewicz, Rationality and Language*; K. Szaniawski, *Ajdukiewicz on Non-deductive Inference*; A. Varzi, *Variable-Binders as Functors*; V. Vasyukov, *Categorial Semantics for Ajdukiewicz-Lambek Calculus*; H. Wassing, *On the Expressiveness of Categorial Grammar*; J. Woleński, *On Ajdukiewicz's Refutation of Scepticism*; *Bibliography of Ajdukiewicz; Bibliography on Ajdukiewicz (writings in western languages).*

VOLUME 41 (1994)

HISTORIOGRAPHY BETWEEN MODERNISM AND POSTMODERNISM. CONTRIBUTIONS TO THE METHODOLOGY OF THE HISTORICAL RESEARCH
(Edited by Jerzy Topolski)

Editor's Introduction; J. Topolski, *A Non-postmodernist Analysis of Historical Narratives*; F.R. Ankersmit, *Historism, Postmodernism and Historiography*; D. Carr, *Getting the Story Straight: Narrative and Historical Knowledge*; W. Wrzosek, *The Problem of Cultural Imputation in History. Culture versus History*; J. Tacq, *Causality and Virtual Finality*; G. Zalejko, *Soviet Historiography as a "Normal Science"*; H. Mamzer and J. Ostoja-Zagórski, *Deconstruction of Evolutionist Paradigm in Archaeology*; N. Lautier, *At the Crossroad of Epistemology and Psychology: Prospects of a Didactic of History*; Teresa Kostyrko, *Remarks on "Aesthetization" in Science on the Basis of History.*

VOLUME 42 (1995)

IDEALIZATION VII: IDEALIZATION, STRUCTURALISM, AND APPROXIMATION
(Edited by Martti Kuokkanen)

Idealization, Approximation and Counterfactuals in the Structuralist Framework – T.A.F. Kuipers, *The Refined Structure of Theories*; C.U. Moulines and R. Straub, *Approximation and Idealization from the Structuralist Point of View*; I.A. Kieseppä, *A Note on the Structuralist Account of Approximation*; C.U. Moulines and R. Straub, *A Reply to Kieseppä*; W. Balzer and G. Zoubek, *Structuralist Aspects of Idealization*; A. Ibarra and T.Mormann, *Counterfactual Deformation and Idealization in a Structuralist Framework*; I.A. Kieseppä, *Assessing the Structuralist Theory of Verisimilitude*. **Idealization, Approximation and Theory Formation** – L. Nowak, *Remarks on the Nature of Galileo's Methodological Revolution*; I. Niiniluoto, *Approximation in Applied Science*; E. Heise, P. Gerjets and R. Westermann, *Idealized Action Phases. A Concise Rubicon Theory*; K.G. Troitzsch, *Modelling, Simulation, and Structuralism*; V. Rantala and T.Vadén, *Idealization in Cognitive Science. A Study in Counterfactual Correspondence*; M. Sintonen and M. Kiikeri, *Idealization in Evolutionary Biology*; T. Tuomivaara, *On Idealizations in Ecology*; M. Kuokkanen and M. Häyry, *Early Utilitarianism and Its Idealizations from a Systematic Point of View*. **Idealization, Approximation and Measurement** – R. Westermann, *Measurement-Theoretical Idealizations and Empirical Research Practice*; U. Konerding, *Probability as an Idealization of Relative Frequency. A Case Study by Means of the BTL-Model*; R. Suck and J. Wienöbst, *The Empirical Claim of Probality Statements, Idealized Bernoulli Experiments and their Approximate Version*; P.J. Lahti, *Idealizations in Quantum Theory of Measurement.*

VOLUME 43 (1995)

Witold Marciszewski and Roman Murawski
MECHANIZATION OF REASONING IN A HISTORICAL PERSPECTIVE

Chapter 1: *From the Mechanization of Reasoning to a Study of Human Intelligence*; Chapter 2: *The Formalization of Arguments in the Middle Ages*; Chapter 3: *Leibniz's Idea of Mechanical Reasoning at the Historical Background*; Chapter 4: *Between Leibniz and Boole: Towards the Algebraization of Logic*; Chapter 5: *The English Algebra of Logic in the 19th Century*; Chapter 6: *The 20th Century Way to Formalization and Mechanization*; Chapter 7: *Mechanized Deduction Systems.*

VOLUME 44 (1995)

THEORIES AND MODELS IN SCIENTIFIC PROCESSES
(Edited by William Herfel, Władysław Krajewski,
Ilkka Niiniluoto and Ryszard Wójcicki)

Introduction; **Part 1. Models in Scientific Processes** – J. Agassi, *Why there is no Theory of Models?*; M. Czarnocka, *Models and Symbolic Nature of Knowledge*; A. Grobler, *The Representational and the Non-Representational in Models of Scientific Theories*; S. Hartmann, *Models as a Tool for the Theory Construction: Some Strategies of Preliminary Physics*; W. Herfel, *Nonlinear Dynamical Models as Concrete Construction*; E. Kałuszyńska, *Styles of Thinking*; S. Psillos, *The Cognitive Interplay Between Theories and Models: the Case of 19th Century Optics*. **Part 2. Tools of Science** – N.D. Cartwright, T. Shomar, M. Suarez, *The Tool-Box of Science*; J. Echeverria, *The Four Contexts of Scienctific Activity*; K. Havas, *Continuity and Change; Kinds of Negation in Scientific Progress*; M. Kaiser, *The Independence of Scientific Phenomena*; W. Krajewski, *Scientific Meta-Philosophy*; I. Niiniluoto, *The Emergence of Scientific Specialties: Six Models*; L. Nowak, *Antirealism, (Supra-) Realism and Idealization*; R.M. Nugayev, *Classic, Modern and Postmodern Scientific Unification*; V. Rantala, *Translation and Scientific Change*; G. Schurz, *Theories and Their Applications – a Case of Nonmonotonic Reasoning*; W. Strawiński, *The Unity of Science Today*; V. Torosian, *Are the Ethics and Logic of Science Compatible?* **Part 3. Unsharp Approaches in Science** – E.W. Adams, *Problems and Prospects in a Theory of Inexact First-Order Theories*; W. Balzer, G. Zoubek, *On the Comparision of Approximative Empirical Claims*; G. Cattaneo, M. Luisa Dalla Chiara, R. Giuntini, *The Unsharp Approaches to Quantum Theory*; T.A.F. Kuipers, *Falsification Versus Effcient Truth Approximation*; B. Lauth, *Limiting Decidability and Probability*; J. Pykacz, *Many-Valued Logics in Foundations of Quantum Mechanics*; R.R. Zapatrin, *Logico-Algebraic Approach to Spacetime Quantization*.

VOLUME 45 (1995)

COGNITIVE PATTERNS IN SCIENCE AND COMMON SENSE.
GRONINGEN STUDIES IN PHILOSOPHY OF SCIENCE,
LOGIC, AND EPISTEMOLOGY
(Edited by Theo A.F. Kuipers and Anne Ruth Mackor)

L. Nowak, *Foreword*; **General Introduction** – T.A.F. Kuipers and A.R. Mackor, *Cognitive Studies of Science and Common Sense*; **Part I: Conceptual Analysis in Service of Various Research Programmes** – H. Zandvoort, *Concepts of Interdisciplinarity and Environmental Science*; R. Vos, *The Logic and Epistemology of the Concept of Drug and Disease Profile*; R.C. Looijen, *On the Distinction Between Habitat and Niché, and Some Implications for Species' Differentiation*; G.J. Stavenga, *Cognition, Irreversibility and the Direction of Time*; R. Dalitz, *Knowledge, Gender and Social Bias*; **Part II: The Logic of the Evaluation of Arguments, Hypotheses, Rules, and Interesting Theorems** – E.G.W. Krabbe, *Can We Ever Pin One Down to a Formal Fallacy?*; T.A.F. Kuipers, *Explicating the Falsificationist and Instrumentalist Methodology by Decomposing the Hypothetico-Deductive Method*; A. Keupink, *Causal Modelling and Misspecification: Theory and Econometric Historical Practice*; M.C.W. Janssen and Y.-H. Tan, *Default Reasoning and Some Applications in Economics*; B. Hamminga, *Interesting Theorems in Economics*; **Part III: Three Challenges to the Truth Approximation Programme** – S.D. Zwart, *A Hidden Variable in the Discussion About 'Language Dependency' of Truthlikeness*; H. Hettema and T.A.F. Kuipers, *Sommerfeld's* Atombau: *A Case Study in Potential Truth Approximation*; R. Festa, *Verisimilitude, Disorder, and Optimum Prior Probabilities*; **Part IV: Explicating Psychological Intuitions** – A.R. Mackor, *Intentional Psychology is a Biological Discipline*; J. Peijnenburg, *Hempel's Rationality. On the Empty Nature of Being a Rational Agent*; L. Guichard, *The Causal Efficacy of Propositional Attitudes*; M. ter Hark, *Connectionism, Behaviourism and the Language of Thought.*

VOLUME 46 (1996)

POLITICAL DIALOGUE: THEORIES AND PRACTICE
(Edited by Stephen L. Esquith)

Introduction – S.L. Esquith, *Political Dialogue and Political Virtue.* **Part I: The Modern Clasics** – A.J. Damico, *Reason's Reach: Liberal Tolerance and Political Discourse*; T.R. Machan, *Individualism and Political Dialogue*; R. Kukla, *The Coupling of Human Souls: Rousseau and the Problem of Gender Relations*; D.F. Koch, *Dialogue: An Essay in the Instrumentalist Tradition.* **Part II: Toward a Democratic Synthesis** – E. Simpson, *Forms of Political Thinking and the Persistence of Practical Philosophy*; J.B. Sauer, *Discoursee, Consensus, and Value: Conversations about the Intelligible Relation Between the Private and Public Spheres*; M. Kingwell, *Phronesis and Political Dialogue*, R.T. Peterson, *Democracy and Intellectual Mediation – After Liberalism and Socialism.* **Part III: Dialogue in Practice** – S. Rohr Scaff, L.A. Scaff, *Political Dialogue in the New Germany: The Burdens of Culture and an Asymmetrical Past*; J.H. Read, *Participation, Power, and Democracy*; S.E. Bennett, B. Fisher, D. Resnick, *Speaking of Politics in the United States: Who Talks to Whom, Why, and Why Not*; J. Forester, *Beyond Dialogue to Transformative Learning: How Deliberative Rituals Encourage Political Judgement in Community Planning Processes*; A. Fatić, *Retribution in Democracy.*

VOLUME 47 (1996)

EPISTEMOLOGY AND HISTORY. HUMANITIES AS A PHILOSOPHICAL PROBLEM AND JERZY KMITA'S APPROACH TO IT
(Edited by Anna Zeidler-Janiszewska)

VOLUME 48 (1996)

THE SOCIAL PHILOSOPHY OF ERNEST GELLNER
(Edited by John A. Hall and Ian Jarvie)

Ernest Gellner in Soviet East. **Part 2: Nations and Nationalism** – B. O'Leary, *On the Nature of Nationalism: An Appraisal of Ernest Gellner's Writings on Nationalism*; K. Minogue, *Ernest Gellner and the Dangers of Theorising Nationalism*; A.D. Smith, *History and Modernity: Reflection on the Theory of Nationalism*; M. Mann, *The Emergence of Modern European Nationalism*; N. Stagardt, *Gellner's Nationalism: The Spirit of Modernisation?* **Part 3: Patterns of Development** – P. Burke, *Reflections on the History of Encyclopaedias*; A. MacFarlane, *Ernest Gellner and the Escape to Modernity*; R. Dore, *Soverein Individuals*; S. Eisenstadt, *Japan: Non Axial Modernity*; M. Ferro, *L'Indépendance Telescopée: De la Décolonisation a L'Impérialisme Multinational.* **Part 4: Islam** – A. Hammoudi, *Segmentarity, Social stratification, Political Power and Sainthood: Reflections on Gellner's Theses*; H. Munson, Jr., *Rethinking Gellner's Segmentary Analysis of Morocco's Ait^cAtta*; J. Baechler, *Sur le charisme*; Ch. Lindholm, *Despotism and Democracy: State and Society in PreModern Middle East*; H. Munson, Jr. *Muslism and Jew in Morocco: Reflections on the Distinction between Belief and Behaviour*; T. Asad, *The Idea of an Anthropology of Islam.* **Part 5: Science and Disenchantment** – P. Anderson, *Science, Politics, Enchantment*; R. Schroeder, *From the Big Divide to the Rubber Cage: Gellner's Conception of Science and Technology*; J. Davis, *Irrationality in Social Life.* **Part 6: Relativism and Universals** – J. Skorupski, *The Post-Modern Hume: Ernest Gellner's 'Enlightenment Fundamentalism'*; J. Wettersten, *Ernest Gellner: A Wittgensteinian Rationalist*; I. Jarvie, *Gellner's Positivism*; R. Boudon, *Relativising Relativism: When Sociology Refutes the Sociology of Science*; R. Aya, *The Empiricist Exorcist.* **Part 7: Philosophy of History** – W. McNeill, *A Swang Song for British Liberalism?*; A. Park, *Gellner and the Long Trends of History*; E. Leone, *Marx, Gellner, Power*; R. Langlois, *Coercion, Cognition and Production: Gellner's Challenge to Historical Materialism and Post-Modernism*; E. Gellner, *Reply to my Critics*; I. Jarvie, *Complete Bibliography of Gellner's Work.*

VOLUME 49 (1996)

THE SIGNIFICANCE OF POPPER'S THOUGHT
(Edited by Stefan Amsterdamski)

Karl Popper's Three Worlds – J. Watkins, *World 1, World 2 and the Theory of Evolution*; A. Grobler, *World 3 and the Cunning of Reason.* **The Scientific Method as Ethics** – J. Agassi, *Towards Honest Public Relations of Science*; S. Amsterdamski, *Between Relativism and Absolutism: the Popperian Ideal of Knowledge.* **The Open Society and its Prospects** – E. Gellner, *Karl Popper – The Thinker and the Man*; J. Woleński, *Popper on Prophecies and Predictions.*

VOLUME 50 (1996)

THE IDEA OF THE UNIVERSITY
(Edited by Jerzy Brzeziński and Leszek Nowak)

Introduction – K. Twardowski, *The Majesty of the University.* **I** – Z. Ziembiński, *What Can be Saved of the Idea of the University?*; L. Kołakowski, *What Are Universities for?*; L. Gumański, *The Ideal University and Reality*; Z. Bauman, *The Present Crisis of the Universities.* **II** – K. Ajdukiewicz, *On Freedom of Science*; H. Samsonowicz, *Universities and Democracy*; J. Topolski, *The Commonwealth of Scholars and New Conceptions of Truth*; K. Szaniawski, *Plus ratio quam vis.* **III** – L. Koj, *Science, Teaching and Values*; K. Szaniawski, *The Ethics of Scientific Criticism*; J. Brzeziński, *Ethical Problems of Research Work of*

Psychologists. **IV** – J. Goćkowski, *Tradition in Science*; J. Kmita, *Is a "Creative Man of Knowledge" Needed in University Teaching?*; L. Nowak, *The Personality of Researchers and the Necessity of Schools in Science.* **Recapitulation** – J. Brzeziński, *Reflections on the University.*

UNIVERSITY
OF BRISTOL
LIBRARY